Mathematics for scientific and technical students
Second edition

Mathematics for scientific and technical students

Second edition

H. G. Davies and G. A. Hicks

 LONGMAN

Addison Wesley Longman Limited
Edinburgh Gate, Harlow
Essex CM20 2JE
England
and Associated Companies throughout the World

©Longman Group Limited 1975
©Addison Wesley Longman Limited 1998

First published 1975
Second edition 1998

ISBN 0 582 41388 5

British Library Cataloguing-in-Publication Data
A catalogue record for this book is
available from the British Library.

Set by MCS Ltd, Salisbury
Produced through Longman Malaysia, VVP

Contents

subsit   190
204
206
215
510

Preface

The first edition, written twenty years or so ago, contains several sections which are now dated, for example, logarithm tables for calculations and angles expressed in degrees and minutes. In the second edition, logarithm tables have been deleted, although the theory of logarithms remains an integral part. A logarithm of a number on base *10* is written as *log*, whilst a logarithm of a number on the base *e* is written as *ln*. Angles are now expressed in degrees and decimals of degrees.

Much of the first edition however, is still valid as a foundation for students preparing for higher education. Additions have been made with chapters in Differential Equations; Determinants and Matrices; and Vectors.

Whilst the book provides a foundation in Mathematics for those aspiring to degrees in Engineering and Physical Science, it also meets the requirements of the three GNVQ syllabuses for additional and further Mathematics for which, a study guide is included.

Most of the exercises in the first edition have been retained. Additional exercises have been included but after many years of teaching, their original sources have been lost. The authors apologise unreservedly should any have been used without proper acknowledgement. Exercises and worked examples have been constructed to illustrate the application of Mathematics in Engineering and Physics.

As always, we are indebted to the staff of Addison Wesley Longman for their support and forbearance during the preparatory stages of the book, and in particular to James Newall, Chris Leeding and Kate Henderson.

H G Davies
G A Hicks

July 1997

Chapter 1

Algebra

1.1 Quadratic equations

The general form of a simple quadratic equation is

$$ax^2 + bx + c = 0$$

where a, b, and c are numbers called **coefficients**. Such an equation may be solved by

1. factorisation;
2. drawing a graph – see section 9.27;
3. using the formula.

(a) Solution by factorisation

To test whether the equation will factorise the following procedure may be adopted:

1. Multiply the coefficients a and c together.
2. Write down the factor pairs of the product ac.
3. When the sign of ac is $+$ then adding together one of the pairs will give the coefficient b if the equation factorises

    or

    When the sign of ac is $-$ then subtracting one of the pairs will produce b if the equation factorises.
4. The equation factorises using this factor pair.

This procedure is illustrated using the equation

$$12x^2 + x - 6 = 0$$

1. The product $ac = -72$.
2. The factor pairs of 72 are (1, 72) (2, 36) (3, 24) (4, 18) (6, 12) (8, 9).
3. Since ac is $-$ one of these pairs must subtract to give the coefficient of x. The required pair is $(-8, +9)$.

4. The equation is written using this pair,

$$12x^2 + 9x - 8x - 6 = 0$$
$$3x(4x + 3) - 2(4x + 3) = 0$$
$$(4x + 3)(3x - 2) = 0$$

so that $(4x + 3) = 0$ or $(3x - 2) = 0$

Therefore the solutions are

$$x = -\frac{3}{4} \quad \text{or} \quad x = -\frac{2}{3}$$

The procedure when ac is $+$ is shown in the factorisation of the equation

$$6x^2 - 19x + 10 = 0$$

1. The product $ac = +60$.
2. The factor pairs of 60 are (1, 60), (2, 30), (3, 20), (4, 15), (5, 12), (6, 10).
3. Since ac is $+$ one of the pairs must add to give the coefficient of x, which is -19. The required pair is $(-4, -15)$, both being negative because the coefficient b is negative.
4. Write the quadratic equation using this pair:

$$6x^2 - 4x - 15x + 10 = 0$$
$$2x(3x - 2) - 5(3x - 2) = 0$$
$$(3x - 2)(2x - 5) = 0$$
$$x = 2/3 \quad \text{or} \quad x = 5/2$$

(b) Solution using the formula

If an equation does not factorise the formula must be used, and this can be obtained as follows. Let the quadratic be

$$ax^2 + bx + c = 0 \tag{i}$$

Divide throughout by a:

$$x^2 + \frac{b}{a}x + \frac{c}{a} = 0$$

so that

$$x^2 + \frac{b}{a}x = -\frac{c}{a}$$

Add $\dfrac{b^2}{4a^2}$ to both sides: $\quad x^2 + \dfrac{b}{a}x + \dfrac{b^2}{4a^2} = \dfrac{b^2}{4a^2} - \dfrac{c}{a}$

The left-hand side factorises: $\quad \left(x + \dfrac{b}{2a}\right)^2 = \dfrac{b^2 - 4ac}{4a^2}$

Taking square roots on both sides $\quad x + \dfrac{b}{2a} = \dfrac{\pm\sqrt{(b^2 - 4ac)}}{2a}$

so that

$$\boxed{x = \dfrac{-b \pm \sqrt{(b^2 - 4ac)}}{2a}}$$

(1.1)

Example 1.1 shows how the formula is used to solve an equation.

Example 1.1 Solve the equation $3x^2 - 4x - 1 = 0$ correct to two decimal places.

The equation does not factorise. In the above formula, by comparison with equation (i),

$$a = 3, \qquad b = -4, \qquad c = -1$$

Therefore

$$x = \frac{-(-4) \pm \sqrt{\{(-4)^2 - 4(3)(-1)\}}}{2(3)}$$

$$= \frac{4 \pm \sqrt{\{16 + 12\}}}{6} = \frac{4 \pm \sqrt{28}}{6}$$

$$= \frac{4 \pm 5.291}{6}$$

Hence $\quad x = \dfrac{9.291}{6} = 1.55$

or $\quad x = \dfrac{-1.291}{6} = -0.22$

Formula (1.1) for the solution of a quadratic equation provides three conditions:

1. $b^2 - 4ac > 0$. This means that the value under the square root is positive and this condition produces two real solutions.
2. $b^2 - 4ac = 0$. In this case the quadratic equation has two identical solutions, $x = -b/2a$. The quadratic equation in this case is a perfect square.
3. $b^2 - 4ac < 0$. In this case the value inside the square root is negative. It is not possible to have real values of the square roots of negative numbers, so that under this condition, the solutions of the quadratic equation are complex numbers in that they contain both real and imaginary numbers. This is further discussed in Chapter 14.

1.2   Equations reducible to quadratic form

(a) **Type 1: Equations of the form $ax^{2n} + bx^n + c = 0$ where n may be $+$ or $-$**
A substitution $z = x^n$ made which reduces the equation to

$$az^2 + bz + c = 0$$

Example 1.2   Solve the equation $\dfrac{2}{y^2} - \dfrac{3}{y} - 3 = 0$ correct to two decimal places.

The equation may be written in the form

$$2y^{-2} - 3y^{-1} - 3 = 0$$

which is type 1, with $n = -1$. Substitute $z = y^{-1}$ so that the equation becomes $2z^2 - 3z - 3 = 0$. Since the equation does not factorise the formula is used.

$$z = \frac{-(-3) \pm \sqrt{\{9 + 24\}}}{4}$$

$$= \frac{3 \pm 5.744}{4}$$

Therefore $z = \dfrac{3 + 5.744}{4}$ or $z = \dfrac{3 - 5.744}{4}$

Hence $\dfrac{1}{y} = 2.186$ or $\dfrac{1}{y} = -0.686$

$y = \dfrac{1}{2.186}$ or $= \dfrac{1}{-0.686}$

that is $y = 0.46$ or $= -1.46$

(b) **Type 2: Equations of the form $ak^{2x} + bk^x + c = 0$, where k is a constant**

In this type of equation the variable x is in the index, and the substitution $z = k^x$ reduces the equation to

$$az^2 + bz + c = 0$$

Example 1.3   Solve $2^{2x+4} - 5 \times 2^{x+1} + 1 = 0$

The equation must first be written as

$$2^4 \times 2^{2x} - 5 \times 2 \times 2^x + 1 = 0$$
$$16 \times 2^{2x} - 10 \times 2^x + 1 = 0$$

Substitute $z = 2^x$
The equation becomes

$$16z^2 - 10z + 1 = 0$$
$$(8z - 1)(2z - 1) = 0$$

Therefore

$$z = \frac{1}{8} \quad \text{or} \quad \frac{1}{2^3} \quad \text{or} \quad z = \frac{1}{2}$$

Hence $\qquad\qquad\qquad 2^x = 2^{-3} \quad$ or $\quad 2^x = 2^{-1}$

Comparing indices: $\qquad x = -3 \quad$ or $\quad x = -1$

EXERCISE 1.1

Solve the following equations:

1. $2x^2 - 11x + 5 = 0$ 2. $4x^2 + 12x - 7 = 0$

3. $7x^2 - 4x - 1 = 0$ 4. $5x^2 - 3x - 4 = 0$

5. $x^4 - 13x^2 + 36 = 0$ 6. $100x^4 - 229x^2 + 9 = 0$

7. $\dfrac{8}{y^6} - \dfrac{217}{y^3} + 27 = 0$ 8. $\dfrac{4}{y^2} - \dfrac{6}{y} - 3 = 0$

9. $2^{2x} - 6 \times 2^x + 8 = 0$ 10. $5 \times 5^{2x} - 26 \times 5^x + 5 = 0$

11. $2^{2x+2} - 17 \times 2^x + 4 = 0$ 12. $4^{2x} - 17 \times 4^x + 16 = 0$

13. $18\left(x - \dfrac{1}{x}\right)^2 + 33\left(x - \dfrac{1}{x}\right) - 40 = 0$

14. $4\left(x - \dfrac{2}{x}\right)^2 + \left(x - \dfrac{2}{x}\right) - 5 = 0$ 15. $2^{2x+1} - 9(2^x) + 4 = 0$

1.3   Simultaneous equations

(a) Linear simultaneous equations in two unknowns

An equation such as

$$y + x = 3$$

has an infinite number of pairs of x and y values that satisfy it. Some of
these are shown in Table 1.1.
   Similarly, pairs of values for the equation

$$y - x = 1$$

Table 1.1				*Table* 1.2	
x	$y = -x + 3$			x	$y = x + 1$
0	3			0	1
1	2	$\longleftrightarrow$		1	2
2	1			2	3
3	0			3	4
4	−1			4	5

are shown in Table 1.2. Only one pair of values, $x = 1$, $y = 2$, satisfies both equations. The two equations together form a simultaneous pair of equations, both containing two variables. $x = 1$, $y = 2$ is the solution of both equations, meaning that it is the only pair of numbers which is common to both equations. Graphically, the solution is represented by the coordinates of the point of intersection P of the straight line graphs corresponding to the two equations (Fig. 1.1).

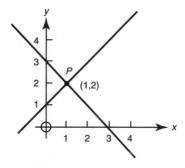

Fig. 1.1

(b) Linear simultaneous equations containing three unknowns

The above discussion is extended to equations containing three variables or unknowns, such as

$$x + y + z = 8$$

There is an infinite number of x, y and z values which satisfy this equation, such as $x = 0$, $y = 2$, $z = 6$.

In combination with two other equations of the same type only one set of values of x, y and z will satisfy all three equations. This single set of values will be the solution of the three simultaneous equations. The solution is obtained as shown below with the three equations.

	Equation number	Current sum
$x + y + z = 8$	(i)	11
$3x - 3y + 2z = 2$	(ii)	4
$x + 4y - 3z = 1$	(iii)	3

Step 1
Eliminate z from (i) and (ii) by making the coefficient the same.
Multiply (i) by 2:

$2x + 2y + 2z = 16$	(iv)	22
$3x - 3y + 2z = 2$	(ii)	4
$-x + 5y \quad = 14$	(v)	18

Subtract (ii) from (iv):

Step 2
Eliminate z from (ii) and (iii) by making the coefficients the same.
Multiply (ii) by 3:
Multiply (iii) by 2:
Add (vi) and (vii):

$9x - 9y + 6z = 6$	(vi)	12
$2x + 8y - 6z = 2$	(vii)	6
$11x - y \quad = 8$	(viii)	18

Step 3
Eliminate y from (v) and (viii) by making the coefficients the same.

Multiply (viii) by 5:
Add (v) and (ix):

$-x + 5y \quad = 14$	(v)	18
$55x - 5y \quad = 40$	(ix)	90
$54x \quad\quad = 54$		108
$x \quad\quad = 1$		

Step 4
Substitute for x in (v): $\quad\quad y \quad = 3$
Substitute for x, y in (i): $\quad z \quad = 4$

The required solution is $x = 1$, $y = 3$, $z = 4$.

In order to check the calculation as we proceed the **current sum** is taken at each line. The current sum is the sum of coefficients and the constant term, and is shown in the column on the right. For equations (i), (ii) and (iii) these sums are 11, 4 and 3 respectively. In step 1, equ. (i) is multiplied by 2, so that the current sum should also be doubled, i.e. 22. Again, equ. (iv) is obtained by subtraction, so that its current sum is the difference between the two numbers above it. In this way a check can be kept on the calculation. This becomes extremely important when the coefficients are not integers, and the calculation is carried out using an electronic calculator.

The solution of the three equations can be interpreted graphically as shown in Fig. 1.2.

Each of the linear equations represents a plane in the three-dimensional coordinate system.

The three planes intersect at a single point P. The coordinates of P (in this case $x = 1$, $y = 3$, $z = 4$) are the values of the solution of the three equations.

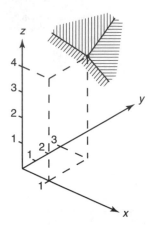

Fig. 1.2

EXERCISE 1.2

Solve the following equations:

1. $x + y + z = 6$
   $x - y + 2z = 5$
   $x + y + 3z = 12$

2. $x + 2y + 3z = -7$
   $3x - y + 4z = -4$
   $-2x + 2y - z = -2$

3. $4x + y - z = -5$
   $x - 3y + z = -2$
   $2x + 5y + 2z = 7$

4. $i_1 - i_2 + i_3 = 6$
   $2i_1 + 3i_2 + 2i_3 = 7$
   $3i_1 - 3i_2 - 2i_3 = 3$

5. $4u - v + 3w = -2$
   $3u + 2v - 2w = -14$
   $u + 3v - w = -12$

6. $a + 2b + 3c = 12$
   $2a + 3b + c = 1$
   $3a + b + 2c = -7$

7. In an electrical network of an out-of-balance Wheatstone Bridge, on applying Kirchoff's law, the following equations were obtained:

$$2i_1 - 3i_2 + 3i_3 = 0$$
$$10(i_1 - i_3) - (i_2 + i_3) - 2i_3 = 0$$
$$2i_1 + 9(i_1 - i_3) = 15$$

Solve the equations and find the currents i_1, i_2, i_3.

8. The curve $y = a + bx + cx^2$ passes through the points (0, 2), (2, 4) and (−1, 4). Find the values of a, b and c.

9. Forces in a framework are specified by the following three equations. Find the value of each force:

$$F_1 + F_2 - F_3 = 1$$
$$3F_1 + 4F_2 - 2F_3 = 3$$
$$-F_1 + F_2 + 4F_3 = 2$$

(c) Simultaneous equations in two unknowns involving one linear equation and one of degree 2

Degree 2 equations contain terms in x^2, y^2 and xy. The equations are solved by substituting for x or y from the linear equation into the degree 2 equation.
  Consider the following pair of equations.

$$2x + 4y = 9 \tag{i}$$
$$4x^2 + 16y^2 - 20x - 4y = -19 \tag{ii}$$

From equation (i)

$$x = \frac{9 - 4y}{2}$$

Substitute for x in (ii)

$$4 \times \left(\frac{9 - 4y}{2}\right)^2 + 16y^2 - 20 \times \left(\frac{9 - 4y}{2}\right) - 4y = -19$$
$$(9 - 4y)^2 + 16y^2 - 10(9 - 4y) - 4y = -19$$

which reduces to the quadratic equation

$$32y^2 - 36y + 10 = 0$$
$$(2y - 1)(8y - 5) = 0$$
$$y = 0.5 \quad \text{or} \quad y = 0.625$$

Substitute for y in equation (i)

$$x = 3.5 \quad \text{or} \quad x = 3.25$$

Therefore the solutions are (3.5, 0.5) and (3.25, 0.625)

Example 1.4   Solve the equations

$$4x - y = 6 \tag{i}$$
$$xy = 10 \tag{ii}$$

From equation (i) $y = 4x - 6$
Substituting for y into equation (ii)

$$x(4x - 6) = 10$$
$$4x^2 - 6x - 10 = 0$$
$$2x^2 - 3x - 5 = 0$$
$$(2x - 5)(x + 1) = 0$$
$$x = -1 \quad \text{or} \quad x = 2.5$$

Substitute for x into equation (ii)

$$y = -10 \quad \text{or} \quad y = 4$$

Therefore the solutions are $(-1, -10)$ and $(2.5, 4)$.

(d) Graphical interpretation of the solution of simultaneous equations of type c

In the above two examples equations of degree 2 are equations of curves, such as the circle, ellipse, hyperbola, as described in Chapter 9. Equations of degree 1 are equations of straight lines. The solutions of the two simultaneous equations in each case are the coordinates of the points of intersection of the straight line with the curve. This is illustrated in Fig 1.3(a).

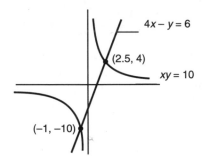

Fig. 1.3(a)

Example 1.5   Solve the two equations

$$3y + 4x = 11 \tag{i}$$

$$2x^2 + 3y^2 = 11 \tag{ii}$$

From equation (i) $x = (11 - 3y)/4$
Substitute for x into equation (ii)

$$\frac{2(11 - 3y)^2}{16} + 3y^2 = 11$$

which reduces to

$$33y^2 - 66y + 33 = 0$$
$$y^2 - 2y + 1 = 0$$
$$(y - 1)^2 = 0$$
$$y = 1$$

Substitute $y = 1$ in to equation (i): $3 + 4x = 11$

$$x = 2$$

Therefore the solutions are identical as $(2, 1)$

When the two solutions are identical the straight line is a tangent to the curve as shown in Fig 1.3(b).

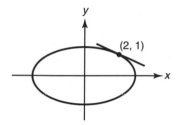

Fig. 1.3(b)

Example 1.6   Solve the equations

$$y = x + 5$$

$$x^2 + y^2 = 5$$

Substitute $y = x + 5$ into $x^2 + y^2 = 5$

$$x^2 + (x + 5)^2 = 5$$

$$x^2 + 5x + 10 = 0$$

Using the formula we obtain

$$x = \frac{-5 \mp \sqrt{25 - 40}}{2} = \frac{-5 \mp \sqrt{-15}}{2}$$

The solutions involve the square root of -15, which is an imaginary quantity. The graphical interpretation in this case is that the straight line does not intersect with the curve as shown in Fig. 1.3(c).

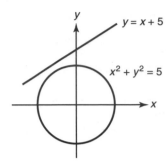

Fig. 1.3(c)

EXERCISE 1.3

Solve the following simultaneous equations.

1. $y = 3x - 2$
    $y^2 = 8x$

2. $6x^2 + 5y^2 = 11$
    $y = 2x + 1$

3. $x + y = 2$
    $x^2 + y^2 = 20$

4. $y = 2x + 5$
    $x^2 + y^2 = 5$

5. $4x - y = 4$
    $\dfrac{1}{x} + \dfrac{2}{y} = 1$

6. $x + 2y = 0$
    $x^2 + 3xy + 4y^2 = 18$

7. $4x^2 - 5y^2 = 65$
    $2x - 5y = 5$

8. $xy = 4$
    $x + y = 5$

1.4 Indices

An algebraic number such as $3.x.x.x.x$ is written as $3x^4$ where

    3 is the coefficient

    x is the base

    4 is the index

Definitions

An algebraic number such as $3x^4$ is called a **term** when it occurs in a more complicated expression.

An expression containing a sum of such terms is called a **polynomial**, e.g. $3x^4 - 2x + 4$.

The highest index in the polynomial containing a single variable is called the **degree** of the polynomial. In the above example the degree is 4.

Rules applying to indices

1. When numbers with the same base are multiplied, indices are added.

   $$x^m \times x^n = x^{m+n}$$

2. When numbers with the same base are divided, indices are subtracted.

   $$x^m \div x^n = x^{m-n}$$

3. When a number in index form is raised to another power, the indices are multiplied.

   $$(x^m)^n = x^{mn}$$

4. A number to an index 0 is 1.

   $$x^0 = 1$$

5. A number with a negative index can be converted into a number with a positive index by inversion.

   $$x^{-m} = \frac{1}{x^m} \quad \text{or} \quad \frac{1}{x^{-n}} = x^n$$

6. A fractional index is a root.

   $$x^{1/m} = \sqrt[m]{x}$$

Example 1.7 Simplify the following:.

(i) $\left(\dfrac{16}{81}\right)^{\frac{3}{4}}$ (ii) $(125)^{-\frac{2}{3}}$ (iii) $3x^{\frac{1}{2}} . 4x^{\frac{3}{2}} \div 2x^3$

(i) Take the fourth root first: $\left(\dfrac{16}{81}\right)^{\frac{3}{4}} = \left(\dfrac{2}{3}\right)^3 = \dfrac{8}{27}$

(ii) Convert to positive index: $(125)^{-\frac{2}{3}} = \left(\dfrac{1}{125}\right)^{\frac{2}{3}} = \left(\dfrac{1}{5}\right)^2 = \dfrac{1}{25}$

(iii) Multiply and divide coefficients,

Add and subtract indices: $\dfrac{3 \times 4}{2} x^{\frac{1}{2}+\frac{3}{2}-3} = 6x^{-1}$

The first example illustrates an important method of evaluating numbers; namely, **take the roots first**. In this way the numbers handled are smaller and more manageable.

EXERCISE 1.4

Simplify the following using the rules of indices:

1. $\left(\dfrac{8}{27}\right)^{-\frac{2}{3}}$

2. $\left(\dfrac{625}{16}\right)^{-\frac{1}{4}}$

3. $\left(\dfrac{32}{50}\right)^{\frac{1}{2}}$

4. $3x^2 . (3x)^2$

5. $2x^{-\frac{1}{4}} . (4x^{\frac{1}{2}})^{\frac{1}{2}}$

6. $(2x^{-\frac{1}{4}})^2 . 4x^{\frac{1}{2}}$

7. $(25y^2)^{\frac{1}{2}} . 5(y^2)^{\frac{1}{2}}$

8. $6z^{\frac{1}{2}} . 3z^{-\frac{3}{4}} \div 2z^{-\frac{1}{4}}$

9. $7x^0$

10. $(7x)^0$

11. $(3x - 4y)^0$

12. $t^{\frac{1}{2}} . t^{\frac{1}{3}} . t^{\frac{1}{10}}$

1.5   Logarithms

(a) The meaning of a logarithm

A number such as 64 can be written as 2^6. The index 6 is called the logarithm of 64 on a base 2, i.e.

$$6 = \log_2 64$$

This equation and

$$64 = 2^6$$

are equivalent equations. The same applies to the following pairs of equations

$$125 = 5^3 \quad \text{and} \quad 3 = \log_5 \ 125$$
$$121 = 11^2 \quad \text{and} \quad 2 = \log_{11} \ 121$$
$$1000 = 10^3 \quad \text{and} \quad 3 = \log_{10} \ 1000$$

EXERCISE 1.5

Change the following to equivalent form:

1. $27 = 3^3$

2. $81 = 9^2$

3. $100 = 10^2$

4. $2 = 4^{\frac{1}{2}}$

5. $N = n^t$

6. $5 = \log_2 32$

7. $1 = \log_6 6$ **8.** $3 = \log_7 343$

9. $\frac{1}{3} = \log_{27} 3$ **10.** $m = \log_x M$

(b) Common logarithms

Since $100 = 10^2$ and $1000 = 10^3$
a number such as 642 must have an index of value somewhere between 2 and 3. The index is actually 2.807 5,

$$642 = 10^{2.807\ 5}$$

The index 2.807 5 is the logarithm of 642 taken on the base of 10. In the same way $6\,420 = 10^{3.8075}$ so that the logarithm of $6\,420 = 3.8075$. Thus, when the number 642 is increased by a factor of 10, the logarithm increases by 1. It is seen that the logarithm is cyclic, so that the logarithm of 64 200 will be 4.807 5.

The logarithm of a number N on a base of 10 is written as log N.

1. To find the logarithm of a number on the base of 10 using the electronic calculator.

   The sequence of key operations on the calculator is

   enter $\boxed{642}$ $\rightarrow$ *press* $\boxed{\text{log}}$ *key* $\rightarrow$ *display* $\boxed{2.8075}$

   log $642 = 2.81$

2. To find a number given that its logarithm on the base 10 is known.

   If log $N = 1.395$, N can be determined using an inverse process on the calculator, as follows:

   enter $\boxed{1.395}$ $\rightarrow$ *press* $\boxed{\text{INV}}$ *key* $\rightarrow$ *press* $\boxed{\text{log}}$ *key* $\rightarrow$ *display* $\boxed{24.83}$

   $N = 24.83$

Example 1.8   (a) Find (i) log 12.46 (ii) log 0.00727
                (b) If (i) log $N = 2.06$ (ii) log $N = -2.772$
                find the value of N.

(a) Using the electronic calculator, the sequence is

(i)   *enter* $\boxed{12.46}$ *press* $\boxed{\text{log}}$ *display* $\boxed{1.0955}$: log 12.46   =   1.10

(ii) *enter* $\boxed{0.00727}$ *press* $\boxed{\text{log}}$ *display* $\boxed{-2.138}$:   log $0.00727 = -2.14$

(b) Using the the electronic calculator, the sequence is

(i)   *enter* $\boxed{2.06}$ *press* $\boxed{\text{INV}}$ $\boxed{\text{log}}$ *display* $\boxed{114.8}$ $N = 115$

(ii) *enter* $\boxed{2.772}$ *press* $\boxed{+/-}$ $\boxed{\text{INV}}$ $\boxed{\text{log}}$ *display* $\boxed{1.69 \ -03}$

                                         $N = 1.69 \times 10^{-3}$

(c) Napierian logarithms

Definition

Besides common logarithms to the base 10 there is another type of logarithm which is extremely important in science and engineering, called **natural** or **Napierian** logarithms. The base of these logarithms is the number e, where e is defined by the series

$$e = 1 + \frac{1}{1} + \frac{1}{1 \times 2} + \frac{1}{1 \times 2 \times 3} + \frac{1}{1 \times 2 \times 3 \times 4} + \cdots$$

$$= 1 + 1 + 0.5 + 0.166\,67 + 0.041\,67$$

The terms decrease sufficiently rapidly for the sum of the series to reach a limiting value as the number of terms becomes very large. The limiting value correct to three places of decimals is 2.718. A series such as this, whose sum reaches a limiting value as the number of terms extends to infinity, is called a **convergent series**. In other words the sum of the series converges to a limiting value.

The logarithm of a number on the base e is written as $\ln N$.

Note: Apart from $\log N$ and $\ln N$ all other bases must be recorded, e.g $\log_5 N$.

(i) To calculate the logarithm of a number on the base of e.

To determine $\ln 63.77$ the following sequence is carried out on the calculator:

enter $\boxed{63.77}$ *press* $\boxed{\ln}$ *display* $\boxed{4.155}$

$\ln 63.77 = 4.15$

(ii) To find a number, given that its logarithm on the base e is known.

Given $\ln N = 3.016$ to find N the sequence on the calculator is

enter $\boxed{3.016}$ *press* $\boxed{\text{INV}}$ *press* $\boxed{\ln}$ *display* $\boxed{20.409}$

$N = 20.4$

Example 1.9   (a) Find (i) $\ln 29.22$ (ii) $\ln = 0.0717$
(b) If $\ln N = 1.33$, (ii) $\ln = -3.159$, find N in each case

The sequence on the calculator is

(a) (i) *enter* $\boxed{29.22}$ *press* $\boxed{\ln}$ *display* $\boxed{3.37}$: $\ln 29.22 = 3.37$

(ii) *enter* $\boxed{0.0717}$ *press* $\boxed{\ln}$ *display* $\boxed{-2.64}$: $\ln 0.0717 = -2.64$

(b) (i) *enter* $\boxed{1.33}$ *press* $\boxed{\text{INV}}\boxed{\text{ln}}$ *display* $\boxed{3.78}$ $: N = 3.78$

   (ii) *enter* $\boxed{3.159}$ *press* $\boxed{+/-}\boxed{\text{INV}}\boxed{\text{ln}}$ *display* $\boxed{0.0425}: N = 0.0425$

EXERCISE 1.6

1. Using the electronic calculator evaluate the following correct to 4 significant figures:

   (a) log 74.31   (b) ln 8.871   (c) ln 1.632   (d) ln 476.3

   (e) ln 0.005721   (f) ln 0.8276   (g) log 0.0003712

2. Find the value of N in each of the following:

   (a) log $N = 3.216$ (b) ln $N = 1.6172$ (c) ln $N = 0.7135$

   (d) ln $N = -2.144$ (e) ln $N = -1.3636$ (f) log $N = -3.0017$

(d) Simple logarithmic equations

Examples of simple logarithmic equations are given in section (b) above. Thus we found that if

$$\log N = 3.4, \qquad N = 2511.$$

The value of N is the solution of the logarithmic equation. This can be extended to an equation such as

$$2 \log(x + 3) = 2.962$$

First we have to divide throughout by 2,

$$\log(x + 3) = 1.481$$

To find the value of $(x + 3)$ from the calculator

     enter $\boxed{1.481}$ *press* $\boxed{\text{INV}}\boxed{\text{log}}$ *display* $\boxed{30.27}$

$$(x + 3) = 30.27$$

$$x = 27.27$$

Example 1.10 In the equation $\log(3x + 7) = \log(x^2 + 3)$ solve for x.

Since both sides have the same base the equation can be solved by inverting both sides:

$$3x + 7 = x^2 + 3$$
$$x^2 - 3x - 4 = 0$$
$$(x - 4)(x + 1) = 0$$
$$x = 4 \quad \text{or} \quad z = -1$$

The same result would be obtained if the logarithms were on any other base. The base, however, must be the same on either side of the equation.

EXERCISE 1.7

Solve the following equations.

1. $3 \log(x - 2) = 4.712\,4$
2. $2 \log(x^2 + 4) = 2.691\,7$
3. $\log(t^2 - 9) = -1.316\,4$
4. $4 \log(3t - 6) = 1.792\,4$
5. $\log(2x - 5) = \log(3x - 9)$
6. $\log(x^2 - 4) = \log 21$
7. $\log(x^2 - 1) = \log(3x - 3)$
8. $5 \log(3x - 2) = 17.61$
9. $3 \ln(5x + 3) = 7.921$
10. $4 \log(5x + 1) + 6.1 = 13.51$

1.6   The laws of logarithms

Since the logarithm of any number is the index when the number is written on a particular base, then the laws of logarithms must be the same as the laws of indices. The laws are derived for logarithms on the base 10, but are applicable for any base.

LAW I

$$\log M . N = \log M + \log N$$

Let $\log M = s$ (i)

$\log N = t$ (ii)

The equivalent forms are

$$M = 10^s \qquad \text{(iii)}$$
$$N = 10^t \qquad \text{(iv)}$$
$$M \times N = 10^s \times 10^t = 10^{s+t} \qquad \text{(v)}$$

Rewriting (v) in log form

$$\log M . N = s + t$$

Using (i) and (ii) for s and t

$$\boxed{\log M . N = \log M + \log N}$$

LAW II

$$\log \frac{M}{N} = \log M - \log N$$

Using equations (i), (ii), (iii) and (iv) above and dividing (iii) by (iv)

$$\frac{M}{N} = 10^s \div 10^t = 10^{s-t} \qquad \text{(vi)}$$

Rewriting (vi) in log form

$$\log \frac{M}{N} = s - t$$

Using (i) and (ii) for s and t

$$\boxed{\log \frac{M}{N} = \log M - \log N}$$

LAW III

$$\log M^n = n \log M$$

From equations (i) and (iii) above, we have

$$M^n = (10^s)^n = 10^{ns} \qquad \text{(vii)}$$

Rewriting (vii) in index form

$$\log M^n = ns$$

Substituting for s from (i)

$$\boxed{\log M^n = n \log M}$$

These laws are used in Examples 1.11 and 1.12.

Example 1.11 Simplifying, show that

$$\frac{\log_x 64 + \log_x 4 - \log_x 8}{\log_x 1024} = \frac{1}{2}$$

On the left-hand side, each of the numbers is expressed as a power of 2, that is

$$\frac{\log_x 2^6 + \log_x 2^2 - \log_x 2^3}{\log_x 2^{10}}$$

$$= \frac{6 \log_x 2 + 2 \log_x 2 - 3 \log_x 2}{10 \log_x 2}, \text{ using law III}$$

$$= \frac{5 \log_x 2}{10 \log_x 2}$$

$$= \tfrac{1}{2}$$

Example 1.12   Solve the equation

$$\log(x+4) + \log(x-3) = 2 \log(x-1)$$

Using law I and law III the equation becomes

$$\log(x+4)(x-3) = \log(x-1)^2$$

Invert both sides

$$(x+4)(x-3) = (x-1)^2$$
$$x^2 + x - 12 = x^2 - 2x + 1$$
$$3x = 13$$
$$x = 13/3$$

Properties of logarithms

1. $\log_x 1 = 0$

   Let $\log_x 1 = s$ (i)

   The equivalent equation is $x^s = 1$

   For this to be true $s = 0$

   Hence in (i) $\log_x 1 = 0$

2. $\log_x x = 1$

   Let $\log_x x = t$ (ii)

   The equivalent equation is $x^t = x$

   This means that $t = 1$

   Hence in equation (ii) $\log_x x = 1$

3. $\log_a b = \dfrac{1}{\log_b a}$

Let $\qquad\qquad\qquad\qquad \log_b a = r \qquad\qquad\qquad$ (iii)

The equivalent equation is $\qquad a = b^r$

Take the rth root $\qquad\qquad a^{1/r} = b$

In log form this equation

becomes $\qquad\qquad\qquad \log_a b = 1/r$

which from (iii) reduces to $\qquad \log_a b = 1/\log_b a$

4. $e^{\ln x} = x$

Let $\qquad\qquad\qquad\qquad\qquad N = e^{\ln x}$

Transpose into logarithm form:

$$\ln N = \ln x$$
$$N = x$$

Therefore $\qquad\qquad\qquad e^{\ln x} = x$

5. $\ln e^x = x$

Let $\qquad\qquad\qquad\qquad\qquad N = \ln e^x$

Transpose into index form:

$$e^N = e^x$$
$$N = x$$

Therefore $\qquad\qquad\qquad \ln e^x = x$

EXERCISE 1.8

Solve the following equations:

1. $\log 3x^4 - \log x^2 = \log 81 - \log x$
2. $\log(2x - 1) + \log(x + 2) = 2\log(x - 1)$
3. $3\log 2 - \log(x - 1) = \log(x - 3)$

Simplify the following:

4. $\log 81 + \log 3 - \log 27$
5. $\log 625 - 5\log 5 + \log 25$
6. $\dfrac{\frac{1}{2}\log 81 + \frac{1}{3}\log 27}{\log 9}$

1.7   Changing the base of a logarithm

It is often necessary to change the base of a logarithm, that is express $\log N$ in terms of $\ln N$. For example, in later chapters it will be seen that differentiation and integration can only be carried out on the ln function, so that any other logarithm function which needs differentiation or integration must be converted into the ln function first.

To express log N in terms of ln N, let

ln $N = s$

Therefore

$N = e^s$

Take logarithms on base 10 on both sides:

$$\log N = \log e^s$$
$$= s \log e \quad \text{(using law III)}$$
$$= \ln N \times \log e$$
$$= 0.4343 \ln N \tag{1.2}$$

1.8   Solution of indicial equations

There are two types of equations to consider as shown below.

1.   *The variable is in the index*, as in

$$(3.4)^x = 16.1$$

Take logs on base 10 on both sides of the equation:

$$\log(3.4)^x = \log 16.1$$
$$x \log 3.4 = \log 16.1$$
$$x = \frac{\log 16.1}{\log 3.4}$$
$$= \frac{1.206\,8}{0.531\,5}$$
$$= 2.271$$

2.   *The variable is in the base*, as in

$$x^{4.1} = 96$$

Take logs to the base 10 on both sides of the equation:

$$\log x^{4.1} = \log 96$$
$$4.1 \log x = \log 96$$
$$\log x = \frac{\log 96}{4.1}$$
$$= \frac{1.982\,3}{4.1}$$
$$= 0.483\,4$$
$$x = 3.004$$

Example 1.13 Solve the equation $4^{2x} = 7^{x-1}$

Take logs to base 10 on both sides of the equation:

$$\log 4^{2x} = \log 7^{x-1}$$
$$2x \log 4 = (x-1) \log 7$$
$$2x \times 0.602\,1 = (x-1)0.845\,1$$
$$1.204\,2x - 0.845\,1x = -0.845\,1$$
$$0.359\,1x = -0.845\,1$$
$$x = -\frac{0.845\,1}{0.359\,1}$$
$$= -2.354$$

EXERCISE 1.9

Solve the following equations correct to three decimal places.

1. $4^x = 27$ 2. $1.96^{2x} = 13.14$
3. $14^x = 0.017\,62$ 4. $0.057\,13^x = 1.816$
5. $y^7 = 91$ 6. $z^{0.176\,3} = 14$
7. $3(7)^{2x} = 4(6)^{x-3}$ 8. $(3)^{2x}(2)^{3y} = 17$
                                    $(4)^x(5)^y = 37$

1.9   Transposition of equations containing logarithms and indices

(a) The variable in the log function is required as the subject

The transposition can be broken down into two steps:

1. Rearrange the equation so that the log function appears alone on one side of the equation.
2. Rewrite the equation in its equivalent form.

Example 1.14 Given that $V = V_0 \ln(M + N)$, express N in terms of the other quantities.

Step 1. Rearrange the equation with the log function on its own.

$$\frac{V}{V_0} = \ln(M + N)$$

Step 2. Rewrite in equivalent form.

$$(M + N) = e^{V/V_0}$$
$$N = e^{V/V_0} - M$$

(b) The variable in the index is required as the subject

Again the transposition can be broken into two steps:

1. Rearrange the equation so that the term containing the index is on its own on one side of the equation.
2. Rewrite the equation in the equivalent form containing the log function.

Example 1.15   Given that $I = I_0 e^{-\lambda t}$, express t in terms of the other variables.

Step 1. Rearrange the equation with the term containing the index on its own.

$$\frac{I}{I_0} = e^{-\lambda t}$$

Step 2. Rewrite in equivalent form:

$$-\lambda t = \ln \frac{I}{I_0}$$
$$t = -\frac{1}{\lambda} \ln(I/I_0)$$

EXERCISE 1.10

1. Given that the tensions in a belt drive are given by $T_1 = T_2 e^{\mu r \theta}$, express θ in terms of the other variables.
2. Given that $L = 2k[1 + \ln(b/a)]$, express b in terms of the other quantities.
3. In a study of the growth of population Malthus used the equation

$$N = N_0 e^{Rt}$$

where N_0 is the population at some starting point and N is the population at some time t later.

Find the expression for t when the population has (a) doubled, (b) increased by a factor of 10, (c) increased by a factor of 100, from a given starting point.

4.   The flow of current from the filament of a vacuum tube is

$$i = kT^2 . e^{(W - E)/kT}$$

Find the equation for E.

5.   The electric potential near a long cylinder due to charges on its surface is given by

$$V = A - \frac{2q}{K} . \ln r$$

Make r the subject of the equation.

6.   The heat conducted radially in a cylindrical tube with inner and outer radii a, b is given by

$$Q = \frac{2 . \pi \kappa \theta}{\ln(b/a)}$$

Find the expression for b in terms of the other quantities.

7.   The work done on a gas in an isothermal expansion from v_1 to v_2 is

$$W = RT \ln \left(\frac{v_2}{v_1} \right)$$

Express v_2 in terms of the other quantities.

8.   The radioactive decay of radium A is given by

$$N_t = N_0 e^{-\lambda t}$$

Find the expression for t. Hence find t when $N_t = 0.1 N_0$, given that $\lambda = 3.1$.

9.   Planck's law of radiation is given by

$$E_\lambda = \frac{a}{\lambda^5 . (e^{b/\lambda T} - 1)}$$

Find the expression for T.

10.   The deflection x in a spring, in a damped system is given by

$$x = 8(1 - e^{-t/T})$$

Rearrange to find the formula for t.

1.10   Partial fractions

Before dealing with partial fractions it is necessary to consider the following points.

1. In an identity such as

$$ax^2 + bx + c = -3x^2 + 4$$

the left hand is identical with the right, so that the respective coefficients and constants on both sides are equal, that is

$$a = -3$$
$$b = 0$$
$$c = 4$$

2. Fractions are added in algebra by first finding the LCM of the denominator:

$$\frac{1}{(x+2)} + \frac{1}{(x+3)} = \frac{(x+3) + (x+2)}{(x+2)(x+3)}$$

$$= \frac{2x+5}{(x+2)(x+3)}$$

3. The reverse process of (2), that is splitting up a single fraction into a sum of two or more fractions, is called the method of **partial fractions**. Three types of problems are considered below.

(a) Type 1: A fraction containing two simple binomial factors in the denominator, each of degree 1

The procedure for obtaining partial fractions is shown in Example 1.16.

Example 1.16   Convert $\dfrac{2x+5}{(x+2)(x+3)}$ into partial fractions.

(i)   Let

$$\frac{2x+5}{(x+2)(x+3)} = \frac{A}{(x+2)} + \frac{B}{(x+3)}$$

$$\frac{A(x+3) + B(x+2)}{(x+2)(x+3)}$$

by adding as in (2) above.

(ii)   The numerators on both sides of the equation are now identical

$$(2x+5) = A(x+3) + B(x+2)$$

Let $x = -3$: $-1 = B \times -1$

$B = 1$

Let $x = -2$: $1 = A$

that is $A = 1$ $B = 1$

Hence $\dfrac{2x + 5}{(x + 2)(x + 3)} = \dfrac{1}{(x + 2)} + \dfrac{1}{(x + 3)}$

In this type of fraction each of the partial fractions contains one of the factors in the denominator. Step (i) is the starting point for this type, which can be extended to more than two factors.

(b) Type 2: A fraction containing a power of a simple binomial in the denominator

The method of obtaining the partial fractions for this type is shown in Example 1.17

Example 1.17   Convert $\dfrac{x + 2}{(x - 2)^2}$ into partial fractions.

(i)   Let

$$\frac{x + 2}{(x - 2)^2} \equiv \frac{A}{(x - 2)} + \frac{B}{(x - 2)^2}$$

$$\equiv \frac{(x - 2)A + B}{(x - 2)^2}$$

by adding together as in (2).

(ii)   The numerators on both sides are now identical

$$x + 2 \equiv (x - 2)A + B$$

Let $x = 2$: $4 = B$

Let $x = 0$: $2 = -2A + B = -2A + 4$

$A = 1$, $B = 4$

$$\frac{x + 2}{(x - 2)^2} = \frac{1}{(x - 2)} + \frac{4}{(x - 2)^2}$$

(c) Type 3: A fraction containing a simple binomial and a quadratic in the denominator

The partial fractions contain the factors that appear in the original fraction. In the case of the partial fraction containing the quadratic factor the numerator must always be a polynomial of degree one less than the denominator. The procedure is shown in Example 1.18.

Example 1.18 Convert $\dfrac{x-4}{(x-1)(x^2+x+1)}$ into partial fractions.

(i) Let

$$\frac{x-4}{(x-1)(x^2+x+1)} = \frac{A}{(x-1)} + \frac{Bx+C}{x^2+x+1}$$

$$= \frac{A(x^2+x+1) + (Bx+C)(x-1)}{(x-1)(x^2+x+1)}$$

(ii) The numerators on both sides are identical:

$$x - 4 \equiv A(x^2+x+1) + (Bx+C)(x-1)$$

$$\begin{array}{llll} \text{Let} & x=1, & -3 = 3A, & A=-1 \\ & x=0, & -4 = A-C, & C=3 \\ & x=2, & -2 = 7A + (2B+C) = -7 + 2B + 3, & B=1 \end{array}$$

$$A = -1, \qquad B = 1, \qquad C = 3$$

Therefore

$$\frac{x-4}{(x-1)(x^2+x+1)} = -\frac{1}{(x-1)} + \frac{x+3}{x^2+x+1}$$

Note: If a fraction is such that the numerator is of higher or equal degree to the denominator, the fraction must first be divided out before partial fractions can be obtained. For example,

$$\frac{x^3 + 3x^2 + 7x + 4}{x^2 + 3x + 2} = x + \frac{5x+4}{x^2+3x+2}$$

$$= x + \frac{5x+4}{(x+1)(x+2)}$$

The second fraction can now be expressed in partial fractions. It should be pointed out that partial fractions can be obtained only if the denominator contains factors.

EXERCISE 1.11

Express the following in partial fractions

1. $\dfrac{3x+2}{(x-4)(x+2)}$

2. $\dfrac{1}{(x+2)(2x+5)}$

3. $\dfrac{4x}{(x-3)(3x+1)}$

4. $\dfrac{2x+3}{(x+1)(5x-1)}$

5. $\dfrac{x-3}{(x+1)^2}$

6. $\dfrac{x}{(2x-1)^2}$

7. $\dfrac{3x-2}{(x-1)(x^2+2x-1)}$

8. $\dfrac{4x-3}{(x+2)(3x^2-2x+1)}$

9. $\dfrac{x^2-x+4}{x^2-x-2}$

10. $\dfrac{4x^3-3x^2+2x-5}{(x+2)(x-3)}$

MISCELLANEOUS EXERCISES 1

1. (a) Solve the simultaneous equations

$$\frac{1}{x}+\frac{1}{y}=2;\qquad x+7y=8.$$

(b) Solve the equation:

$$2^{2x}-9(2^x)+8=0$$

2. Solve the equation $3^{x^2+6}=9^{2.5x}$

3. (a) The expression ax^2+bx+c has a value 4 when $x=1.13$, when $x=2$ and 26 when $x=3$. Find the numerical values of a, b and c.

(b) Solve the following equation for x:

$$5^{(3x+1)}=25^{(x+2)}$$

4. (a) Given that $T=A+(B-A)\mathrm{e}^{-kt}$, where A, B and k are constants, express t in terms of the other symbols.

(b) The sides of a triangle are a, b and c metres long. Given that

$$a+\ b-\ c=\ \ 8$$
$$3a-2b+6c=124$$
$$5a+3b+2c=148$$

find the values of a, b and c, and hence prove that one of the angles is a right angle.

5. Solve the following equations, giving the results, where appropriate, correct to three significant figures:
   (a) $\ln 3x = 2 \log 3$
   (b) $25^x - 5^{x+2} + 100 = 0$
   (c) $5^{2x} = 4(3^{x-1})$

6. Solve the following equations for x and y:

   $\ln y^3 + \ln x^2 = 1.163\,2$

   $\ln y^2 + \ln x = 0.47$

7. (a) Evaluate $\ln \dfrac{7.6}{3.3}$
   (b) Given $\ln(x + 7) = 2.995\,7$, find x.

8. Express in simpler form using positive indices only:

   (a) $\dfrac{(2a^{-3}b)^{-2}}{(3a^{-2}b^{-4})^3}$ (b) $\left(\dfrac{216a^6b^{-5}}{b^4c^{\frac{1}{4}}}\right)^{-\frac{1}{3}}$

9. Express $\dfrac{x - 4}{(x - 1)(x - 2)(x - 3)}$ in partial fractions.

10. Calculate the value of x in each of the equations
    (a) $x = 13.74^{-0.2}$ (b) $3.555^x = 17.8$

Chapter 2

Computation

2.1   Introduction

Computation, as its name suggests, involves numerical calculations.

Numerical calculations can be in error due to ordinary arithmetical mistakes and/or due to the use of approximate values in the calculation.

Arithmetical mistakes can be largely eliminated by a systematic approach to numerical work and by carrying out rough checks of the arithmetical results wherever possible.

Approximate quantities are frequently used in computation. Scientific data such as length, mass, time, etc., can only be as accurate as the measuring equipment used and these quantities are therefore usually in error to some degree. Also, the results of arithmetic operations on some numbers can only be expressed as approximate values. For example, 3.333 and 1.414 are approximate values of $\frac{10}{3}$ and $\sqrt{2}$ respectively, since only the first four figures are quoted in the answers, the remaining figures being ignored.

2.2   Rounding off numbers

Numbers that have been terminated after a predetermined number of digits are approximate.

The terminating process may be carried out by (i) a 'chopping' process or (ii) a 'rounding off' process. The chopping process does not take account of the size of the digits being removed. For example 34.9 becomes 34 when 'chopped', even though it is closer to 35. 'Rounding off' takes into account the size of the digits being removed. For example, 34.9 when rounded to 2 figures becomes 35. 'Rounding off' reduces the errors caused by terminating the number. The rule operating for rounding off is given below.

Apart from those zeros which must be used in a number to locate the decimal point, the more digits retained in a number after rounding off the more accurately that number is expressed.

Rule.   (i) If the digit to be removed is greater than 5, then the immediately preceding digit is increased by 1.

(ii) If the digit to be removed is equal to 5, then the immediately preceding digit is increased by 1 only if it is odd.

For example:

49.68	rounds to 49.7	correct to 1 decimal place
35.2	rounds to 35	correct to the nearest unit
397 500	rounds to 398 000	correct to the nearest thousand
0.0192	rounds to 0.020	correct to 3 decimal places
845	rounds to 840	correct to the nearest ten
63.7	rounds to 64	correct to 2 figures

2.3   Significant figures

It is seen in the above examples that the accuracy of a rounded off number can be expressed in different ways. One of the most commonly used ways is to count the number of digits in the number. For example, 502.6 has 4 digits and is said to be accurate to 4 significant figures.

Other examples are:

36.396	is correct to 5 significant figures
30.90 mm	is correct to 4 significant figures
0.000 26 $(= 2.6 \times 10^{-4})$	is correct to 2 significant figures
0.010 2 $(= 1.02 \times 10^{-2})$	is correct to 3 significant figures

It is not always possible to decide from the number itself how accurate the number is, unless further information is available. For example, the population of a town may be given as 38 400. This may be accurate to 3, 4, or 5 significant figures.
The actual population could be

between 38 350 and 38 450,	i.e. 38 400 (3 significant figures)
between 38 395 and 38 405,	i.e. 38 400 (4 significant figures)
exactly 38 400,	i.e. 38 400 (5 significant figures)

It will be noticed that significant figures are determined irrespective of the position of the decimal point.
If a quantity is quoted say to 3 significant figures such as a length of 18.2 mm, it means that the actual length lies between 18.15 mm and 18.25 mm.

2.4   Computational accuracy

It is misleading for an answer to a calculation to be quoted with more significant figures than in the original data. This would suggest that the answer is more accurate than it really is.

In fact, as a general rule, an answer should contain the same number of significant figures as the least accurate data.

Even so, too much reliance must not be placed upon the accuracy of the last significant figure in the final answer. However, at intermediate stages during a calculation two extra significant figures should be retained as guarding figures. If this is not done, continuous rounding off can affect the last significant digit in the answer.

Example 2.1   A rectangular steel plate has the dimensions 82 mm × 10.1 mm. Calculate the area of the plate.

Using the dimensions as they are gives the area of the plate as $10.1 \times 82 = 828.2 \text{ mm}^2$.

But 10.1 mm represents a length between 10.05 mm and 10.15 mm and 82 mm represents a length between 81.5 mm and 82.5 mm.

$$\text{Minimum possible area} = 10.05 \times 81.5$$
$$= 819.075 \text{ mm}^2$$
$$\text{Maximum possible area} = 10.15 \times 82.5$$
$$= 837.375 \text{ mm}^2$$

Since the difference between the possible maximum and minimum areas is 18.3 mm², it would be absurd to quote the area of the plate as 828.2 mm², because this suggests that the area is correct to four significant figures. The original dimensions were given to 2 and 3 significant figures accuracy. It is therefore sufficient to quote the area correct to two significant figures.

$$\text{Area of plate} = 830 \text{ mm}^2$$

Note: As stated in the text, this answer of 830 mm² suggests that the true area lies between 825 mm² and 835 mm². In fact the area could be as low as 819 mm² or as high as 837 mm². Care must therefore be exercised when interpreting the accuracy of the last significant figure in a result.

2.5   Definition of error

The value of the error involved in using an approximation may be expressed in two ways:

1. Absolute error $= |$ approximate value $-$ correct value $|$

2. Relative error $= \dfrac{\text{absolute error}}{\text{correct value}} \times 100\%$

The modulus is used in (i) because the error may be positive or negative. The relative error is a useful indicator of error because it is important to compare the size of the error to the correct value. For example, an error of 1 mm in a measurement of 20 mm represents a relative error of $\frac{1}{20} \times 100 = 5\%$ which is large. On the other hand the same error of 1 mm in a measurement of 1 m represents a relative error of 0.05%, which is small.

Example 2.2   In Example 2.1 calculate the maximum possible relative errors in

(i)   the length of the plate;
(ii)   the width of the plate;
(iii)   the area of the plate.

Since the correct values of length, width and area are unknown the nominal values of 82 mm, 10.1 mm and 828.2 mm^2 are used.

(i)   Maximum relative error in length $= \dfrac{|81.5 - 82|}{82} \times 100$

$= 0.6098\%$

(ii)   Maximum relative error in width $= \dfrac{|10.05 - 10.1|}{10.1} \times 100$

$= 0.4950\%$

(iii)   Maximum relative error in area $= \dfrac{|819.075 - 828.2|}{828.2} \times 100$

$= 1.102\%$

$\text{or} = \dfrac{|837.375 - 828.2|}{828.2} \times 100$

$= 1.108\%$

There are two possible results to part (iii) but since there is very little difference between them either figure may be used.

Although the range of 819 mm^2 to 837 mm^2 for the value of the area suggests that the area cannot be calculated very accurately it is seen from (iii) that the maximum possible relative error in the answer is only 1%.

EXERCISE 2.1

1.   Round off each of the following numbers:
   (a)   53.5   to the nearest unit
   (b)   4.359 8   to three decimal places

(c) 148.551   to the nearest tenth
(d)   0.037 86 to the nearest thousandth
(e) 631 964   to the nearest hundred.
2. Express each of the following numbers correct to three significant
    figures:
    292 467, 0.312 6, 0.009 158, 0.000 5012
3. Use the equation $F = ma$ to find force F acting on a mass m to give the
    mass an acceleration a if $m = 10.5$ kg and $a = 2.673$ m/s^2. What is the
    maximum possible error in this answer as a result of rounding off in m
    and a?
4. The diameter of a sphere is measured as 154 mm to the nearest mm.
    What is the volume and surface area of this sphere correct to three
    significant figures? How inaccurate could these answers be?
5. If 4643 is rounded off to two significant figures what is (a) the absolute
    error, (b) the relative error?
6. Find the possible minimum and maximum values of $(5.62 - 3.41)/7.84$
    if all the numbers have been rounded off to three significant figures.
    Calculate also the possible absolute and relative errors.
7. Given that $A = 54°$ and $B = 28°$ where both angles have been
    measured to two significant figures find the range of values possible
    for each of the following: (a) $\sin(A - B)$, (b) $\cos 2A$, (c) $\sin A + \sin B$.
    Find the possible relative error in each case.
8. Find the maximum possible errors in the positive root of the equation
    $2.2x^2 - 1.6x - 3.1 = 0$ if all the numbers have been rounded to two
    significant figures.
9. The impedance of an electric circuit is given by

$$Z = \sqrt{\{R^2 + (X_L - X_C)^2\}}$$

If R, X_L and X_C lie within the ranges 996–1032 Ω, 1238–1321 Ω and
1724–797 Ω respectively, determine the range within which Z lies.
10. A rectangular plate has a length of 260 ± 1.2 mm, a breadth of
    120 ± 0.8 mm. Find the greatest and least possible values for the area
    of the plate, correct to the nearest mm^2.
    Calculate the percentage relative error in the answer, correct to two
    significant figures.

2.6   Error prediction

It is essential to be able to predict the effect of using approximate values on
the final result of a calculation. The more complicated the calculation the
more essential it becomes to predict the error range that the final answer
must lie within. It is possible to carry out the calculation using minimum
and maximum values of all the variables, and thus find the maximum and

minimum possible values of the result, as in Example 2.1. This can be tedious. The following rules may be used to predict the accuracy of the final answer.

Rule 1.   If two quantities are to be added or subtracted the maximum possible absolute error in the answer is the sum of the two individual absolute errors.

Rule 2.   If two numbers are to be multiplied or divided the maximum possible relative error in the answer is the algebraic sum of the two individual relative errors.

Example 2.3   The voltages measured across two resistors in series are 8.6 V and 12.3 V, measured to the nearest 0.1 V. What is the maximum possible absolute error for the total voltage across both resistors?

1st voltage: maximum possible absolute error

$$= |8.55 - 8.6| = 0.05 \text{ V}$$

2nd voltage: maximum possible absolute error

$$= |12.25 - 12.3| = 0.05 \text{ V}$$

By rule 1, for the combined voltage:

$$\text{maximum possible absolute error} = 0.05 + 0.05$$
$$= 0.1 \text{ V}$$

Total voltage $= 20.9 \pm 0.1 \text{ V}$

Example 2.4   The speed of an object changes from 26.47 to 34.08 m/s. Find the maximum possible absolute error for the change of speed.

1st speed: maximum possible absolute error

$$= |26.465 - 26.47| = 0.005 \text{ m/s}$$

2nd speed: maximum possible absolute error $= |34.075 - 34.08|$
$$= 0.005 \text{ m/s}$$

By rule 1: maximum possible absolute error for change of speed

$$= 0.005 + 0.005$$
$$= 0.01 \text{ m/s}$$

Change of speed $= 7.61 \pm 0.01 \text{ m/s}$

Example 2.5 Use rule 2 in Example 2.2 to verify the maximum possible relative error in in the calculation of the area.

Since the length and width are multiplied together, using rule 2:

Maximum possible relative error in area

 = maximum possible relative error in length

      + maximum possible relative error in width

 $= 0.6098 + 0.4950\%$

 $= 1.105\%$

This result agrees with (iii) in Example 2.2, showing that rule 2 can be used to predict maximum possible relative error in a final result.

Example 2.6 The torque T in a round bar is given by the formula

$$T = \frac{\pi G D^4 \theta}{32 L}$$

Given that the maximum relative errors in the measurements are $G = \pm 1\%$; $D = \pm 2\%$; $\theta = 0.5\%$; $L = \pm 0.8\%$, find the maximum possible relative error in T.

Using the above rules, maximum possible relative error in
$T = 1 + 4 \times 2 + 0.5 + 0.8 = 10.3\%$ (high)
or $= -1 - 4 \times 2 - 0 - 0.8 = -9.8\%$ (low)

Note:

(i) $D^4 = D \times D \times D \times D$. Therefore, using rule 2, the error in $D^4 = 2 + 2 + 2 + 2 = 8\%$.

(ii)   Since L is in the denominator -0.8% error in L will have a $+0.8\%$ error in T and vice versa.

(iii) 10.3% is a relative large error, most of which is contributed by D^4. It is important therefore that errors in variables with higher powers are made as small as possible. In this example it is imperative that D is measured as accurately as possible.

(iv)   For the low T value the relative error θ is taken as 0%, its lowest value.

The algebraic addition of relative errors by Rule 2 is in agreement with the application of the binomial theorem in Examples 10.19 and 10.20.

EXERCISE 2.2

Use the rules given in Section 2.6 to calculate errors in the following questions.

1. Find the absolute and relative errors in calculating the third angle of a triangle if the other two angles are $54°$ and $38°$, correct to two significant figures.

2. In triangle ABC, $a = 15\,cm$, $b = 20\,cm$ and angle $C = 1.32$ radians, where all the numbers have been rounded. Given that area $= \frac{1}{2}ab\,\sin C$ find the maximum possible relative error and the maximum possible value of the area.

3. Given that $x = 1.76$ correct to three significant figures find the maximum possible relative error in

    (a) $4x$ (b) $\sqrt{x}$ (c) $5x^2$

4. Find the maximum possible relative error in the value of $(7.86 - 3.10)/25.3$ if all the numbers have been rounded to three significant figures. What is the possible range of values for the answer?

5. If $x = 21.2$, $y = 14.31$ and $z = 6.2$ where all the numbers have been rounded, find
    (a) the possible absolute errors in $x - z$, $x + z$, $x^2 - y^2$
    (b) the possible relative errors in xy/z, x^2y^2/z.

6. For two resistors R_1 and R_2 in parallel, the equivalent resistance is

    $$R = \frac{R_1 R_2}{R_1 + R_2}$$

    If $R_1 = 40\,\Omega$ with a possible relative error of $\pm 1.5\%$ and $R_2 = 70\,\Omega$ with a possible relative error of $\pm 2\%$, what is the maximum possible error in the value of R?

7. The time of oscillation of a simple pendulum is given by

    $$T = 2\pi\sqrt{\frac{L}{g}}$$

    If the maximum possible error in the value of L is $\pm 0.8\%$ what is the maximum possible error in the value of T?

8. Find the maximum possible relative error in the calculated volume of a cylinder given that the maximum possible errors in the radius and height are $\pm 2\%$ and 1% respectively.

9. The resonant frequency f in a circuit is given by

    $$f = \frac{1}{2\pi\sqrt{LC}}$$

If the possible relative errors in L and C are 2% and -0.5% what is the possible relative error in f? If $L = 160$ mH and $C = 30$ μF calculate the possible absolute error in f (L and C must be in H and F).

10. The deflection of a cantilever is given by

$$d = \frac{FL^3}{3EI}$$

Given that the maximum possible relative errors are $F = -1.2\%$, $L = \pm 0.8\%$, $I = \pm 1.4\%$ find the possible relative error in d.

11. Given that $a = 5.234, b = 3.671, c = 8.436$ and $d = 6.372$, with all numbers rounded to four significant figures find the possible absolute and relative error in the values of

(a) $\dfrac{a+b}{cd}$ (b) $\dfrac{abc}{c-d}$

2.7 Standard form

Very large or very small numbers cannot be entered into a calculator if they contain more digits than are available in the display. For example neither 6 817 000 000 nor 0.000 000 000 025 can be displayed on an 8-digit calculator. Such numbers can be rearranged in standard form, in which there is only one digit before the decimal point. Therefore

$$6\,817\,000\,000 = 6.187 \times 10^9$$
$$0.000\,000\,000\,025 = 2.5 \times 10^{-11}$$

Standard form numbers can be entered into a calculator in the following sequence:

$$\boxed{6.187}\ \boxed{\text{EXP}}\ \boxed{9}\ \boxed{=}$$

$$\boxed{2.5}\ \boxed{\text{EXP}}\ \boxed{11}\ \ \ \boxed{+/-}\ \boxed{=}$$

EXERCISE 2.3

1. Evaluate the following, correct to three significant figures

(a) $\sqrt{\left(\dfrac{31.29 \times 1.294}{10.29}\right)}$

(b) $\dfrac{17.19 \times 0.095\,4}{0.369\,4}$

(c) $\dfrac{11.64(22.95 - 20.37)}{0.002\,951 \times 0.951}$

(d) $\dfrac{51.34 \times 0.915\,4^3}{\sqrt{29.26} + 243.6}$

2. Find the value of

(a) $\sqrt[4]{0.003\,179)}$

(b) $0.233\,6^{\frac{3}{4}}$

(c) $48.2^{1.6}$

(d) $36.5^{-0.52}$

(e) $0.006\,27^{2.9}$

(f) $0.1369^{-1.35}$

(g) $\dfrac{16.29}{40.29^{1.2}}$

(h) $0.295^{0.64} \times 31.38^{0.15}$

(j) $21.92^{-0.85} \times \log e$

(k) $e^{-1.6}$

3. The pressure p and volume v of a gas are connected by the formula $pv^n = c$.

    (a) Given that $p = 94.2$, $v = 2.14$ and $n = 1.27$ find c.

    (b) Given that $v = 2.95$, $n = 1.30$ and $c = 220$ find p.

    (c) Given that $p = 104$, $n = 1.35$ and $c = 283$ find v.

4. The formula for the current i in an a.c. circuit is given by $i = Ie^{-Rt/L} \sin 1800t°$. Find i when $I = 30$, $t = 0.0050$, $R = 10$, $L = 0.030$.

5. The time t of one complete oscillation of a compound pendulum is given by

$$t = 2\pi \sqrt{\left(\frac{k^2 + h^2}{gh} \right)}$$

Calculate t if $k = 43.86 \times 10^{-3}$, $h = 28.29 \times 10^{-3}$ and $g = 9.81$.

6. Find the force F in a bar which is given by the formula

$$F = \frac{AEx}{L}$$

where $E = 200 \times 10^9 \, \text{N/m}^2$, $A = 3.0 \times 10^{-6} \, \text{m}^2$, $x = 0.90 \times 10^{-3} \, \text{m}$ and $L = 1.2 \, \text{m}$.

7. The formula for the current i (A) flowing in an RC circuit is

$$i = -\frac{E}{R} e^{-t/RC}$$

Find the current flowing when $t = 0.010 \, \text{s}$, if $C = 10 \times 10^{-6} \, \text{F}$, $R = 1000 \, \Omega$, and $E = 100 \, \text{V}$.

8. The deflection y (m) of a rod can be found from the formula

$$y = \frac{mgL^3}{3EI}$$

Given that $m = 12 \, \text{kg}$, $L = 0.25 \, \text{m}$, $E = 120 \times 10^9 \, \text{N/m}^2$, $I = 1.11 \times 10^{-9} \, \text{m}^4$, $g = 9.81 \, \text{m/s}^2$ find y.

9. Find the stress σ from the formula

$$\sigma = \sqrt{\frac{mv^2 E}{LA}}$$

given that $m = 21$, $v = 0.40$, $E = 200 \times 10^9$, $L = 10$, $A = 400 \times 10^{-6}$

10.   The strain energy U (J) in a wire rope is given by

$$U = \frac{\rho^2 g^2 A L^3}{6E}$$

Find U if $\rho = 8.0 \times 10^3 \, \text{kg/m}^3$, $g = 9.81 \, \text{m/s}^2$, $A = 1100 \times 10^{-6} \, \text{m}^2$, $L = 8.0 \, \text{m}$, $E = 200 \times 10^9 \, N/\text{m}^2$.

11.   Find the stress which is given by the formula

$$\sigma = \sqrt{\left(\frac{M^2 g^2}{A^2} + \frac{2hEMg}{AL} \right)}$$

given that $M = 20$, $g = 9.81$, $A = 0.5 \times 10^{-3}$, $h = 0.070$, $L = 2.5$, $E = 200 \times 10^9$.

Chapter 3

Trigonometry

3.1 Introduction

Trigonometry is the branch of Mathematics that deals with the relationships between the sides and the angles of a triangle. Provided that a minimum of

(i) 3 sides,   (ii) 2 sides and 1 angle or   (iii) 1 side and 2 angles

are known about any triangle, the other unknown sides or angles can be calculated using trigonometrical methods.

Trigonometry is based on the trigonometrical ratios of sine, cosine and tangent, which are the ratios between the sides of a right-angled triangle. The values of these ratios depend upon the size of the angles, and do not depend on the size of the triangle.

3.2 Trigonometric ratios

In Fig. 3.1 sides are given names with respect to the angle θ. The side opposite the right-angle (90°) is called the hypotenuse. The other two sides are named according to their position relative to the angle. The ratios are defined as

$$\text{sine of } \theta = \frac{\text{opposite}}{\text{hypotenuse}} \quad \text{or} \quad \boxed{\sin \theta = \frac{o}{h}}$$

$$\text{cosine of } \theta = \frac{\text{adjacent}}{\text{hypotenuse}} \quad \text{or} \quad \boxed{\cos \theta = \frac{a}{h}}$$

$$\text{tangent of } \theta = \frac{\text{opposite}}{\text{adjacent}} \quad \text{or} \quad \boxed{\tan \theta = \frac{o}{a}}$$

For a known angle the values of these ratios can be obtained from an electronic calculator. For example $\sin 40° = 0.6428$ can be obtained by entering 40 and pressing the $\boxed{\sin}$ key.

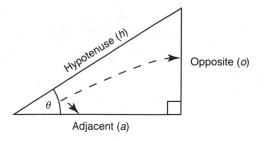

Fig. 3.1

Example 3.1 Find the lengths of the unknown sides in the steel bracket *ABC* shown in Fig 3.2.

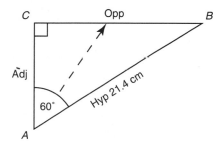

Fig. 3.2

Using $\quad \sin \theta = \dfrac{o}{h} \quad$ with $\theta = 60°$ $\quad$ and $\quad h = 21.4$ cm

$$\sin 60° = \frac{o}{21.4}$$

Multiply both sides by 21.4:

$$o = 21.4 \times \sin 60$$
$$= 21.4 \times 0.866$$
$$BC = 18.5 \text{ cm}$$

Using $\quad \cos \theta = \dfrac{a}{h} \quad$ again with $\theta = 60°$ $\quad$ and $\quad h = 21.4$ cm

$$\cos 60° = \frac{a}{21.4}$$
$$a = 21.4 \times \cos 60 = 21.4 \times 0.5$$
$$AC = 10.7 \text{ cm}$$

Example 3.2   Fig 3.3 shows a voltage diagram for an electronic circuit. Calculate the phase angle ϕ between the two voltages.

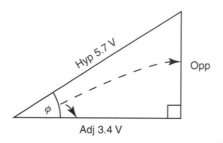

Fig. 3.3

Using $\cos\phi = \dfrac{a}{h}$ with a = 3.4 V   and   h = 5.7 V

$$= \frac{3.4}{5.7} = 0.5965$$

$$\phi = \cos^{-1} 0.5965 = 53.4°$$

Note: $\phi = \cos^{-1} 0.5965$ means that ϕ is an angle which has a cos of 0.5965. The value of ϕ can be obtained from the calculator by entering 0.5965 and pressing the $\boxed{\cos^{-1}}$ key. $\cos^{-1}$ is the inverse of the cosine (see Section 3.15).

3.3   Theorem of Pythagoras

This is an important theorem and is a useful alternative method of finding the third side of a right-angled triangle when the other two sides are known. In the right-angled triangle in Fig 3.4 the sides are labelled *a*, *b*, *c* according to the angles that they are opposite.

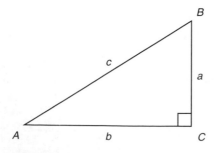

Fig. 3.4

Pythagoras's theorem states that the square of the hypotenuse is equal to the sum of the squares of the other two sides, that is

$$c^2 = a^2 + b^2$$

Example 3.3 Find the unknown force F in the force diagram shown in Fig 3.5

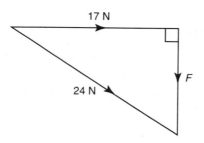

17 N

24 N

F

Fig. 3.5

Using Pythagoras's theorem $\qquad c^2 = a^2 + b^2$

$$24^2 = 17^2 + F^2$$

$$576 = 289 + F^2$$

Subtract 289 from both sides $\qquad F^2 = 576 - 289 = 287$

Take square roots on both sides $\quad F = 17\,\text{N}$

EXERCISE 3.1

1. Fig. 3.6 shows a crank-connecting rod mechanism. Calculate, for the position shown, the distance AC and the angle BAC.

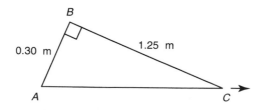

B

0.30 m

1.25 m

A

C

Fig. 3.6

2. A tie wire 6.0 m long is fixed at one end to the top of a vertical radio mast, 4.2 m tall, and the other end is bolted to the horizontal ground. Calculate the angle of inclination of the wire and the distance of the bolted joint from the foot of the mast.

3. In the vector diagram shown in Fig 3.7 calculate the voltage V_R across the resistance and the phase angle ϕ.

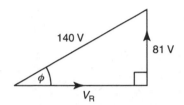

Fig. 3.7

4. For the taper roller shown in Fig 3.8 calculate the taper angle θ and the length of the bearing surface AB.

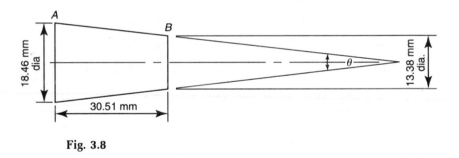

Fig. 3.8

3.4   Trigonometric ratios for 0°, 30°, 45°, 60°, 90°

These angles occur frequently in trigonometrical work and for this reason they deserve special mention.

(a) 30° and 60°

Consider the equilateral triangle ABC with sides of length 2 units (Fig. 3.9). The perpendicular AD will bisect the base.

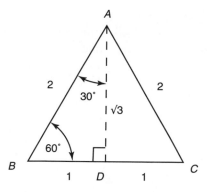

Fig. 3.9

In $\triangle ABD$, by Pyhagoras' Theorem, $AD = \sqrt{3}$. Therefore

$$\sin 60° = \frac{\sqrt{3}}{2} \qquad \sin 30° = \frac{1}{2}$$

$$\cos 60° = \frac{1}{2} \qquad \cos 30° = \frac{\sqrt{3}}{2}$$

$$\tan 60° = \sqrt{3} \qquad \tan 30° = \frac{1}{\sqrt{3}}$$

(b) 45°

Consider the isosceles triangle PQR with equal sides of length 1 unit (Fig. 3.10). By Pythagoras' Theorem $PQ = \sqrt{2}$.

$$\sin 45° = \frac{1}{\sqrt{2}}$$

$$\cos 45° = \frac{1}{\sqrt{2}}$$

$$\tan 45° = 1$$

Note: Numbers such as $\sqrt{2}$ and $\sqrt{3}$ etc. which cannot be evaluated exactly are known as surds. In some calculations it can be more convenient to use the trigonometric ratios for 30° and 60° in surd form rather than the more normal decimal form.

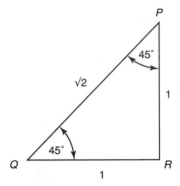

Fig. 3.10

(c) 0°

Consider the triangle ABC where θ is very small (Fig. 3.11). As θ gets smaller, B moves towards C. When B reaches C then $\theta = 0°$, $a = 0$ and $c = b$. Therefore

$$\sin 0° = \frac{a}{c} = \frac{0}{c} = 0$$

$$\cos 0° = \frac{b}{c} = \frac{b}{b} = 1$$

$$\tan 0° = \frac{a}{b} = \frac{0}{b} = 0$$

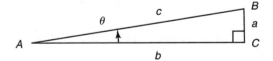

Fig. 3.11

(d) 90°

In triangle ABC, θ is very nearly 90° (Fig. 3.12). As θ gets closer to 90°, A moves closer to C. When A reaches C then $\theta = 90$, $b = 0$, $c = a$. Therefore

$$\sin 90° = \frac{a}{c} = \frac{a}{a} = 1$$

$$\cos 90° = \frac{b}{c} = \frac{0}{c} = 0$$

$$\tan 90° = \frac{a}{b} = \frac{a}{0} = \infty$$

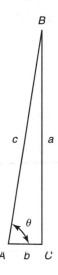

Fig. 3.12

Note: A fraction such as $\frac{a}{0}$ is an extremely large number, and is written as infinity (∞).

These results are summarised in Table 3.1.

Table 3.1

Angle θ	0°	30°	45°	60°	90°
$\sin \theta$	0	$\frac{1}{2}$	$\frac{1}{\sqrt{2}}$	$\frac{\sqrt{3}}{2}$	1
$\cos \theta$	1	$\frac{\sqrt{3}}{2}$	$\frac{1}{\sqrt{2}}$	$\frac{1}{2}$	0
$\tan \theta$	0	$\frac{1}{\sqrt{3}}$	1	$\sqrt{3}$	∞

3.5   Trigonometric ratios of angles greater than 90°

Since angles greater than 90° do not exist in a right-angled triangle a method must be devised for finding the trigonometric ratios for the angles greater than 90°. In Fig 3.13(a–d) angles θ are generated by a line OP rotating in an anticlockwise direction, θ being the angle between the positive x-axis OX and the line OP. Since OP is a rotating line it is always positive. *The trigonometric ratios for θ are always taken to be numerically equal to those for the angle POQ in the right-angled triangle OPQ.* These ratios may be positive or negative depending upon the signs of the sides PQ and OQ in each quadrant, as explained in sections (a) to (d) below.

(a) Angles between 0° and 90°

$$\theta = \text{angle } POQ$$

As shown in Fig. 3.13a the ratios are all positive, that is,

$$\sin \theta = \frac{PQ}{OP} = \frac{+}{+} = \qquad + \text{ value}$$

$$\cos \theta = \frac{OQ}{OP} = \frac{+}{+} = \qquad + \text{ value}$$

$$\tan \theta = \frac{PQ}{OQ} = \frac{+}{+} = \qquad + \text{ value}$$

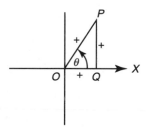

Fig. 3.13(a)

(b) Angles between 90° and 180°

$$\theta_2 = 180° - \text{angle } POQ$$

The ratios are defined as in (a) but the side OQ (Fig. 3.13b) is now negative

$$\sin \theta_2 = \frac{PQ}{OP} = \frac{+}{+} = \qquad + \text{ value}$$

$$\cos \theta_2 = \frac{OQ}{OP} = \frac{-}{+} = \qquad - \text{ value}$$

$$\tan \theta_2 = \frac{PO}{OQ} = \frac{+}{-} = \qquad - \text{ value}$$

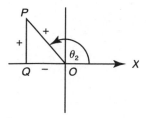

Fig. 3.13(b)

(c) Angles between 180° and 270°

$$\theta_3 = 180° + \text{angle } POQ$$

The ratios are defined as in (a) but the sides OQ and PQ (Fig. 3.13c) are both negative in the ratios so that

$$\sin \theta_3 = \frac{PQ}{OP} = \frac{-}{+} = \quad - \text{ value}$$

$$\cos \theta_3 = \frac{OQ}{OP} = \frac{-}{+} = \quad - \text{ value}$$

$$\tan \theta_3 = \frac{PO}{OQ} = \frac{-}{-} = \quad + \text{ value}$$

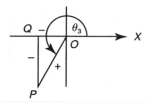

Fig. 3.13(c)

(d) Angles between 270° and 360°

$$\theta_4 = 360° - \text{angle } POQ$$

The ratios are defined as in (a) but the side PQ (Fig. 3.13d) is now negative in the ratios so that

$$\sin \theta_4 = \frac{PQ}{OP} = \frac{-}{+} = \quad - \text{ value}$$

$$\cos \theta_4 = \frac{OQ}{OP} = \frac{+}{+} = \quad + \text{ value}$$

$$\tan \theta_4 = \frac{PO}{OQ} = \frac{-}{+} = \quad - \text{ value}$$

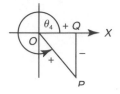

Fig. 3.13(d)

Fig. 3.13e identifies which trigonometric ratios are positive in each quadrant. Example 3.4 illustrates the theory for finding the trigonometric ratio of an angle greater than 90°.

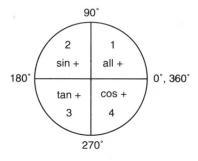

Fig. 3.13(e)

Example 3.4   Determine the value of sin 225°.

In Fig. 3.14 the angle 225° is in quadrant 3, where the sine is negative (see Fig. 3.13e).

Since
$$\theta_3 = 180 + POQ$$
$$225 = 180 + POQ$$
$$POQ = 45°,$$

so that   sin 225 = −0.7071.

However, the values of the trigonometric ratios can be determined directly using the electronic calculation In the above example

enter 225°,   *press* ⌐sin⌐ *key*   *to display* = −0.7071

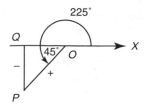

Fig. 3.14

Note: If *OP* rotates in a clockwise direction the angle moved through is a negative angle. In Fig. 3.15 the rotation of *OP* in a clockwise direction from *OX* produces an angle of −50°, which is seen to be the same as a rotation in an anticlockwise direction of 310°.

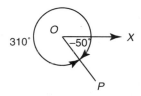

Fig. 3.15

(e) To determine the angle from a given trigonometrical ratio

There are always two values of θ in the range 0° to 360° with the same trigonometrical ratio. Using the calculator to determine the angle will only give one of these values. To find both values of θ first of all enter the positive value of the trigonometric ratio into the calculator to find the angle *POQ*. The two values of θ are then found using the method given in sections (a) to (d) above, as shown by the following examples.

Example 3.5 Find the values of θ in the range 0° to 360° for which tan θ = −1.4.
    Using the calculator enter +1.4 and press the $\boxed{tan^{-1}}$ key to display the angle *POQ* = 54.5°. Fig. 3.13e shows that the tangent is negative in the second and fourth quadrants. Therefore the

angles θ_2 and θ_4 are required, as shown in Fig. 3.16.

$$\theta_2 = 180 - \text{angle } POQ \quad \text{and} \quad \theta_4 = 360 - \text{angle } POQ$$
$$= 180 - 54.5 \qquad\qquad\qquad = 360 - 54.5$$
$$= 125.5° \qquad\qquad\qquad\quad = 305.5°$$

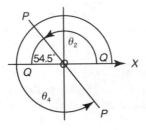

Fig. 3.16

Example 3.6 Find the values of θ in the range $0°$ to $360°$ for which $\sin\theta = 0.722$.

Using the calculator we obtain $POQ = 46.2°$ (the ratio is positive). Fig. 3.13e shows that sine is positive in quadrant 1 and 2. Therefore, the two angles $\theta = POQ$ and θ_2 are required, as shown in Fig. 3.17.

$$\theta = \text{Angle } POQ = 46.2° \quad \text{and} \quad \theta_2 = 180 - \text{angle } POQ$$
$$= 180 - 46.2$$
$$= 133.8°$$

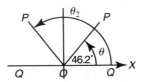

Fig. 3.17

Example 3.7 Find the values of θ in the range $0°$ to $360°$ for which $\cos\theta = -0.61$.

From the calculator we find the angle corresponding to $\cos\theta = +0.61$, which is $POQ = 52.4°$. Fig 3.13e shows that the

cosine is negative in the second and third quadrants. Therefore
the angles θ_2 and θ_3 are required, as shown in Fig 3.18.

$$\theta_2 = 180° - POQ \quad \text{and} \quad \theta_3 = 180° + POQ$$
$$= 180° - 52.4° \qquad\qquad = 180° + 52.4$$
$$= 127.6° \qquad\qquad\qquad = 232.4°$$

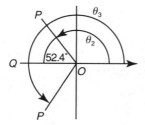

Fig. 3.18

EXERCISE 3.2

Determine the two values of θ between $0°$ and $360°$ in each of the following
ratios:

1. $\sin \theta = 0.64$ **2.** $\sin \theta = -0.317$
3. $\cos \theta = 0.3$ **4.** $\cos \theta = -0.542$
5. $\tan \theta = 0.21$ **6.** $\tan \theta = -2.1$

3.6   The sine and cosine rules

The trigonometrical ratios defined in section 3.2 are based on a right-angled
triangle. Consequently they cannot be used for triangles without a right
angle. In order to calculate sides and angles in such triangles the sine and
cosine rules are used. The two rules may be derived by dividing any triangle,
such as ABC, in Fig. 3.19 into two right-angled triangles.

The sine rule states that,

$$\boxed{\frac{a}{\sin A} = \frac{b}{\sin B} = \frac{c}{\sin C}}$$

The cosine rule states that

$$\boxed{\begin{aligned} a^2 &= b^2 + c^2 - 2bc \cos A \\ b^2 &= a^2 + c^2 - 2ac \cos B \\ c^2 &= a^2 + b^2 - 2ab \cos C \end{aligned}}$$

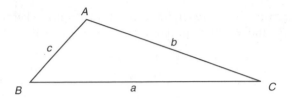

Fig. 3.19

In any triangle there are six unknowns: three sides and three angles. In order to be able to determine all six values at least three of these must be known.
   The cosine rule is used if the following are known:

(i)   3 sides;
(ii) 2 sides and the angle in between them.

The sine rule is used if the following are known

(i)   1 side and 2 angles;
(ii) 2 sides and an angle which is *not* in between.

Note: At least one side must always be specified. The application of the two rules is shown in Examples 3.8 to 3.11.

Example 3.8   Find the length of the base of the template shown in Fig. 3.20.

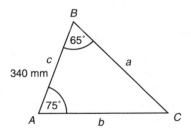

Fig. 3.20

Since one side and two angles are known the sine rule is used.
Now angle $C = 180 - (65 + 75) = 40°$ (because the angles of a triangle always total $180°$)

$$\frac{b}{\sin B} = \frac{c}{\sin C}$$

$$\frac{b}{\sin 65} = \frac{340}{\sin 40}$$

Multiply by sin 65: $b = \dfrac{340 \sin 65}{\sin 40} = 340 \times \dfrac{0.9063}{0.6428}$

$\qquad\qquad\qquad = 479.4$ mm

$\qquad\qquad\qquad = 480$ mm (correct to 2 significant figures)

Example 3.9 Fig. 3.21 shows an assembly for a crane. Find the angle between the girders AB and BC.

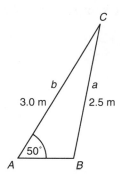

Fig. 3.21

Two sides and an angle *not* between them are known. Therefore the sine rule is used. For ease of working we use the inverted form of the sine rule.

$$\frac{\sin A}{a} = \frac{\sin b}{b}$$

$$\frac{\sin 50}{2.5} = \frac{\sin B}{3.0}$$

Multiply by 3.0 $\sin B = \dfrac{3.0 \sin 50}{2.5} = 0.9193$

$$B = \sin^{-1} 0.9193 = 66.8°$$

Referring to the Fig. 3.21, the angle ABC is greater than 90°. This other solution, following Example 3.6, is in the second quadrant, that is $180 - 66.8 = 113.2°$ which is 110° correct to two significant figures, that is $B = 110°$.

Example 3.10 The total current I taken by an a.c. circuit is given by the line BC in Fig. 3.22. Calculate the value of this total current and the phase angle ϕ.

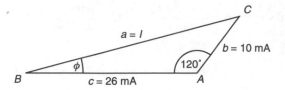

Fig. 3.22

Two sides and the angle in between are known so that the cosine rule is used.

$$a^2 = b^2 + c^2 - 2bc \cos A$$

$$= 26^2 + 10^2 - 2 \times 26 \times 10 \cos 120$$

$$= 676 + 100 - 520(-0.5)$$

$$= 1036$$

$$a = \sqrt{1036} = 32.2$$

$$= 32 \text{ mA (correct to two significant figures)}$$

Now that four quantities are known either rule may be used to find ϕ. Using the inverted sine rule

$$\frac{\sin B}{b} = \frac{\sin A}{a}$$

$$\frac{\sin \phi}{10} = \frac{\sin 120}{32.2}$$

$$\sin \phi = \frac{10 \sin 120}{32.2} = 0.2690$$

$$\phi = \sin^{-1} 0.2690 = 15.6°$$

$$= 15.6° \text{ (correct to two significant figures)}$$

Example 3.11   Fig. 3.23a shows three forces acting at a point. Fig 3.23b is the triangle of forces for this loading arrangement. Calculate the angle θ between the forces F_3 and F_2.

Since the three sides are known the cosine rule is used to find the angle θ at C.

$$c^2 = a^2 + b^2 - 2ab \cos C$$

Transposing $2ab \cos C = a^2 + b^2 - c^2$

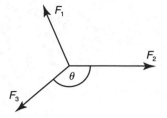

Fig. 3.23(a)

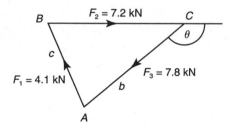

Fig. 3.23(b)

Dividing by $2ab$

$$\cos C = \frac{a^2 + b^2 - c^2}{2ab}$$

$$= \frac{7.2^2 + 7.8^2 - 4.1^2}{2 \times 7.2 \times 7.8} = 0.8535$$

$$C = \cos^{-1} 0.8535$$

$$= 31.4°$$

Therefore

$$\theta = 180 - 31.4 = 148.6$$

$$= 150° \text{ (correct to two significant figures)}$$

EXERCISE 3.3

1. Fig. 3.24 shows a wall crane. Find the length of the tie and the angle the tie makes with the wall.
2. An electric cable is to run due east from a substation A for 800 m, and then north-east to a factory B. Given that the factory B is on a bearing

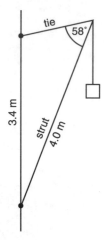

Fig. 3.24

of 055° from *A* calculate the total length of cable required and the direct distance from *A* to *B*.

3. The vector diagram for an electrical circuit is shown in Fig 3.25. Calculate the voltage *V* and the angle *θ* it makes with the 3.2 V vector.

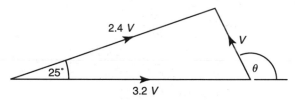

Fig. 3.25

4. Fig. 3.26 shows a parallelogram of vectors. Calculate the resultant vector *R* and the inclination it makes with the horizontal vector.

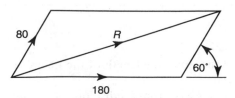

Fig. 3.26

5. The installation of a new motor involves suspending it by two wire ropes as shown in Fig 3.27. If the angle that the longer rope makes with the ceiling must not be less than 30°, what is the minimum length of the smaller rope? For this arrangement what will be the distance of the point A below the ceiling?

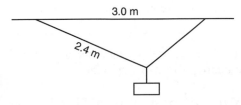

3.0 m

2.4 m

Fig. 3.27

3.7 Area of a triangle

1. In Fig. 3.28 the area of a triangle is $= \dfrac{1}{2}$ base × height $= \dfrac{1}{2} ah$

In triangle AXC $\quad \dfrac{h}{b} = \sin C \quad$ so that $\quad h = b \sin C$

Area of triangle $ABC = \dfrac{1}{2} ab \sin C$ $\qquad\qquad$ (3.1)

$$= \dfrac{1}{2} ac \sin B$$

$$= \dfrac{1}{2} bc \sin A$$

by a similar manner.

In order to use these formulae two sides and the angle in between must be known.

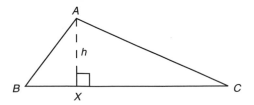

Fig. 3.28

2. If three sides are known and no angles are known then the following formula is suitable.

$$\text{Area of triangle} = \sqrt{s(s-a)(s-b)(s-c)}$$
where $2s = a + b + c$

3. If one side and two angles are known the sine rule is used to find a second side; then equation 3.1 may be used to find the area.

Example 3.12 Calculate the area of sheet metal required to make the machine guard shown in Fig. 3.29.

$$\text{Area of triangle ADE} = \frac{1}{2}\,ad\,\sin E = \frac{1}{2}\times 1.0 \times 1.2 \times \sin 70$$

$$= 0.564\,\text{m}^2$$

$$\text{Area of triangle ABC} = \sqrt{s(s-a)(s-b)(s-c)}$$

where $2s = a + b + c = 0.9 + 1.1 + 0.8 = 2.8$

$$s = 1.4$$

$$\text{Area of triangle ABC} = \sqrt{1.4(1.4 - 0.9)(1.4 - 1.1)(1.4 - 0.8)}$$

$$= \sqrt{0.126}$$

$$= 0.355\,\text{m}^2$$

Therefore area of guard $= 0.564 + 0.355$

$$= 0.919$$

$$= 0.92\,\text{m}^2\,(\text{correct to 2 significant figures})$$

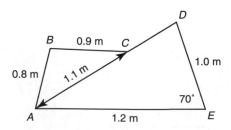

Fig. 3.29

EXERCISE 3.4

1. A support plate is in the form of a triangle *ABC* with
   $AC = 10.4$ cm, $BC = 13.1$ cm, and the angle $C = 60°$. Calculate the
   area of the plate.

2. An electric heat sink is triangular in shape and to lose the required
   amount of heat it must have an area of at least 2400 mm^2. If the two
   sides of this triangle must have lengths of 80 mm and 160 mm what is
   the minimum value of the angle between these two sides?

3. A mesh guard for a robot is to be in the form of a trapezium as shown
   in Fig. 3.30. Calculate the area of metal required for the guard.

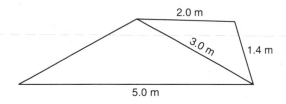

2.0 m

3.0 m 1.4 m

5.0 m

Fig. 3.30

4. An instrument panel must have the dimensions shown in Fig 3.31.
   Find the area of the panel.

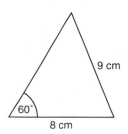

9 cm

60°

8 cm

Fig. 3.31

3.8   Trigonometric graphs

(a) Sine and cosine graphs

Physical quantities such as a.c. currents and voltages, moving parts such as
pistons etc. can be modelled with the trigonometric functions sin θ and cos θ.
The graphs of both these functions are shown in Fig. 3.32.

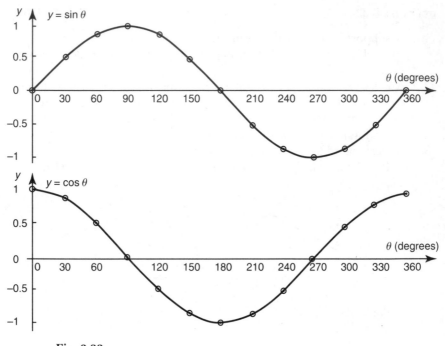

Fig. 3.32

The values of both functions are obtained in intervals of θ of $30°$ over a range of $0°$ to $360°$ (Table 3.2).

Table 3.2

$\theta°$	0	30	60	90	120	150	180	210	240	270	300	330	360
$y = \sin \theta$	0	0.5	0.87	1.0	0.87	0.5	0	−0.5	−0.87	−1.0	−0.87	−0.5	0
$y = \cos \theta$	1.0	0.87	0.5	0	−0.5	−0.87	−1.0	−0.87	−0.5	0	0.5	0.87	1.0

These trigonometric graphs have the following significant features:

1.   Both graphs will repeat themselves after every $360°$. The **period** of each graph is $360°$.
2.   The maximum and minimum value of each function, is $+1$ and $−1$. This is called the **amplitude**. For the graph of $y = A \sin \theta$ the maximum value of y is A, that is, A is the amplitude.
3.   The graph of $\cos \theta$ is seen to have the same form as the graph of $\sin \theta$ but displaced to the left by $90°$, that is, it is $90°$ ahead of $\sin \theta$, and hence can be written as $y = \sin(\theta + 90°)$, where $90°$ is the **phase angle**.

(b) The tangent graph

Table 3.3

$\theta°$	0	30	60	90	120	150	180	210	240	270	300	330	360
tan θ	0	0.577	1.732	∞	−1.732	−0.577	0	0.577	1.732	∞	−1.732	−0.577	0

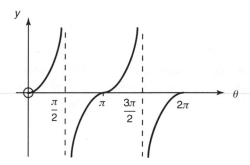

Fig. 3.33

The graph of tan θ has the following significant features:

1. The graph is periodic and will repeat itself after 180°.
2. As the angle approaches 90° and 270° the function tan θ approaches $\pm\infty$.

(c) Radian measure of an angle

So far we have measured angles in degrees. Another measure used widely in engineering is radians. Refering to Fig. 3.34a, 1 radian is defined as the angle *AOB* which cuts the circumference of a circle of radius 1 with an arc of length 1.

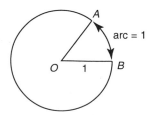

Fig. 3.34(a)

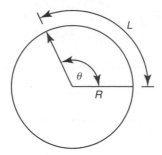

Fig. 3.34(b)

To convert radians to degrees we see that a complete revolution of this circle has an arc length, that is its circumference, of $2\pi l$. The complete revolution is 2π radians, which is the same as $360°$. Therefore

$$2\pi \text{ radians} = 360°$$
$$\pi \text{ radians} = 180°$$
$$1 \text{ radian} = 180/\pi = 57.30 \text{ radians correct to 4 significant figures.}$$

In general, as shown in Fig. 3.34(b),

$$\text{angle in radians } (\theta) = \frac{\text{length of arc}(L)}{\text{radius }(R)}$$

Since this definition involves the division of two lengths the radian measure has no units which is a major advantage in engineering calculations.

Conversions are carried out as follows:

(i)   Degrees to radians

$$40° = 40 \times \frac{\pi}{180} = 0.70 \text{ radians}$$

$$60° = 60 \times \frac{\pi}{180} = \frac{\pi}{3} \text{ radians}$$

(ii)   Radians to degrees

$$1.6 \text{ radians} = 1.6 \times \frac{180}{\pi} = 92°$$

$$\frac{\pi}{4} \text{ radians} = \frac{\pi}{4} \times \frac{180}{\pi} = 45°$$

(d) The general equation of a sine wave $y = A \sin(\theta \pm \alpha)$

In the equation $y = A \sin(\theta + \alpha)$ the phase angle α is positive and the sine wave is said to **lead** $A \sin \theta$.

In the equation $y = A \sin(\theta - \alpha)$ the phase angle is negative and the sine wave is said to lag $A \sin \theta$. The two cases are shown in Fig. 3.35.

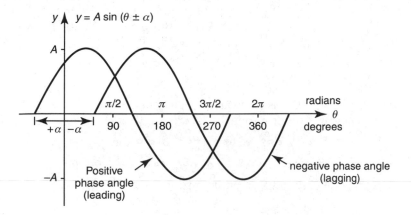

Fig. 3.35

(e) The rotating vector method of constructing a sine wave

A sine wave can be constructed by using a rotating line equal in length to the amplitude. This rotating vector is called a **phasor**. The construction is shown in Fig. 3.36.

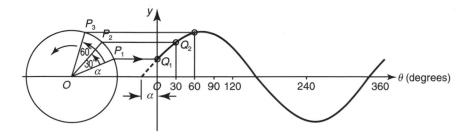

Fig. 3.36

The line OP is the phasor, which rotates in an anticlockwise direction, and has length equal to the amplitude A of $y = A \sin(\theta + \alpha)$. OP_1 is drawn at an angle α to the horizontal in Fig. 3.36, indicating a positive value for the phase angle. If α were negative then OP would start below the horizontal axis, with an angle $-\alpha$. In the position OP^1 the angle $\theta = 0°$. The point P_1 is projected to Q_1 on the y-axis, that is, the $\theta = 0°$ axis. The phasor OP is then rotated $30°$ in an anticlockwise direction to the position OP_2. The point P_2 is now projected to Q_2 on the $\theta = 30°$ ordinate, to give the second point on the sine wave. The process is repeated until the phasor returns to the position OP_1, when the sine wave will have completed one cycle. This rotating method illustrates the relationship between the sine wave and the process of generating a.c. voltages using a rotating coil.

(f) Angular velocity ω

Let the phasor OP rotate at a constant angular velocity, ω rad/s. Every second the angle swept through is ω radians. Then in t seconds OP will have rotated through an angle

$$\theta = \omega t \tag{3.2}$$

The general equation will then become

$$y = A \sin(\omega t + \alpha)$$

Note: In this form α must be expressed in radians.

(g) Periodic time and frequency

(i) Periodic Time T

The periodic time is the time taken in seconds to complete one cycle of the sine wave, which is the same as the time for the phasor to complete one revolution. Therefore, when $t = T$, $\theta = 2\pi$ so that equation (3.2) becomes

$$2\pi = \omega T$$

$$T = \frac{2\pi}{\omega} \tag{3.3}$$

(ii) Frequency f

Frequency f is the number of sine waves completed in one second. It is measured in cycles per second or hertz (Hz) where Hz=1 cycle/s. From the above we see that

in T s the number of cycles generated = 1

in 1 s the number of cycles generated = $1/T$ so that

$$f = \frac{1}{T}$$

From equation (3.3)

$$f = \frac{\omega}{2\pi}$$

Example 3.13   An a.c. voltage has the formula
$v = 340 \sin(314t - 0.7)$. Determine

(i)   the maximum voltage;
(ii)   the frequency of the voltage;
(iii)   the phase angle in degrees.

From the information make a rough sketch of the waveform by plotting the voltage against the angle of rotation in degrees, showing the main features.

Comparing the formula with the general equation
$y = A \sin(\omega t - \alpha)$ we obtain

(i)   maximum voltage $= A = 340 \, \text{V}$
(ii)   frequency $f = \omega/2\pi = 314/2\pi = 50 \, \text{Hz}$
(iii)   phase angle $\alpha = -0.7 \, \text{rad} = -0.7 \times 180/\pi = -40°$ which
shows that the wave is lagging.

The sketch of the sine wave is shown in Fig. 3.37.

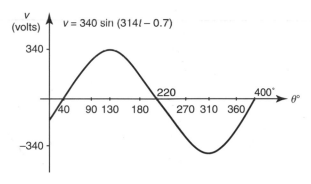

Fig. 3.37

(h) Addition of sine waves

In electrical engineering electric circuits a.c. voltages and currents arise, and need to be combined. In mechanical engineering the sinusoidal movements of mechanical mechanisms may need to be combined. Adding together two sine waves with the same frequency will produce another sine wave of the same frequency. Adding together two sine waves of different frequencies will produce a complicated waveform, which will not be in the form of a sine wave. Sine waves may be added together in two ways:

● *Method 1.* Both sine waves are drawn and the ordinates (that is, the vertical values at each value of the angle) added together algebraically, to produce the combined waveform. This method may be used to combine sine waves with the same or different frequencies.

- *Method 2*. The rotating vector may be used after first obtaining the resultant of the two sine wave phasors. This method can only be used to combine sine waves with the same frequencies.

Both methods are described in the following examples.

Example 3.14   The movement of part of a packaging machine is made up of two movements given by $y_1 = 50 \sin \theta$ and $y_2 = 100 \sin(\theta + 50°)$.

Both movements are measured in mm. Draw graphs of the two sine waves and find the resultant movement by adding the two sine waves together. Hence find the formula for the resultant movement.

Using method 1 the graphs of y_1 and y_2 are plotted for θ over one cycle of 360° as shown in Fig. 3.38, using the data from Table 3.4.

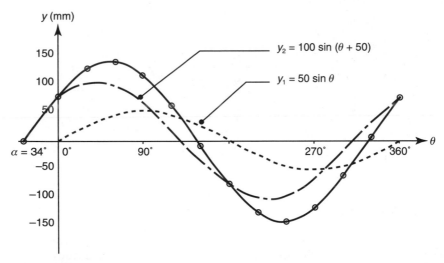

Fig. 3.38

The combined graph has ordinates $y_1 + y_2$ and these can be found either from the table or from the graphs, using algebraic addition, to take into account the signs. Because the two sine waves have the same frequency, the combined graph will be a sine

Table 3.4

$\theta°$	0	30	60	90	120	150	180	210	240	270	300	330	360
$\sin \theta$	0	0.5	0.87	1.0	0.87	0.5	0	-0.5	-0.87	-1.0	-0.87	-0.5	0
$y_1 = 50 \sin \theta$	0	25	43.5	50	43.5	25	0	-25	-43.5	-50	-43.5	-25	0
$(\theta + 50)°$	50	80	110	140	170	200	230	260	290	320	350	380	410
$\sin(\theta + 50)$	0.766	0.985	0.940	0.643	0.174	-0.342	-0.766	-0.985	-0.940	-0.642	-0.174	0.342	0.766
$y_2 = 100 \sin(\theta + 50)$	76.6	98.5	94.0	64.3	17.4	-34.2	-76.6	-98.5	-94.0	-64.2	-17.4	34.2	76.6

wave with the same frequency. If we examine the combined graph we see that its phase angle is 34° and its peak value is 137 mm, so that its formula is

$$Y = 137 \sin(\theta + 34°)$$

Example 3.15   The sine waves $y_1 = 10 \sin \theta$ and $y_2 = 15 \sin 2\theta$ are to be added together to create the combined waveform $y = 10 \sin \theta + 15 \sin 2\theta$. Draw the wave form for one complete cycle.

With $y_1 = 10 \sin \theta$ one complete cycle is 360°. With $y_2 = 15 \sin \theta$ one complete cycle is $360°/2 = 180°$. The combined wave form will complete one cycle in the LCM of the two cycles, that is the LCM of 180° and 360°, which is 360°.

A table of values is completed for the range $\theta = 0°$ to 360° in intervals of 15, this interval ensuring an adequate number of points for the plot of the higher frequency wave. Part of the table is shown as Table 3.5

The combined waveform is shown in Fig. 3.39, which is seen to be non-sinusoidal and periodic with a 360° cycle.

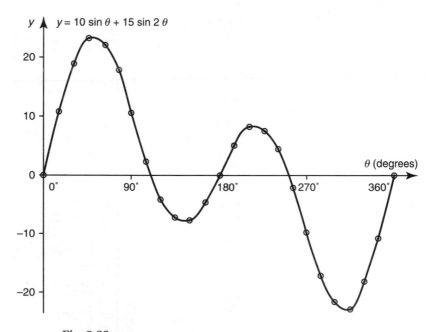

Fig. 3.39

Table 3.5

$\theta°$	0	15	30	45	60	75	90	105	120	135	360
$y_1 = 10 \sin \theta$	0	2.59	5.0	7.071	8.66	9.66	10.0	9.66	8.66	7.071	0
2θ	0	30	60	90	120	150	180	210	240	270	720
$y_2 = 15 \sin 2\theta$	0	7.5	12.99	15.0	12.99	7.5	0	-7.5	-12.99	-15.0	0
$y_1 + y_2$	0	10.09	17.99	22.071	21.65	17.16	10	2.16	-4.33	-7.93	0

Example 3.16   Using the rotating vector method draw a graph of the output current in amperes from an electrical circuit which is given by

$$i = 5 \sin \omega t + 7.5 \sin(\omega t - \pi/3)$$

Referring to Fig. 3.40 we represent $5 \sin \omega t$ with a phasor $OA = 5$ units long drawn horizontally, since the phase angle is zero. The

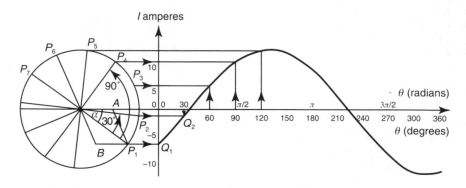

Fig. 3.40

sine wave $7.5 \sin(\omega t - \pi/3)$ is represented by a phasor $OB = 7.5$ units long, drawn at $-\pi/3$ rad, that is, $60°$ below the axis because the phase angle is negative. OP_1, the resultant of these two phasors is now obtained by drawing a parallelogram. P_1 is projected across to give the first point Q_1 on the combined graph at $\theta = 0°$. The resultant OP_1 is now rotated $30°$ to OP_2 and P_2 is projected across to fix the next point Q_2 on the graph at $\theta = 30°$. The process is continued until the whole cycle has been completed. Because we are adding two sine waves of equal frequency the output waveform is sinusoidal. From the graph it is seen that

Amplitude A $= 11.2\ A$

Phase angle α $= -35° = -0.61$ rad.

Output formula is $i = 11.2 \sin(\omega t - 0.61)$

EXERCISE 3.5

1. Use π radians $= 180°$ to convert the following angles:
    (a) from radians to degrees: $\dfrac{\pi}{6}$, $\dfrac{3\pi}{2}$, 5π, $\dfrac{7\pi}{3}$
    (b) from degrees to radians: $45°$, $18°$, $720°$, $150°$, $135°$.
2. Convert the following:
    (a) radians to degrees: 0.62, 1.37, 3.91
    (b) degrees to radians: $16°$, $50.45°$, $309.2°$.
3. Draw the graphs of $y = \sin\theta$, $y = \sin 2\theta$, $y = 2\sin\theta$ and $y = \sin\frac{1}{2}\theta$, on the same axes, for $0° \leqslant \theta \leqslant 360°$.
4. Draw the graphs of $\cos\theta$, $\cos(\theta + 30°)$ and $\cos(\theta - 30°)$ for values of θ from $0°$ to $180°$.
5. Draw the graph of $y = 3\sin(2\theta + 40°)$ for $0° \leqslant \theta \leqslant 180°$. Hence find the value of θ which makes $y = \pm 1.8$.
6. Determine the amplitude, period and phase in the following:
    (a) $3\cos 2x$ (b) $5\sin(6\theta + 27°)$,   (c) $10\sin(6\pi t - 0.4)$,
    (d) $v = 250\sin 80\pi t$,   (e) $16\sin(2\pi t + \frac{1}{3}\pi)$.
7. Draw the graphs of $y = \sin 2x$ and $y = \sin(x - 30°)$ for values of x between $0°$ and $360°$. Hence draw the graph $y = \sin 2x + \sin(x - 30°)$.
8. Draw the graphs (a) $y = 3\sin x + \sin 2x$,
    (b) $y = 2\sin x - \sin(x + 30°)$ for one complete cycle. Find the general equation for the combined wave in (b).
9. The displacement x metres of an oscillating mechanism after a time t seconds is given by $x = 2.4\sin(5t - 0.1)$. Find the maximum displacement, the time at which this occurs and the displacement after 0.2 s. Draw the displacement time graph.
10. Two voltages $v_1 = 50\sin 100\pi t$ and $v_2 = 80\sin\left(100\pi t + \dfrac{\pi}{6}\right)$ are combined in a circuit. Plot a graph showing the resultant voltage. What is the periodic time of this resultant voltage? What is the general equation of the combined voltage?
11. Draw the graphs of $y_1 = 20\sin\theta$ and $y_2 = 30\sin(\theta - 40°)$ on the same axes. By adding together the ordinates (that is, the heights) of both graphs draw the combined graph $y = 20\sin\theta + 30\sin(\theta - 40°)$. By examining the combined graph give its formula in the form $y = A\sin(\theta + \alpha)$.
12. Use the rotating vector method to draw the graph

    $$v = 4\sin\omega t + 6\sin(\omega t - \pi/4)$$

    What is the general formula for this graph?

3.9   Other trigonometric ratios

In addition to the three ratios sine, cosine and tangent other trigonometric ratios are used which are reciprocals of these three. They are

$$\text{cosec(ant) } \theta = \frac{1}{\sin \theta}$$

$$\text{sec(ant) } \theta = \frac{1}{\cos \theta}$$

$$\text{cot(angent) } \theta = \frac{1}{\tan \theta}$$

These ratios can be used in preference to sine, cosine and tangent to simplify mathematical expressions.

Example 3.17   Evaluate $\sec \theta$, $\text{cosec } \theta$ and $\cot \theta$ if $\theta = 130°$.

Using the rules corresponding to sin, cos and tan:

$$\sec 130° = \frac{1}{\cos 130°} = \frac{1}{-0.6425} = -1.5557$$

$$\text{cosec } 130° = \frac{1}{\sin 130°} = \frac{1}{0.7660} = 1.3054$$

$$\cot 130° = \frac{1}{\tan 130°} = \frac{1}{-1.192} = -0.8391$$

3.10   Trigonometric identities

Identities differ from equations in that they are true for any value of the unknown variable that they contain, whereas equations are true for only certain values of the variable. There are certain basic identities in

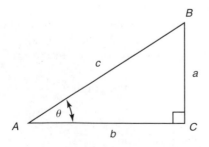

Fig. 3.41

trigonometry which can be established using right-angled triangles as shown in Fig. 3.41. Note first that a power of a trigonometric ratio is written in a special way, e.g.

$$(\sin \theta)^2 = \sin^2 \theta$$

$$(\tan x)^4 = \tan^4 x$$

Using the definitions

(a) $\sin^2 \theta + \cos^2 \theta = \dfrac{a^2}{c^2} + \dfrac{b^2}{c^2}$

$$= \dfrac{a^2 + b^2}{c^2}$$

$$= \dfrac{c^2}{c^2}$$

since $a^2 + b^2 = c^2$ by Pythagoras' theorem.

Therefore $\boxed{\sin^2 \theta + \cos^2 \theta = 1}$ (3.4)

(b) $\dfrac{\sin \theta}{\cos \theta} = \dfrac{a/c}{b/c} = \dfrac{a}{\not{c}} \times \dfrac{\not{c}}{b} = \dfrac{a}{b}$

that is $\boxed{\dfrac{\sin \theta}{\cos \theta} = \tan \theta}$ (3.5)

Similarly it can be shown that

$$\boxed{\begin{array}{l} 1 + \tan^2 \theta = \sec^2 \theta \\ 1 + \cot^2 \theta = \operatorname{cosec}^2 \theta \end{array}}$$

and   (3.6)
                                                                                                (3.7)

These identities can be used to simplify and manipulate trigonometric expressions and equations.

Example 3.18   Prove the identity $(1 - \operatorname{cosec}^2 \theta) \times \tan^2 \theta = -1$

$(1 - \operatorname{cosec}^2 \theta) \times \tan^2 \theta = - \cot^2 \theta \times \tan^2 \theta$, using equation (3.7)

$$= -\dfrac{1}{\tan^2 \theta} \times \tan^2 \theta = -1$$

Example 3.19   Show that $\dfrac{1 + \tan^2 \theta}{1 - \tan^2 \theta} = \dfrac{1}{2\cos^2 \theta - 1}$

$$\frac{1 + \tan^2 \theta}{1 - \tan^2 \theta} = \frac{\left\{ 1 + \dfrac{\sin^2 \theta}{\cos^2 \theta} \right\}}{\left\{ 1 - \dfrac{\sin^2 \theta}{\cos^2 \theta} \right\}} = \frac{\left\{ \dfrac{\cos^2 \theta + \sin^2 \theta}{\cos^2 \theta} \right\}}{\left\{ \dfrac{\cos^2 \theta - \sin^2 \theta}{\cos^2 \theta} \right\}}$$

$$= \frac{\cos^2 \theta + \sin^2 \theta}{\cos^2 \theta - \sin^2 \theta} = \frac{1}{\cos^2 \theta - (1 - \cos^2 \theta)}$$

$$= \frac{1}{2\cos^2 \theta - 1}$$

EXERCISE 3.6

1.   Find the values of sec θ, cot θ and cosec θ if θ is

   (a)   125.32°,   (b)   240.45°,   (c)   345.87°,   (d)   900°,   (e) $-\dfrac{19\pi}{3}$

2.   Prove the following identities
   (a) $2\cos^2 \theta - 1 = 1 - 2\sin^2 \theta$
   (b) $(\sin \theta + \cos \theta)^2 - 1 = 2\sin \theta \times \cos \theta$
   (c) $\dfrac{1}{\tan^2 \theta}\left(\dfrac{1}{\cos^2 \theta} - 1\right) = 1$
   (d) $\dfrac{\cot A - \tan B}{\cot B - \tan A} = \cot A \times \tan B$
   (e) $\dfrac{\tan A + \cot A}{\sec A} = \operatorname{cosec} A$
   (f) $\dfrac{1 + \cot^2 A}{\cot A \operatorname{cosec} A} = \sec A$
   (g) $\sec x(\cot x - 1) + \operatorname{cosec} x(1 - \tan x) = 2(\operatorname{cosec} x - \sec x)$
   (h) $(\sec B + \operatorname{cosec} B)(\sec B - \operatorname{cosec} B) = \tan^2 B - \cot^2 B$
   (j) $(\cot \theta + \cos \theta)^2 - (\cot \theta - \cos \theta)^2 = 4\cos^2 \theta \times \operatorname{cosec} \theta$
3.   Simplify: $\operatorname{cosec}(180 + A) \times \sec(90 + A) - 1$.

4.   Given that $\tan \theta = \dfrac{a}{h}$ where θ is acute, find expressions for $\sin \theta$ and $\sec^2 \theta$.

3.11   Trigonometric ratios of compound angles

In the ratios such as $\sin(A + B)$, $\tan(A - B)$, etc., the angles such as $(A + B)$ and $(A - B)$ are known as compound angles. They occur frequently when trigonometry is used in science and technology, and it is therefore necessary to express trigonometric ratios containing such angles in terms of the simple angles A and B.

Consider Fig. 3.42 containing the angles A and B.

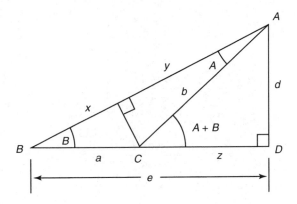

Fig. 3.42

Now angle $ACD = A + B$ (external angle of a triangle).

1. In triangle ACD:

    $$\sin(A + B) = \frac{d}{b}$$

    But in triangle ABD,

    $$d = (x + y) \sin B$$

    Therefore

    $$\sin(A + B) = \frac{(x + y) \sin B}{b}$$

    $$= x \frac{\sin B}{b} + \frac{y}{b} \sin B$$

    But in triangle ABC, by the sine rule

    $$\frac{\sin B}{b} = \frac{\sin A}{a}$$

    Therefore

    $$\sin(A + B) = \frac{x}{a} \sin A + \frac{y}{b} \sin B$$

    But from the diagram $\frac{x}{a} = \cos B$, $\frac{y}{b} = \cos A$.

    $$\boxed{\sin(A + B) = \sin A \cos B + \cos A \sin B}$$ (3.8)

2. In triangle ACD:

$$\cos(A + B) = \frac{z}{b} = \frac{e - a}{b}$$

But in triangle ABD,

$$e = (y + x) \cos B$$

Therefore

$$\cos(A + B) = \frac{y \cos B + x \cos B - a}{b}$$

Since $x = a \cos B$, and $y = b \cos A$,

$$\cos(A + B) = \frac{y \cos B + a \cos B \cos B - a}{b}$$

$$= \frac{y}{b} \cos B + \frac{a}{b}(\cos^2 B - 1)$$

$$= \cos A \cos B - \frac{a}{b} \sin^2 B$$

Again from the sine rule

$$\frac{\sin B}{b} = \frac{\sin A}{a}$$

Therefore

$$\cos(A + B) = \cos A \cos B - a \sin B \frac{\sin A}{a}$$

$$\boxed{\cos(A + B) = \cos A \cos B - \sin A \sin B} \qquad (3.9)$$

3. $\tan(A + B) = \dfrac{\sin(A + B)}{\cos(A + B)}$

$$= \frac{\sin A \cos B + \cos A \sin B}{\cos A \cos B - \sin A \sin B}$$

Dividing top and bottom by $\cos A \cos B$

$$\tan(A + B) = \frac{\left\{\dfrac{\sin A \cos B}{\cos A \cos B} + \dfrac{\cos A \sin B}{\cos A \cos B}\right\}}{\left\{\dfrac{\cos A \cos B}{\cos A \cos B} - \dfrac{\sin A \sin B}{\cos A \cos B}\right\}}$$

$$\boxed{\tan(A + B) = \frac{\tan A + \tan B}{1 - \tan A \tan B}} \qquad (3.10)$$

4. Replacing B by $-B$ in the three equations

$$\sin(A - B) = \sin A \cos B - \cos A \sin B \qquad (3.11)$$
$$\cos(A - B) = \cos A \cos B + \sin A \sin B \qquad (3.12)$$
$$\tan(A - B) = \frac{\tan A - \tan B}{1 + \tan A \tan B} \qquad (3.13)$$

To obtain these last three equations we used the results

$$\sin(-B) = -\sin B, \ \cos(-B) = \cos B, \ \tan(-B) = -\tan B$$

The six formulae are called the compound angle formulae.

3.12 Double angle formulae

These can be established by putting $B = A$ in the three compound angle formulae (3.8) to (3.10):

(a) $\sin(A + A) = \sin A \cos A + \sin A \cos A$

$$\boxed{\sin 2A = 2 \sin A \cos A} \qquad (3.14)$$

(b) $\cos(A + A) = \cos A \cos A - \sin A \sin A$

$$\boxed{\cos 2A = \cos^2 A - \sin^2 A} \qquad (3.15)$$

If we substitute for $\cos^2 A$ in equation (3.15) using equation (3.4), that is substitute

$$\cos^2 A = 1 - \sin^2 A.$$

then

$$\cos 2A = 1 - \sin^2 A - \sin^2 A$$

that is, $\boxed{\cos 2A = 1 - 2 \sin^2 A} \qquad (3.16)$

Again, if we substitute $\sin^2 A = 1 - \cos^2 A$, then

$$\cos 2A = \cos^2 A - (1 - \cos^2 A)$$

that is, $\boxed{\cos 2A = 2 \cos^2 A - 1} \qquad (3.17)$

(c) $\tan(A + A) = \dfrac{\tan A + \tan A}{1 - \tan A \cdot \tan A}$

$$\boxed{\tan 2A = \frac{2 \tan A}{1 - \tan^2 A}} \qquad (3.18)$$

Example 3.20   Rearrange the expression $2 \sin(A + 20°)$ in simple angles.

Taking B as $20°$ and using the compound angle formula for $\sin(A + B)$ we obtain

$$2 \sin(A + 20°) = 2 \sin A \cos 20 + 2 \cos A \sin 20$$

$$= 1.88 \sin A + 0.68 \cos A$$

Example 3.21   Prove the identities

(i) $\sin(A - B) + \sin(A + B) = 2 \sin A \cos B$

(ii) $\sec 2\theta(\cos \theta + \sin \theta) = \dfrac{1}{\cos \theta - \sin \theta}$

(i) $\sin(A - B) + \sin(A + B) = \sin A \cos B - \cos A \sin B$

$$+ \sin A \cos B + \cos A \sin B$$

$$= 2 \sin A \cos B$$

(ii) $\sec 2\theta(\cos \theta + \sin \theta) = \dfrac{\cos \theta + \sin \theta}{\cos 2\theta}$

$$= \dfrac{\cos \theta + \sin \theta}{\cos^2 \theta - \sin^2 \theta}$$

$$= \dfrac{(\cos \theta + \sin \theta)}{(\cos \theta - \sin \theta)(\cos \theta + \sin \theta)}$$

$$= \dfrac{1}{(\cos \theta - \sin \theta)}$$

Example 3.22   Use the compound angle formulae to find the general equation for the resultant sine wave when the two sine waves $y_1 = 30 \sin \theta$ and $y_2 = 40 \sin(\theta + 40°)$ are added together.

The combined sine wave is

$$y = 30 \sin \theta + 40 \sin(\theta + 40)$$

$$= 30 \sin \theta + 40[\sin \theta \cos 40 + \cos \theta \sin 40) \text{ using equation (3.8)}$$

$$= 30 \sin \theta + 30.64 \sin \theta + 25.71 \cos \theta$$

$$= 60.64 \sin \theta + 25.71 \cos \theta \qquad\qquad\qquad (i)$$

We now express this equation in the form $y = R \sin(\theta + \alpha)$ using the $\sin(A + B)$ formula

$$R \sin(\theta + \alpha) = [R \cos \alpha] \sin \theta + [R \sin \alpha] \cos \theta \qquad \text{(ii)}$$

Compare the coefficients of $\sin \theta$ and $\cos \theta$ in equation (i) and (ii):

$$R \cos \alpha = 60.64 \qquad \text{(iii)}$$
$$R \sin \alpha = 25.71 \qquad \text{(iv)}$$

Divide (iv) by (iii):

$$\tan \alpha = 0.4240$$
$$\alpha = 23.0°$$

Substitute into (iii):

$$R \cos 23.0 = 60.64$$
$$R = 65.87$$

If we substitute the values of R and α into $y = R \sin(\theta + \alpha)$ we obtain the general equation of the resultant

$$y = 65.9 \sin(\theta + 23)$$

Since this formula is that of a sine wave it can be analysed easily. It has a maximum value of 65.9 when $\theta + 23° = 90°$, that is when $\theta = 67°$. It has a minimum value of -65.9 when $\theta + 23° = 270°$, that is, when $\theta = 247°$.

EXERCISE 3.7

1. By substituting $B = 2A$ in the compound angle formulae show that
    (a) $\sin 3A = 3 \sin A - 4 \sin^3 A$
    (b) $\cos 3A = 4 \cos^3 A - 3 \cos A$
    (c) $\tan 3A = \dfrac{3 \tan A - \tan^3 A}{1 - 3 \tan^2 A}.$

2. Prove that
    (a) $\sin(A + B) - \sin(A - B) = 2 \cos A \sin B$
    (b) $\cos(A + B) + \cos(A - B) = 2 \cos A \cos B$
    (c) $\cos(\frac{1}{3} \pi + \theta) - \cos(\frac{1}{3} \pi - \theta) = -\sqrt{3} \sin \theta$
    (d) $\dfrac{\sin(A - B)}{\sin A \sin B} = \cot B - \cot A$

(e) $\dfrac{\cos 2A}{\sin 4A} = \tfrac{1}{2}\operatorname{cosec} 2A$

(f) $\dfrac{1 + \cos A}{\sin A} = \cot \tfrac{1}{2}A$

(g) $\dfrac{1 + \tan^2 A}{1 - \tan^2 A} = \sec 2A$

(h) $\cos^4 A - \sin^4 A = \cos 2A.$

3. Taking $A = 45°$ and $B = 30°$ use a compound angle formula to show that $\sin 15° = (\sqrt{6} - \sqrt{2})/4$ and $\cos 15° = (\sqrt{6} + \sqrt{2})/4$. Use these expressions to find $\tan 15°$ in surd form.

3.13   Sum and difference formulae

The compound angle formulae can be used to obtain expressions for the sum or difference of sines or cosines. These expressions are useful for the solution of some trigonometric equations, the integration of trigonometric functions and the simplification of trigonometric expressions. Using the compound angles from section 3.11, let

$$A + B = C \qquad \text{and} \qquad A - B = D$$

Therefore $A = \tfrac{1}{2}(C + D)$ $\qquad$ and $\qquad$ $B = \tfrac{1}{2}(C - D)$
From section 3.11,

$$\sin(A + B) = \sin A \cos B + \cos A \sin B \tag{i}$$
$$\sin(A - B) = \sin A \cos B - \cos A \sin B \tag{ii}$$

Add (i) and (ii)

$$\sin(A + B) + \sin(A - B) = 2 \sin A \cos B$$

Substituting for A and B in terms of C and D

$$\boxed{\sin C + \sin D = 2 \sin \tfrac{1}{2}(C + D) \cos \tfrac{1}{2}(C - D)} \tag{3.19}$$

Subtracting (ii) from (i), and substituting,

$$\boxed{\sin C - \sin D = 2 \cos \tfrac{1}{2}(C + D) \sin \tfrac{1}{2}(C - D)} \tag{3.20}$$

Again from section 3.11

$$\cos(A + B) = \cos A \cos B - \sin A \sin B \tag{iii}$$
$$\cos(A - B) = \cos A \cos B + \sin A \sin B \tag{iv}$$

Adding (iii) and (iv)

$$\cos(A + B) + \cos(A - B) = 2 \cos A \cos B$$

Substituting for A and B,

$$\boxed{\cos C + \cos D = 2 \cos \tfrac{1}{2}(C + D) \ \cos \tfrac{1}{2}(C - D)}$$ (3.21)

Similarly, subtracting (iv) from (iii) and substituting,

$$\boxed{\cos C - \cos D = -2 \sin \tfrac{1}{2}(C + D) \ \sin \tfrac{1}{2}(C - D)}$$ (3.22)

Example 3.23 Express the following as products:
(a) $\sin 6\theta + \sin 4\theta$, (b) $\cos 18\theta - \cos 2\theta$.

(a) $C = 6\theta, D = 4\theta$
Therefore, $\sin 6\theta + \sin 4\theta = 2 \sin \tfrac{1}{2}(6\theta + 4\theta) \cos \tfrac{1}{2}(6\theta - 4\theta)$
$$= 2 \sin 5\theta \cos \theta$$
(b) $C = 18\theta, D = 2\theta$
$$\cos 18\theta - \cos 2\theta = -2 \sin \tfrac{1}{2}(18\theta + 2\theta) \sin \tfrac{1}{2}(18\theta - 2\theta)$$
$$= -2 \sin 10\theta \sin 8\theta$$

These formulae can be used in the reverse manner to express products as sums or differences, as shown in Example 3.24.

Example 3.24 Express $\cos 7\theta \cos 3\theta$ as a sum or difference.

From the $\cos C + \cos D$ formula the product of two cosines can be expressed as the sum of two cosines.
  Therefore

$$\tfrac{1}{2}(C + D) = 7\theta \qquad \text{and} \qquad \tfrac{1}{2}(C - D) = 3\theta$$
$$C + D = 14\theta \qquad \text{and} \qquad C - D = 6\theta$$

Solving these two simultaneous equations gives

$$C = 10\theta \qquad \text{and} \qquad D = 4\theta$$

so that

$$\cos 7\theta \cos 3\theta = \tfrac{1}{2}(\cos 10\theta + \cos 4\theta)$$

EXERCISE 3.8

1. Express as products
   (a) $\sin 4\theta + \sin \theta$ (b) $\sin 8\theta - \sin 6\theta$
   (c) $\cos 12\theta + \cos 10\theta$ (d) $\cos \theta - \cos 2\theta$
   (e) $\sin \tfrac{1}{3}\pi + \sin \tfrac{1}{4}\pi$ (f) $\cos 70° - \cos 30°$.

2. Express as sums or differences
    (a) $2 \sin 6\theta \cos 2\theta$ (b) $2 \cos 5\theta \cos 3\theta$
    (c) $2 \cos 4\theta \sin \theta$ (d) $2 \sin 7\theta \sin 5\theta$
    (e) $\sin \frac{1}{3}\pi \cos \frac{1}{6}\pi$ (f) $\cos 150° \sin 70°$
    (g) $2 \cos(A + B)\cos(A - B)$ (h) $\sin \pi \sin \frac{1}{4}\pi$.

3. Simplify:

    (a) $\dfrac{\cos 4\theta + \cos 2\theta}{\sin 4\theta - \sin 2\theta}$ (b) $\dfrac{\sin 2A + \sin 2B}{\cos(A - B)}$ (c) $\dfrac{\sin \frac{1}{2}\theta - \sin \frac{3}{2}\theta}{\sin 2\theta}$.

4. Prove the following identities:

    (a) $\dfrac{\sin(4x + \theta) + \sin(4x - \theta)}{\cos(2x + \theta) + \cos(2x - \theta)} = 2 \sin 2x$

    (b) $\dfrac{\sin \theta - \sin \alpha}{\sin \theta + \sin \alpha} = \tan \frac{1}{2}(\theta - \alpha)\cot \frac{1}{2}(\theta + \alpha)$

    (c) $\dfrac{\sin 15\theta + \sin 11\theta}{\cos 15\theta - \cos 11\theta} = \tan \theta - \dfrac{1}{\sin 2\theta}$.

3.14   Trigonometric equations

It has been seen that trigonometric functions are periodic and for this reason trigonometric equations will have an unlimited number of solutions. There are many types of trigonometric equations, but it is possible to classify them into basic types for the purpose of solving them. All equations require the procedure developed in Section 3.5.

(a) Type 1: Equations containing a single trigonometric function

---

Example 3.25   Solve the equation $\cos x = 0.3420$

From the calculator angle $POQ = \cos^{-1} 0.3420 = 70°$. Since the cosine is positive the solutions must be in quadrants 1 and 4. Therefore

$$x = POQ = 70° \quad \text{and} \quad x = 360° - POQ = 290°$$

Since $\cos x$ has a period of $360°$ further solutions will be at intervals of $360°$. The solutions are therefore

$$x = 70°, 290°, 430°, 650°, \text{etc.}$$

Example 3.26 Solve the equation $\sin 3x = 0.5$ for x values in the range $0°$ to $180°$.

Replace $3x$ by X so that the equation now reads

$$\sin X = 0.5$$

From the calculator angle $POQ = \sin^{-1} 0.5 = 30°$. Since the sine is positive the solutions must be in quadrants 1 and 2. Therefore

$$X = POQ = 30° \quad \text{and} \quad X = 180° - POQ = 150°$$

so that $\quad x = 10°, 50°$

Since $\sin 3x$ has a period of $360°/3 = 120°$ there will be further solutions at $120°$ intervals. Therefore the solutions in the $180°$ range are

$$x = 10°, 50°, 130°, 170°$$

Example 3.27 Find the values of θ in the range $0°$ to $180°$ which satisfy the equation

$$3 \tan(2\theta + 10) + 5.595 = 0$$

Let $X = 2\theta + 10$

Therefore

$$3 \tan X + 5.595 = 0$$

$$\tan X = -1.865$$

From the calculator angle $POQ = \tan^{-1} -1.865 = 61.8°$. Since the tangent is negative the solutions must lie in quadrants 2 and 4. Therefore

$$X = 180° - POQ \quad \text{and} \quad X = 360° - POQ$$

$$= 180° - 61.8° \quad \text{and} \quad X = 360° - 61.8°$$

that is $2\theta + 10 = 118.2°$ $\quad$ and $\quad = 298.2°$

$$\theta = 54.1° \quad \text{and} \quad 144.1°$$

Since the $\tan 2\theta$ function is periodic in $180°/2$ there will be further solutions at intervals of $90°$. However there will be no further solutions in the range $0°$ to $180°$.

(b) Type 2: Equations reducible to quadratic form

A trigonometric equation is of quadratic form if it contains one type of function of degree 2 and a common angle throughout. Trigonometric identities can be used sometimes to bring about this condition.

Example 3.28   What values of ϕ in the range of $0° \leqslant \phi \leqslant 360°$ satisfy the equation?

$3 \cos 2\phi + \cos \phi + 1 = 0$

From equation (3.17) $\cos 2\phi = 2 \cos^2 \phi - 1$

Therefore $3(2 \cos^2 \phi - 1) + \cos \phi + 1 = 0$

$6 \cos^2 \phi + \cos \phi - 2 = 0$

This is now a quadratic equation of the form $ax^2 + bx + c = 0$, where $x = \cos \phi$. Factorising:

$(3 \cos \phi + 2)(2 \cos \phi - 1) = 0$

Either $3 \cos \phi + 2 = 0$ | or $2 \cos \phi - 1 = 0$
$\cos \phi = -0.6667$ | $\cos \phi = 0.5$
$POQ = \cos^{-1} -0.6667$ | or $POQ = \cos^{-1} 0.5$
$= 48.2°$ | $= 60°$
Since cosine is negative ϕ must | Since cosine is positive ϕ must
lie in quadrants 2 and 3 | lie in quadrants 1 and 4
$\phi = 180° - 48.2° = 131.8°$ | $\phi = 60°$
$\phi = 180° + 48.2° = 228.3°$ | $\phi = 360° - 60° = 300°$

The solutions are $60°$, $131.2°$, $228.2°$, $300°$.

(c) Type 3: Equations of the type $\sin C \pm \sin D = 0$ or $\cos C \pm \cos D = 0$

These equations can be solved by using the appropriate sum and difference formulae to express the left-hand side as products.

Example 3.29   Find the roots of the equation $\sin 7\theta = \sin \theta$ for $0° \leqslant \theta \leqslant 90°$.

The equation can be written

$$\sin 7\theta - \sin \theta = 0$$
$$2 \cos \tfrac{1}{2}(7\theta + \theta) \sin \tfrac{1}{2}(7\theta - \theta) = 0$$
$$2 \cos 4\theta \sin 3\theta = 0$$

Either cos $4\theta = 0$ or sin $3\theta = 0$

$4\theta = 90°, 270°$ $3\theta = 0°, 180°$

$\theta = 22.5°, 67.5°$ $\theta = 0°, 60°$

Solutions are $0°, 22.5°, 60°, 67.5°$.

(d) Type 4: Equations for the form $a \sin \theta \pm b \cos \theta = c$

Such equations can be solved by first putting the left-hand side in the form $R \sin(\theta \pm \alpha)$ or $R \cos(\theta \pm \alpha)$. Any of these four variations can be used, but to avoid confusion it is best to use the one which matches the equation most closely. Once the equation is in this form it becomes an equation of type 1.

Example 3.30   Solve the equation $3 \sin \theta + 4 \cos \theta = 3$, for $0° \leqslant \theta \leqslant 360°$.

Since $\sin(\theta + \alpha) = \sin \theta \cos \alpha + \cos \theta \sin \alpha$

then $R \sin(\theta + \alpha) = [R \cos \alpha] \sin \theta + [R \sin \alpha] \cos \theta$

Now if $R \sin(\theta + \alpha) = [3] \sin \theta + [4] \cos \theta$

the coefficients of sin θ and cos θ on the right-hand sides in these two latter expressions must be identical, that is,

$R \cos \alpha = 3$ (i)

$R \sin \alpha = 4$ (ii)

Dividing (ii) by (i)

tan $\alpha = 1.3333$ (Note: smallest positive value of α is used)

$\alpha = 53.1°$

R can be used found by substitution or by squaring and adding (i) and (ii).

$R^2 \cos^2 \alpha + R^2 \sin^2 \alpha = 3^2 + 4^2$

$R^2(\cos^2 \alpha + \sin^2 \alpha) = 25$

$R = \sqrt{25}$ (since $\sin^2 \alpha + \cos^2 \alpha = 1$)

$R = 5$ (Note: positive value used)

The original equation can now be written

$5 \sin(\theta + 53.1°) = 3$

Let $\qquad X = \theta + 53.1°$

then $\qquad \sin X = \frac{3}{5} = 0.6$

The angle $POQ = \sin^{-1} 0.6 = 36.9°$.

Since the sine is positive the values of X must lie in quadrants 1 and 2.

$$X = 36.9°, \ 180° - 36.9°$$
$$\theta + 53.1° = 36.9°, \ 143.1°, \ 360 + 36.9°$$
$$\theta = -16.2°, \ 90°, \ 343.8°$$

(*Note*: $-16.2°$ is equivalent to $343.8°$. The solutions are $90°$ and $343°$).

The process of putting expressions such as a $a \sin \theta \pm b \cos \theta$ in the form $R \sin(\theta \pm \alpha)$ may be used to calculate maximum and minimum values of such expressions. In its new form it is a single sinusoidal function and therefore it is easily analysed.

EXERCISE 3.9

1. Find the values of θ in the range $0° \leqslant \theta \leqslant 360°$ which satisfy the following equations:
    (a) $\sin \theta = 0.875\,1$ (b) $\tan \theta = 1.391\,2$
    (c) $\cos \theta = 0.219\,5$ (d) $\sin \theta = -0.226\,4$
    (e) $\operatorname{cosec} \theta = 1.312\,7$ (f) $\cot \theta = -0.362\,7$
    (g) $\sec \theta = 1.917$ (h) $\sin 2\theta = -0.814\,1$
    (j) $\tan \frac{1}{2}\theta = 0.626\,7$ (k) $\cos 3\theta = -0.494\,4$
    (l) $2 \sin(\theta + 30°) - 1 = 0$ (m) $3 \sin(2\theta - 10.45°) + 0.669\,2 = 0$
    (n) $\tan(\theta + 81.17°) = 1.217$ (o) $\cos(3\theta - 27.9°) + 0.0927 = 0$.
2. What values of x in the range $0° \leqslant x \leqslant 180°$ are roots of the following equations?
    (a) $\sin 2x = \sin 60°$ (b) $\cos 3x = \cos 120°$
    (c) $\tan^2 x = \frac{1}{3}$ (d) $\tan \frac{3}{2}\theta = \sqrt{3}$
3. Find the roots of the following equations in the range $0°$ to $360°$ inclusive, unless otherwise stated:
    (a) $2 \sin^2 \theta + \sin \theta - 1 = 0$ (b) $16 \tan^2 \theta - 24 \tan \theta + 9 = 0$
    (c) $3 \tan^2 x - \sec x - 7 = 0$ (d) $12 \sin^2 A - \cos A - 6 = 0$
    (e) $\sin 2\theta = \sin \theta$ (f) $\cos 2\theta = \cos \theta$
    (g) $\cos 6\theta = \cos 4\theta$,   for θ in the range $0° - 180°$
    (h) $\sin 3\theta + \sin \theta = 0$,   for θ in the range $0° - 180°$
    (i) $\cos 4\theta + \cos \theta = 0$,   for θ in the range $0° - 180°$

(j) $\sin 6\theta - \sin 2\theta = 0$,   for θ in the range $0° - 180°$
(k) $40 \sin x + 9 \cos x = 16.1$ (l) $3 \cos x + 4 \sin x = 4.35$
(m) $7 \sin \theta - 8 \cos \theta = 9$ (n) $5 \cos \phi + 12 \sin \phi = 4.7$

3.15   Inverse trigonometric functions

The statement $\sin y = x$ means that the sine of the angle y is equal to x. An alternative statement would be that y is an angle of which the sine is x. The latter statement is written mathematically as

$$y = \sin^{-1} x$$

Similar interpretations can be given to $y - \cos^{-1} x$ and $y = \tan^{-1} x$. These functions ($\sin^{-1} x$, $\cos^{-1} x$, $\tan^{-1} x$) are called the inverse trigonometric functions. The angle is often expressed in radians, and for this reason the inverse functions are also called *arcsin*, *arcos*, and *arctan*.

Inverse functions are periodic so that again it becomes necessary to specify the smallest or principal value of the angle. For $\sin^{-1} x$, $\tan^{-1} x$, $\operatorname{cosec}^{-1} x$, $\cot^{-1} x$ the principal value is in the range $-\frac{1}{2}\pi \leqslant y \leqslant \frac{1}{2}\pi$.

For $\cos^{-1} x$ and $\sec^{-1} x$ the principal value is in the range $0 \leqslant y \leqslant \pi$. All principal values are obtained directly from the calculator by entering the x value *inclusive of its sign* and pressing the appropriate inverse key. If the angle is required in radians the calculator must be preset in the radians mode.

Example 3.31   Find the principal values of

(a) $\sin^{-1}(-\frac{1}{2})$ (b) $\tan^{-1}(\sqrt{3})$ (c) $\cos^{-1}(-1)$
(d) $\operatorname{cosec}^{-1}(\sqrt{2})$

(a)   With the calculator enter -0.5, press $\boxed{\sin^{-1}}$ key

$$\sin^{-1}(-\tfrac{1}{2}) = -30° = -\tfrac{1}{6}\pi \text{ rad}$$

Similarly

(b) $\tan^{-1}(\sqrt{3}) = 60° = \frac{1}{3}\pi \text{ rad}$
(c) $\cos^{-1}(-1) = 180° = \pi \text{ rad}$
(d) $\operatorname{cosec}^{-1}(\sqrt{2}) = 45° = \frac{1}{4}\pi \text{ rad}$

Example 3.32   Find the value of $\sin^{-1}(\frac{1}{4}) + \cos^{-1}(\frac{1}{3})$

$\sin^{-1}(\frac{1}{4}) = \sin^{-1} 0.25 = 0.252\,8 \text{ rad}$

$\cos^{-1}(\frac{1}{3}) = \cos^{-1} 0.333\,3 = 1.231\,1 \text{ rad}$

Therefore $\sin^{-1}(\frac{1}{4}) + \cos^{-1}(\frac{1}{3}) = 0.252\,8 + 1.231\,1$

$$= 1.483\,8 \text{ rad}$$

EXERCISE 3.10

1. Find the principal values of
    (a) $\sin^{-1} 0.369\,5$ (b) $\cos^{-1} 0.385\,1$
    (c) $\tan^{-1} 1.827\,9$ (d) $\operatorname{cosec}^{-1}(-1.559\,2)$
    (e) $\cos^{-1}(-0.281\,5)$ (f) $\sin^{-1} 1$
    (g) $\tan^{-1}\left(\dfrac{1}{\sqrt{3}}\right)$ (h) $\sec^{-1}\sqrt{2}$
    (j) $\cot^{-1}\infty$.

2. Draw the graphs of $y = \sin^{-1} x$, $y = \cos^{-1} x$ and $y = \tan^{-1} x$ for the principal values.

3. Find the value of
    (a) $3\sin^{-1}(\frac{1}{2}) - \cos^{-1}(\frac{1}{2})$ (b) $\tan^{-1}(\frac{1}{3}) + \tan^{-1}(\frac{3}{4})$.

4. Prove that

    $$3\tan^{-1}(1) = \tan^{-1}(-1)$$

5. Show that

    $$3\cos^{-1}(\tfrac{1}{2}) - \tan^{-1}\left(\frac{1}{\sqrt{3}}\right) = \sec^{-1}\left(-\frac{2}{\sqrt{3}}\right)$$

6. Use the formula for $\tan(A + B)$ to show that

    $$\tan^{-1} x + \tan^{-1} y = \tan^{-1}\left\{\frac{x + y}{1 - xy}\right\}$$

MISCELLANEOUS EXERCISE 3

1. (a) Prove that
        (i) $\dfrac{\cos(45^\circ + \theta)}{\cos(45^\circ - \theta)} = \dfrac{1 - \tan\theta}{1 + \tan\theta}$

        (ii) $\dfrac{\sin\theta + \sin 2\theta}{1 + \cos\theta + \cos 2\theta} = \tan\theta$.

    (b) Solve the equation $5\sin 2x + 3\cos 2x = 1$, giving the value of x between 0° and 180°.

2. Expand $4\sin(\theta + 30^\circ) + 8\sin(\theta - 60^\circ)$ and hence express it in the form $R\sin(\theta + \alpha)$. State the frequency, amplitude and phase angle of the new result.

3. Express $4\sin\theta - 5\cos\theta$ in the form $R\sin(\theta - \alpha)$. From the result find
    (a) the greatest and least values of $4\sin\theta - 5\cos\theta$ and the corresponding values of θ in the range 0° to 360°;
    (b) the solution of the equation $4\sin\theta - 5\cos\theta = 3$ for those values of θ in the range 0° to 360°.

4. In a survey of a large field with four corners A, B, C and D, it is known that B is 730 m due north of A, and C is due north of D. Observations from A show that C is N 40 E and D is N 75 E, and an observation from B gives D as S 65 E. Calculate the distance from C to D.

5. (a) In a certain type of ammeter the current I amperes, when the deflection of the needle is θ degrees, is given by the equation

    $$I^2 \sin(\theta + 30°) = 0.735 \cos(\theta + 30°)$$

    Find the deflection θ when the current is 1.65A.

    (b) Express $(\cos\theta - \sin\theta)$ in the form $R\cos(\theta + \alpha)$ and find the values of R and α.

    Hence or otherwise, solve the equation

    $$\cos\theta - \sin = \theta\tfrac{5}{16}$$

    giving all values of θ between 0 and 360.

6. (a) Prove that $(\cot\theta\sin\theta + \tan\theta\cos\theta)^2 = 1 + \sin 2\theta$.

    (b) Given that

    $$\frac{8\sin\theta}{5\sin\theta - \cos\theta} = 1$$

    find the values of the angle θ which lie between 0 and 360.

7. Plot $y = 2\sin x$ and $y = \cos x$ using the same axes, for values of x from 0 to π.

    Use these graphs to plot $y = 2\sin x - \cos x$, and hence find one value of x for which $2\sin x = \cos x$.

8. $ABCD$ is a plot of ground. $AB = 18$ m, $BC = 12$ m, $CD = 21$ m, $DA = 29$ m, and $AC = 22$ m. Find all the internal angles, also the length BD.

9. (a) Using the formula calculate the largest angle in the triangle with sides 136, 84.5 and 76.5 mm.

    $$\tan\tfrac{1}{2}A = \sqrt{\left\{ \frac{(s-b)(s-c)}{s(s-a)} \right\}}$$

10. (a) Express $\cos\theta + \sqrt{3}\sin\theta$ in the form $r\cos(\theta - \alpha)$ and find the value of θ between 0° and 360 for which $\cos\theta + \sqrt{3}\sin\theta = 1.24$.

    (b) Find all the angles between 0 and π radians for which $\sin 7x - \sin x = 0$.

11. (a) Show that
    (i) $\tan\theta + \cot\theta = \sec\theta\operatorname{cosec}\theta$,
    (ii) $\sec^2\theta(\sin 2\theta + \cos 2\theta - 1) = 2\tan\theta(1 - \tan\theta)$.

    (b) Solve the equation $3 \sin \theta - 2 \cos \theta = 1$, giving the values of θ between $0°$ and $180°$.

12. (a) Given that $x = \sin(2t + \frac{1}{4}\pi)$, $y = \cos(2t - \frac{1}{4}\pi)$ express $x + y$ in the form $R \sin(2t + \alpha)$, where R is a positive number and α is in radians.

    (b) Find the smallest positive values for t for which $x + y$ (i) is zero, (ii) has it greatest and least values, and state these values. Hence sketch the graph of $x + y$ over the complete period.

13. Solve the following equations, giving the values of x between 0 and 360 :

    (a) $6 \cos 2x - 5 \sin x + 1 = 0$;

    (b) $4 \cos x - 3 \sin x = 2.5$.

14. Taking the smallest positive angle in each case, show that $\cos^{-1}(-0.8) + \tan^{-1}(0.75) = \pi$.

Chapter 4

Differentiation

4.1 Functional notation

An equation such as

$$y = 3x^2 + 4x - 6 \qquad (4.1)$$

contains two variables. For every value assigned to x there is a corresponding value of y. The variable x is called the **independent variable**, and y is called the **dependent variable**. y is said to be a function of x, that is,

$$y = f(x)$$

where in this particular example

$$f(x) = 3x^2 + 4x - 6 \qquad (4.2)$$

Substitution is indicated by replacing x by the particular value in $f(x)$. If $x = 2$ then from equation (4.2)

$$f(2) = 3(2)^2 + 4(2) - 6$$
$$= 14,$$

that is, at

$$x = 2, \qquad y = 14$$

The meaning of $f(2)$ is shown in Fig. 4.1a. It represents the length of the ordinate at $x = 2$. In general $f(x)$ is the length of the ordinate at any point x.

$f(4)$ and $f(3)$ are shown in Fig 4.1b. The difference $f(4) - f(3)$ is shown as the length AB in the figure.

$$f(4) = 3.4^2 + 4.4 - 6$$
$$= 58$$
$$f(3) = 3.3^2 + 4.3 - 6$$
$$= 33$$
$$AB = f(4) - f(3)$$
$$= 58 - 33$$
$$= 25$$

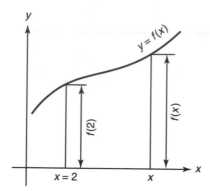

Fig. 4.1(a)

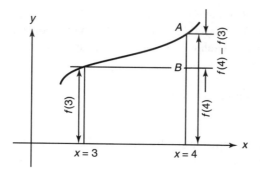

Fig. 4.1(b)

EXERCISE 4.1

1.   Given $f(x) = 3x^2 + 5x - 8$, find

$$f(0), \quad f(2), \quad f(-1), \quad 3f(1), \quad f(2) - f(-1)$$

2.   Given $f(x) = 6x^3$, find

$$3f(1) - 4f(-1), \quad f(x+h) - f(x), \quad \frac{f(h)}{h}$$

3.   Given $f(x) = 4x^4$, find

$$f(3) \div f(1), \quad f(4) \times f(-1)$$

4.2   Incremental values

In this chapter use is made of very small parts of a variable, called increments. For example, a small length of the x axis is often required. Such a length is written as δx, which stands for a small increment of x. Similarly a small increment of y is written as δy.

Small increments of x and y are shown in Fig. 4.2. In this instance δy is seen to be the difference between $f(x + \delta x)$ and $f(x)$.

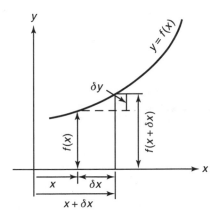

Fig. 4.2

4.3 Limits

(a) The meaning of infinity

A number N which is extremely large is said to approach infinity, and written as $N \to \infty$. For example, the distance between the Earth and the nearest star is very large and approaches infinity. Infinity is not a number in the sense that it can be multiplied or divided. Rather it stands for the statement that a number is extremely large.

(b) The meaning of a limit

The idea of a limit may be understood by considering the equation

$$y = 1 + \frac{1}{x}$$

The values of y are determined for various values of x, and given in Table 4.1. The values are plotted on a graph as in Fig. 4.3.

As x increases, the value of y gets closer to 1, which is the ultimate value of y as $x \to \infty$. This ultimate value is called the limit of y as x approaches infinity and is written as

$$\lim_{x \to \infty} y = 1$$

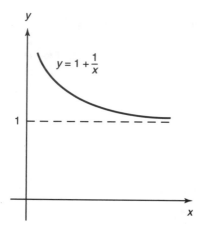

Fig. 4.3

Table 4.1

x	$\frac{1}{2}$	1	10	100	1000
y	3	2	1.1	1.01	1.001

Example 4.1   Find the limit of

$$\left(4 + \frac{1}{x}\right)^{1/x} \quad \text{as} \quad x \to \infty$$

It is obvious that $\dfrac{1}{\infty} = 0$.

For convenience write $1/x$ as n. Thus as $x \to \infty$, $\dfrac{1}{x} \to 0$ and hence $n \to 0$.

Thus we require the limit

$$(4 + n)^n \quad \text{as} \quad n \to 0$$

The inside of the bracket tends to 4, and the index tends to 0. Therefore, the expression tends to 4^0 which is 1.

EXERCISE 4.2

What are the limits of the following expressions?

1. $4 + \dfrac{2}{x^2}$ as $x \to \infty$

2. $\dfrac{1-3^{x^2}}{3^{x^2}}$ as $x \to 0$ and $x \to \infty$.

3. $\dfrac{1-\dfrac{1}{x}}{1+\dfrac{1}{x}}$ as $x \to \infty$ and $x \to 0$.

(c) The area of a trapezium as the width becomes extremely small

In Fig. 4.4a a trapezium is shown which is composed of an isosceles triangle (shaded) on top of a rectangle.
Its area A_t, is

$$A_t = 5h + \tfrac{1}{2}h^2$$

The area A_t is examined as the width h of the strip decreases, keeping the side AC constant at 5 units. For each value of h, the ratio of the area of the trapezium to the area of the rectangle is determined.

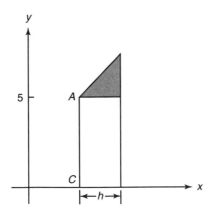

Fig. 4.4(a)

Table 4.2

h	Area of rectangle $A_r = 5h$	Area of trapezium $A_t = 5h + \tfrac{1}{2}h^2$	Ratio A_t/A_r
1	5	5.5	1.1
0.1	0.5	0.505	1.01
0.01	0.05	0.050 05	1.001

As h decreases the ratio $\dfrac{A_t}{A_r}$ gets closer and closer to 1, that is, as $h \to 0$ the ratio approaches a limiting value of 1,

$$\lim_{h \to 0} A_t = A_r \tag{4.3}$$

The area of the trapezium becomes identical with the area of the rectangle as $h \to 0$. This conclusion is extremely important and is used in Chapter 7.

(d) The indeterminate ratio $\dfrac{0}{0}$

In such a ratio the 0's really represent extremely small quantities. More exactly the ratio should be written as $\varepsilon_1/\varepsilon_2$, where the ε's are extremely small. The relative values of the ε's are unknown so that the fraction $\dfrac{0}{0}$ is indeterminate.

An example of such a ratio is shown in Fig. 4.4b. For small distances from A, the gradient is given by $\varepsilon_1/\varepsilon_2$, both ε's being extremely small.

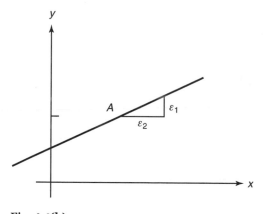

Fig. 4.4(b)

4.4   The gradient of a curve

Consider the graph of $y = x^2$, as shown in Fig. 4.5(a). The gradient of the chord AB is given by

$$\frac{BE}{AE} = \frac{BD - AC}{AE}$$

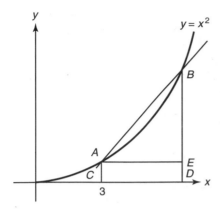

Fig. 4.5(a)

Let AC be the ordinate at $x = 3$ and BD the ordinate at any other point x. Then

$$AC = f(3) = 3^2 = 9$$
$$BD = f(x) = x^2$$

The gradient of the chord is examined as B moves closer to A. The values of the gradient are given in Table 4.3.

Table 4.3

x value of B	$f(x) = x^2$	$f(x) - f(3) = BE$	$x - 3 = AE$	Gradient $\dfrac{BE}{AE}$
4	16	7	1	7
3.1	9.61	0.61	0.1	6.1
3.01	9.060 1	0.060 1	0.01	6.01
3.001	9.006 001	0.006 001	0.001	6.001

(i)   From the table the limiting value of the gradient of the chord is 6.
(ii) Fig. 4.5(b) shows that as B approaches A the chord becomes the tangent at A.
(iii) Hence the limiting value of 6 is the gradient of the tangent at A.
(iv) Since the gradient of the tangent represents the gradient of the curve then the value 6 is the gradient of the curve at A.

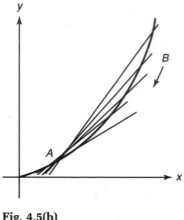

Fig. 4.5(b)

4.5   Differentiation

The mathematical process of finding an expression for the gradient of a curve at any point is called **differentiation**. It is carried out as follows. Fig. 4.6a shows two points A and B very close together on the curve. AE is an increment δx, and hence BE is an increment δy. The gradient of the chord is

$$\frac{\delta y}{\delta x}$$

As B approaches A, i.e. as $\delta x \to 0$, then $\delta y \to 0$, and the above fraction becomes $0/0$, which is indeterminate.

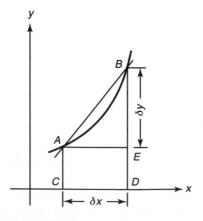

Fig. 4.6(a)

As $\delta x \rightarrow 0$ the chord becomes a tangent, so that the fraction $0/0$ actually corresponds to the gradient of the tangent. The limiting value of $\delta y/\delta x$ is written as dy/dx, i.e.,

$$\frac{dy}{dx} = \lim_{x \rightarrow 0} \frac{\delta y}{\delta x} \tag{4.4}$$

It should be noted that dy/dx is not a fraction although it is derived as the limit of a fraction. dy/dx is called the **derivative or the differential coefficient** of y with respect to x.

This process is now applied to the determination of dy/dx for the function

$$y = x^2$$

Referring to Fig. 4.6b, it is seen that

$$AC = f(x) \qquad = x^2$$
$$BD = f(x + \delta x) = (x + \delta x)^2$$
$$= x^2 + 2x.\delta x + (\delta x)^2$$
$$BE = \delta y \qquad = f(x + \delta x) - f(x)$$
$$= 2x.\delta x + (\delta x)^2$$
$$\frac{BE}{AE} = \frac{\delta y}{\delta x} \qquad = \frac{\delta x(2x + \delta x)}{\delta x}$$
$$= 2x + \delta x$$

As $\delta x \rightarrow 0, \qquad \dfrac{\delta y}{\delta x} \rightarrow \dfrac{dy}{dx}$

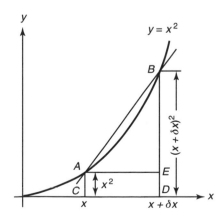

Fig. 4.6(b)

Therefore $\dfrac{dy}{dx} = 2x$

At $x = 3$, the value of the gradient becomes 6 which agrees with the result of Section 4.4. This method of obtaining dy/dx is called **differentiation from first principles**. The stages of obtaining dy/dx are

1. Find $f(x + \delta x)$
2. Find $f(x + \delta x) - f(x)$
3. Divide by δx
4. Let $\delta x \to 0$ to find the limit

Example 4.2   Differentiate from first principles

(i) $y = 4x^3$
(ii) $y = 2x^{\frac{1}{2}}$

Find the value of the gradient of each curve at $x = 4$.

(i) $\begin{aligned} y = f(x) &= 4x^3 \\ y + \delta y = f(x + \delta x) &= 4(x + \delta x)^3 \\ &= 4x^3 + 12x^2\,\delta x + 12x(\delta x)^2 + 4(\delta x)^3 \end{aligned}$

Subtract:

$$\delta y = f(x + \delta x) - f(x) = 12x^2\,\delta x + 12x(\delta x)^2 + 4(\delta x)^3$$

$$\text{Gradient of chord} = \frac{\delta y}{\delta x} = \frac{12x^2\,\delta x + 12x(\delta x)^2 + 4(\delta x)^3}{\delta x}$$

$$= 12x^2 + 12x\,\delta x + 4(\delta x)^2$$

As $\delta x \to 0$, chord $\to$ tangent, and $\dfrac{\delta y}{\delta x} \to \dfrac{dy}{dx}$

$\dfrac{dy}{dx} = 12x^2$

At $x = 4$,

$$\frac{dy}{dx} = 12(4)^2 = 192$$

(ii) $\begin{aligned} y = f(x) &= 2x^{\frac{1}{2}} \\ y + \delta y = f(x + \delta x) &= 2(x + \delta x)^{\frac{1}{2}} \end{aligned}$

The factor $(x + \delta x)^{\frac{1}{2}}$ may be expanded using the Binomial theorem as described in Chapter 10:

$$(x + \delta x)^{\frac{1}{2}} = x^{\frac{1}{2}} + \tfrac{1}{2}x^{-\frac{1}{2}}\,\delta x + \frac{(\frac{1}{2})(-\frac{1}{2})}{1.2}x^{-\frac{3}{2}}(\delta x)^2 + \cdots$$

Therefore

$$y + \delta y = 2x^{\frac{1}{2}} + x^{-\frac{1}{2}}\delta x - \tfrac{1}{4}x^{-\frac{3}{2}}(\delta x)^2 + \cdots$$

$$y = 2x^{\frac{1}{2}}$$

Subtract:

$$\delta y = x^{-\frac{1}{2}}\delta x - \tfrac{1}{4}x^{-\frac{3}{2}}(\delta x)^2$$

Gradient of chord

$$\frac{\delta y}{\delta x} = x^{-\frac{1}{2}} - \tfrac{1}{4}x^{-\frac{3}{2}}\delta x + \cdots$$

As $\delta x \rightarrow 0$

$$\frac{dy}{dx} = x^{-\frac{1}{2}}$$

At $x = 4$,

$$\frac{dy}{dx} = \frac{1}{x^{\frac{1}{2}}} = \frac{1}{4^{\frac{1}{2}}}$$

$$= \tfrac{1}{2}$$

EXERCISE 4.3

Differentiate the following from first principles:

1. x^6
2. $3x^4$
3. $2x^{-1}$
4. $7x^{\frac{3}{4}}$
5. $2x^{-2}$
6. $3x^{-\frac{1}{2}}$.

4.6   Differentiation of $y = px^n$, where p, n are constants

The method is exactly the same as shown for $y = x^2$ in Section 4.5.

$$y = f(x) \qquad = px^n \tag{i}$$
$$y + \delta y = f(x + \delta x) = p(x + \delta x)^n$$

The bracket on the right-hand side is expanded using the binomial theorem:

$$y + \delta y = p\left[x^n + nx^{n-1}\,\delta x + \frac{n(n-1)}{1.2}x^{n-2}(\delta x)^2 + \cdots\right] \tag{ii}$$

Subtracting equation (i) from (ii) gives

$$\delta y = npx^{n-1}\,\delta x + \tfrac{1}{2}n(n-1)px^{n-2}(\delta x)^2 + \cdots.$$

The gradient of the chord is $\delta y/\delta x$. Therefore the last equation is divided by δx,

$$\frac{\delta y}{\delta x} = npx^{n-1} + \tfrac{1}{2}n(n-1)px^{n-2}(\delta x) + \cdots .$$

As $\delta x \to 0$, $\delta y/\delta x \to dy/dx$. Therefore, dy/dx is obtained by making $\delta x \to 0$, on the right-hand side of the above equation. As a result the second and all following terms disappear. Thus

$$\boxed{\text{If } \quad y = px^n, \qquad \frac{dy}{dx} = npx^{n-1}} \tag{4.5}$$

This result can now be used to differentiate any monomial in x, without resorting to the method of first principles. Example 4.3 illustrates the use of the above result as a working rule.

Example 4.3   Differentiate with respect to x (i) $8x^7$, (ii) $3x^{-1/2}$ (iii) $4x^{2/3}$, using rule 4.5.

(i) $\dfrac{dy}{dx} = 8 \times 7x^{7-1}$ (ii) $\dfrac{dy}{dy} = 3(-\tfrac{1}{2})x^{-\frac{1}{2}-1}$ (iii) $\dfrac{dy}{dx} = 4 \times \tfrac{2}{3}x^{\frac{2}{3}-1}$

$\qquad\quad = 56x^6$ $\qquad\quad = -\tfrac{3}{2}x^{-\frac{3}{2}}$ $\qquad\quad = \tfrac{8}{3}x^{-\frac{1}{3}}$

EXERCISE 4.4

Differentiate the following, with respect to x:

1. $6x^7$
2. $7x^{-3}$
3. $4x^{-1}$
4. $3x^{\frac{1}{2}}$
5. $7x^{\frac{7}{2}}$
6. $5x^{-\frac{3}{2}}$
7. $-2x^{-\frac{1}{2}}$
8. $9x^{-\frac{2}{3}}$
9. $-\tfrac{2}{3}x^{\frac{3}{2}}$
10. Find the gradient of the curve $y = 7x^4$ at the points $x = 1$ and $x = 3$.
11. At what point on the curve $y = 4x^3$ is the gradient equal to (a) 0, (b) 108?

4.7   Another notation for the differential coefficient

So far we have written equations in the form $y = f(x)$, such as

$$y = 7x^4$$

and obtained dy/dx. This process can be expressed as

$$\frac{d}{dx}(7x^4) = 28x^3$$

Example 4.4 Evaluate the following:

(i) $\frac{d}{dx}(2x^{-4})$,   (ii) $\frac{d}{d\theta}(5\theta^{1.5})$,   (iii) $\frac{d}{dt}(-2t^{\frac{3}{4}})$.

(i) $\frac{d}{dx}(2x^{-4}) = -8x^{-5}$

(ii) $\frac{d}{d\theta}(5\theta^{1.5}) = 7.5\theta^{0.5}$

(iii) $\frac{d}{dt}(-2t^{\frac{3}{4}}) = -\frac{3}{2}t^{-\frac{1}{4}}$.

4.8 Differentiation of polynomials

In a function such as

$$y = 4x^3 + \frac{2}{x^2}$$

each term must be written in the form px^n, and then the function is differentiated term by term in accordance with rule 4.5.

$$y = 4x^3 + 2x^{-2}$$

$$\frac{dy}{dx} = 12x^2 - 4x^{-3}$$

Some functions are reducible to polynomials as shown in Example 4.5.

Example 4.5 Differentiate with respect to x:

(i) $\dfrac{3x^2 + 4x^5}{x^3}$

(ii) $(2x + 3)^2$

(i) Divide each term in the numerator by x^3

$$y = 3x^{-1} + 4x^2$$

$$\frac{dy}{dx} = -3x^{-2} + 8x$$

(ii)   Multiply out

$$y = 4x^2 + 12x + 9$$

$$\frac{dy}{dx} = 8x + 12$$

EXERCISE 4.5

Differentiate the following, with respect to the appropriate variable:

1. $4x^2 + 2x^4$

2. $3x^4 - \dfrac{2}{x^2} - \dfrac{3}{x^{\frac{1}{2}}}$

3. $5x^3 - \dfrac{7}{x^{\frac{1}{2}}} + \dfrac{2}{x^{\frac{3}{2}}}$

4. $\dfrac{3}{x^7} - \dfrac{3}{x^4}$

5. $\dfrac{7t - 4t^7 - 3}{t^3}$

6. $\dfrac{7\theta - 4\theta^3}{\theta^{\frac{1}{2}}}$

7. $(2x + 1)^2$

8. $\dfrac{(2x - 1)(x + 3)}{x^3}$

9.   Given that the curve $y = ax^2 + bx + c$ passes through the points $(0, 1)$, $(-1, 2)$, and the gradient of the curve at $x = 0$ is 2, find a, b and c.

10.   Given that $y = 3x^2 - 6x + 4$, find the coordinates of the points at which

(a) $\dfrac{dy}{dx} = 0$,   (b) $\dfrac{dy}{dx} = 6$

4.9   Repeated differentiation

If a function is differentiated the differential coefficient is written as $\dfrac{dy}{dx}$ or $f'(x)$. If the expression is differentiated again the second differential coefficient is written as

$$\frac{d^2y}{dx^2} \quad \text{or} \quad f''(x)$$

Similarly the third differentiation is

$$\frac{d^3y}{dx^3} \quad \text{or} \quad f'''(x)$$

Thus if $y = f(x) = 4x^5$

$$\frac{dy}{dx} = f'(x) = 20x^4$$

$$\frac{d^2y}{dx^2} = f''(x) = 80x^3$$

$$\frac{d^3y}{dx^3} = f'''(x) = 240x^2$$

EXERCISE 4.6

1.   Given $y = 4x^3 - 3x^2 + 2$, find $\dfrac{dy}{dx}$ and $\dfrac{d^2y}{dx^2}$. Calculate the value of each
   at $x = 2$.

2.   Given $y = 7x^6$, show that

$$\tfrac{1}{2}x\frac{dy}{dy} + \tfrac{1}{10}x^2\frac{d^2y}{dx^2} = 6y$$

3.   Given $y = 4x^2 - 3x$, show that

$$2x\frac{dy}{dx} - x^2\frac{d^2y}{dx^2} = 2y$$

4.   Given $y = ax^2 + bx$, find a and b if $\dfrac{d^2y}{dx^2} = 6$
   and $\dfrac{dy}{dx} = 7$ when $x = 1$.

4.10   Differentiation of a function of a function

A function of a function can be regarded as a function within a function, such as, for example,

$$y = (x^2 + 1)^8$$

It is inconvenient, and sometimes impossible, to expand such functions into polynomials, so that differentiation as outlined in Section 4.8 cannot be carried out.

   The way to tackle the problem is to substitute u for the inner function; that is, substitute

$$u = x^2 + 1 \qquad\qquad\qquad\text{(i)}$$

so that

$$y = u^8 \qquad\qquad\qquad\text{(ii)}$$

As x increases to $x + \delta x$ there will be corresponding increases of δu and δy in u and y respectively. Since these are finite quantities it is possible to write

$$\frac{\delta y}{\delta x} = \frac{\delta y}{\delta u} \times \frac{\delta u}{\delta x}$$

Both y and u are dependent variables of x. Therefore as $\delta x \to 0$ and $\delta u \to 0$, then

$$\frac{\delta y}{\delta x} \to \frac{dy}{dx}, \qquad \frac{\delta y}{\delta u} \to \frac{dy}{du}, \qquad \frac{\delta u}{\delta x} \to \frac{du}{dx}$$

Therefore, in the limit

Function of a function rule

$$\boxed{\frac{dy}{dx} = \frac{dy}{du} \times \frac{du}{dx}} \tag{4.6}$$

Returning to the example,

$$\frac{du}{dx} = 2x, \text{ from equation (i)}$$

and

$$\frac{dy}{du} = 8u^7, \text{ from equation (ii)}$$

so that

$$\frac{dy}{dx} = 8u^7 2x$$

$$= 8(x^2 + 1)^7 2x$$

$$= 16x(x^2 + 1)^7$$

Example 4.6   Differentiate $y = (x^2 - 6x + 1)^{\frac{1}{2}}$

Let $u = x^2 - 6x + 1$, so that $\dfrac{du}{dx} = 2x - 6$

Then

$$y = u^{\frac{1}{2}}, \qquad \text{and} \qquad \frac{dy}{du} = \tfrac{1}{2}u^{-\frac{1}{2}}$$

$$\frac{dy}{dx} = \frac{dy}{du} \times \frac{du}{dx}$$

$$= \tfrac{1}{2}u^{-\frac{1}{2}} \times (2x - 6)$$

$$= \tfrac{1}{2}(x^2 - 6x + 1)^{-\frac{1}{2}} \times 2(x - 3)$$

$$= (x - 3)(x^2 - 6x + 1)^{-\frac{1}{2}}$$

Differentiation of a function of a function may be carried out without going through this substitution procedure each time. The method can be summarised as follows:

1. Differentiate the bracket as a simple variable
2. Differentiate the inside of the bracket
3. $\dfrac{dy}{dx}$ = product of (i) and (ii)

Example 4.6 now becomes

(i) $d(\quad)^{\frac{1}{2}} = \frac{1}{2}(\quad)^{-\frac{1}{2}}$

(ii) $d(x^2 - 6x + 1) = 2x - 6$

(iii) $\dfrac{dy}{dx} = (x - 3)(x^2 - 6x + 1)^{-\frac{1}{2}}$

EXERCISE 4.7

Differentiate the following with respect to the appropriate variable:

1. $(2x + 3)^4$
2. $(7x^2 - 3)^6$
3. $(-3x^{\frac{1}{2}} - 4)^3$
4. $(4t - 3)^{\frac{1}{2}}$
5. $\left(6t^2 - \dfrac{6}{t^2}\right)^{\frac{3}{4}}$
6. $\left(\theta - \dfrac{1}{\theta}\right)^2$
7. $\dfrac{1}{\sqrt[3]{x^2 - 3}}$
8. $\dfrac{1}{\sqrt{t^2 - 3t + 4}}$

4.11   Differentiation of the product of two functions

Consider the equation

$$y = x^3(x + 1)^4$$

The right-hand side can be considered as the product of two functions, namely, x^3 and $(x + 1)^4$. In order to develop a rule for differentiating such a product let

$$u = x^3 \quad \text{and} \quad v = (x + 1)^4 \tag{i}$$

Therefore

$$y = u \cdot v \tag{ii}$$

If x is replaced by $(x + \delta x)$, in equations (i) then u, v and hence y become $u + \delta u$, $v + \delta v$, $y + \delta y$ so that

$$y + \delta y = (u + \delta v)(v + \delta v)$$
$$= uv + v\,\delta u + u\,\delta v + \delta u\,\delta v \tag{iii}$$

Subtract (ii) from (iii). Then

$$\delta y = v\,\delta u + u\,\delta v + \delta u\,\delta v$$

Divide throughout by δx.

$$\frac{\delta y}{\delta x} = v\frac{\delta u}{\delta x} + u\frac{\delta v}{\delta x} + \frac{\delta u}{\delta x}\delta v \qquad\qquad\text{(iv)}$$

As $\delta x \to 0$ the fractions in equation (iv) become the respective differential coefficients; that is,

$$\frac{\delta y}{\delta x} \to \frac{dy}{dx}, \qquad \frac{\delta v}{\delta x} \to \frac{dv}{dx}, \qquad \frac{\delta v}{\delta x} \to \frac{dv}{dx}, \qquad \delta v \to 0$$

In the limit, equation (iv) becomes

$$\frac{dy}{dx} = v\frac{du}{dx} + u\frac{dv}{dx}$$

Product rule

$$\boxed{\text{If } y = uv \text{ then} \qquad \frac{dy}{dx} = v\frac{du}{dx} + u\frac{dv}{dx}} \qquad\qquad\text{(4.7)}$$

Using rule 4.5 in the original example we differentiate equations (i)

$$\frac{du}{dx} = 3x^2, \qquad \frac{dv}{dx} = 4(x+1)^3$$

dy/dx is now obtained from rule (4.7)

$$\frac{dy}{dx} = (x+1)^4 \cdot 3x^2 + x^3 \cdot 4(x+1)^3$$

Common factors are taken outside a bracket

$$\frac{dy}{dx} = x^2(x+1)^3[3(x+1) + 4x]$$

$$= x^2(x+1)^3(7x+3)$$

EXERCISE 4.8

Differentiate the following products with respect to x:

1. $x^2(4x-3)$
2. $4x^3(x-3)^{\frac{1}{2}}$
3. $7x^2(x^2-1)^4$
4. $x^{\frac{1}{2}}(7x^2-4)^2$
5. $x^{\frac{1}{2}}(1+x^2)^{\frac{1}{2}}$
6. $\sqrt{(x+1)}(x^2-4)$

4.12   Differentiation of a quotient of two functions

Consider the equation

$$y = \frac{x^3}{(x+1)^4}$$

The right-hand side is the quotient of two functions, namely x^3 and $(x^2+1)^4$. In order to differentiate such a quotient let

$$u = x^3 \quad \text{and} \quad v = (x+1)^4 \tag{i}$$

Therefore

$$y = \frac{u}{v} \tag{ii}$$

If x is replaced by $(x+\delta x)$ in equations (i) then u, v and hence y become $u + \delta u$, $v + \delta v$, $y + \delta y$ so that

$$y + \delta y = \frac{u + \delta u}{v + \delta v} \tag{iii}$$

Subtract (ii) from (iii)

$$\delta y = \frac{u + \delta u}{v + \delta v} - \frac{u}{v}$$

The LCM is $v(v + \delta v)$, so that

$$y = \frac{(u + \delta u)v - u(v + \delta v)}{v(v + \delta v)}$$

$$= \frac{v\,\delta u - u\,\delta v}{v(v + \delta v)}$$

Divide both sides by δx.

$$\frac{\delta y}{\delta x} = \frac{v\dfrac{\delta u}{\delta x} - u\dfrac{\delta v}{\delta x}}{v(v + \delta v)} \tag{iv}$$

As $\delta x \to 0$,

$$\frac{\delta y}{\delta x} \to \frac{dy}{dx}, \quad \frac{\delta u}{\delta x} \to \frac{du}{dx}, \quad \frac{\delta v}{\delta x} \to \frac{dv}{dx}, \quad \delta v \to 0$$

In the limit equation (iv) becomes $\dfrac{dy}{dx} = \dfrac{v\dfrac{du}{dx} - u\dfrac{dv}{dx}}{v^2}$

Quotient rule

$$\text{If } y = \frac{u}{v} \quad \text{then} \quad \frac{dy}{dx} = \frac{v\dfrac{du}{dx} - u\dfrac{dv}{dx}}{v^2} \tag{4.8}$$

Returning to the original example, and differentiating equations (i).

$$\frac{du}{dx} = 3x^2, \quad \frac{dv}{dx} = 4(x+1)^3$$

dy/dx for the original equation is obtained by substituting these last results, and equations (i) into rule (4.8).

$$\frac{dy}{dx} = \frac{(x+1)^4.3x^2 - x^3.4(x+1)^3}{(x+1)^8}$$

The numerator is simplified by taking common factors outside the bracket.

$$\frac{dy}{dx} = \frac{x^2(x+1)^3[3(x+1) - 4x]}{(x+1)^8}$$

The terms inside the square bracket are simplified and collected.

$$\frac{dy}{dx} = \frac{x^2[-x+3]}{(x+1)^5}$$

EXERCISE 4.9

Differentiate the following, with respect to x using the quotient rule given by equation (4.8):

1. $\dfrac{x^2}{(2x+1)}$

2. $\dfrac{2x^3}{2x^2+1}$

3. $\dfrac{x^2-1}{x^2+1}$

4. $\dfrac{4x}{\sqrt{(x^2+1)}}$

5. $\dfrac{x^2+1}{\sqrt{(x^2+4)}}$

6. $\dfrac{x^2-3x+1}{(x+1)}$

7. $\dfrac{\sqrt{x}}{x+2}$

4.13   The limit of $\sin\theta/\theta$ as $\theta \to 0$, θ being in radians

When $\theta = 0$, $\sin\theta/\theta = \dfrac{0}{0}$ which is indeterminate. The limiting value of $\sin\theta/\theta$ as $\theta \to 0$ is found by making use of Fig. 4.7.

From triangle AOC $h = R\sin\theta$ $\qquad\qquad\qquad\qquad$ (i)

From triangle DOB $H = R\tan\theta$ $\qquad\qquad\qquad\qquad$ (ii)

Area of sector of circle $OAB = \frac{1}{2}R^2\theta$ (see Appendix 1)

Area of triangle $OAB = \frac{1}{2}Rh = \frac{1}{2}R^2\sin\theta$ (using(i))

Area of triangle $OBD = \frac{1}{2}RH = \frac{1}{2}R^2\tan\theta$ (using(ii))

From Fig. 4.7 it is clear that

area of triangle OAB < area of sector AOB < area triangle OBD

Using the expressions obtained for these areas:

$$\tfrac{1}{2}R^2\sin\theta < \tfrac{1}{2}R^2\theta < \tfrac{1}{2}R^2\tan\theta$$

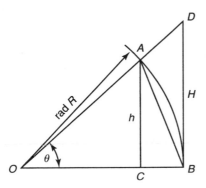

Fig. 4.7

Divide throughout by $\frac{1}{2}R^2\sin\theta$:

$$1 < \frac{\frac{1}{2}R^2\theta}{\frac{1}{2}R^2\sin\theta} < \frac{\frac{1}{2}R^2\tan\theta}{\frac{1}{2}R^2\sin\theta}$$

$$1 < \frac{\theta}{\sin\theta} < \frac{1}{\cos\theta}$$

$\theta/(\sin\theta)$ lies between 1 and $\dfrac{1}{\cos\theta}$. But as $\theta \to 0$, $\cos\theta \to 1$, therefore $\theta/(\sin\theta)$ lies between 1 and 1. This conclusion is illustrated in Fig. 4.8. The value of $\theta/(\sin\theta)$ lies somewhere in the shaded region for each value of θ, where the two boundaries are the graphs of $y = 1$ and $y = 1/(\cos\theta)$. As $\theta \to 0$ the shaded region contracts to a point A having a value 1, so that $\theta/(\sin\theta)$ reaches a value of 1, that is

$$\lim_{\theta \to 0} \frac{\theta}{\sin\theta} = 1$$

Stated in another way, when θ is small,

$$\sin\theta \simeq \theta$$

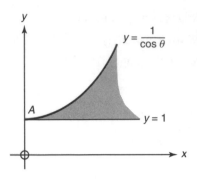

Fig. 4.8

This can be illustrated quite easily by considering a numerical example:

Take $\qquad\qquad\qquad\qquad \theta = 0.5°$

From tables $\qquad\qquad \sin 0.5° = 0.008\ 7$

$$0.5 \text{ in radians} = 0.5 \times \frac{\pi}{180} = 0.008\ 7,$$

i.e. for values of θ of the order $0.5°$, $\sin \theta = \theta$, correct to 4 places of decimals.

4.14   The differential coefficient of the trigonometric functions

(a) y = sin x

Initial note: From equ(3.20) the difference between two sines was expressed as a product. The expression is required in the form

$$\sin(x + \delta x) - \sin x = 2\cos(x + \tfrac{1}{2}\,\delta x)\sin(\tfrac{1}{2}\,\delta x) \qquad\qquad\qquad \text{(i)}$$

Referring to Fig. 4.9:

at A $\qquad\qquad\qquad\qquad y = f(x) \qquad\quad = \sin x$

at B $\qquad\qquad\qquad y + \delta y = f(x + \delta x) = \sin(x + \delta x)$

Subtract $\qquad\qquad\qquad\quad \delta y = \sin(x + \delta x) - \sin x$

$$= 2\cos(x + \tfrac{1}{2}\,\delta x)\sin(\tfrac{1}{2}\delta x) \text{ using equation (i)}$$

Gradient of chord $\qquad \dfrac{\delta y}{\delta x} = \dfrac{2\cos(x + \tfrac{1}{2}\,\delta x)\sin(\tfrac{1}{2}\,\delta x)}{\delta x}$

$$= \cos(x + \tfrac{1}{2}\,\delta x) \cdot \dfrac{\sin(\tfrac{1}{2}\,\delta x)}{(\tfrac{1}{2}\,\delta x)}$$

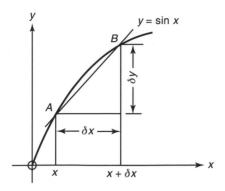

Fig. 4.9

As $\delta x \to 0$

$$\frac{\delta y}{\delta x} \to \frac{dy}{dx}$$

$$x + \tfrac{1}{2}\delta x \to x$$

$$\frac{\sin(\tfrac{1}{2}\delta x)}{(\tfrac{1}{2}\delta x)} \to 1$$

Therefore

$$\frac{dy}{dx} = \cos x$$

$$\boxed{\frac{d(\sin x)}{dx} = \cos x}$$

(4.9)

(b) $y = \cos x$

Initial note: From equ(3.22) the difference between two cosines was expressed as a product. The expression is required in the form

$$\cos(x + \delta x) - \cos x = -2\sin(x + \tfrac{1}{2}\delta x)\sin(\tfrac{1}{2}\delta x) \tag{ii}$$

Following the method for $\sin x$, we have
at any point x

$$y = \cos x$$

at any point $(x + \delta x)$,

$$y + \delta y = \cos(x + \delta x)$$

Subtract

$$\delta y = \cos(x + \delta x) - \cos \delta x$$
$$= -2\sin(x + \tfrac{1}{2}\delta x)\sin(\tfrac{1}{2}\delta x) \text{ using equation (ii)}$$

Gradient of the chord is

$$\frac{\delta y}{\delta x} = \frac{-2 \sin(x + \frac{1}{2} \delta x) \sin(\frac{1}{2} \delta x)}{\delta x}$$

$$= -\sin(x + \tfrac{1}{2} \delta x) \frac{\sin(\frac{1}{2} \delta x)}{\frac{1}{2} \delta x}$$

As $x \to 0$ $\qquad \dfrac{dy}{dx} = -\sin x$

$$\boxed{\frac{d(\cos x)}{dx} = -\sin x}$$

(4.10)

(c) Alternative methods of finding derivatives of sin x and cos x

In Fig. 4.10b OAB is a right-angled triangle with the hypotenuse OA of length 1 unit, and the angle $BOA = x$ rad. With centre O, OA is rotated through an angle AOC, where

angle $AOC = \delta x$ rad.

From Fig. 3.34b length of arc $AC = OA\,\delta x = 1.\delta x = \delta x$

When $\delta x \to 0$ the arc AC and the straight line AC are practically identical, as shown in Fig. 4.10a. In this case the angle $OAC \fallingdotseq 90°$. As seen in the diagram, this means that the angle $ECA = x$ rad.

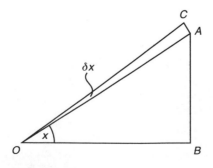

Fig. 4.10(a)

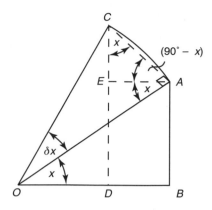

Fig. 4.10(b)

(i) $y = \sin x$

In triangle AOB		$\sin x =$	$\dfrac{AB}{OA}$	$= \quad AB$

In triangle DOC $\qquad \sin(x + \delta x) = \dfrac{CD}{OC} \qquad = \quad CD$

$$\sin(x + \delta x) - \sin x = \quad \delta(\sin x) \quad = CD - AB = CE$$

Divide by δx $\qquad \dfrac{\delta(\sin x)}{\delta x} = \dfrac{\sin(x + \delta x) - \sin x}{\delta x} = \dfrac{CE}{\delta x}$

$$= \dfrac{CE}{\text{arc } AC}$$

As $\delta x \to O$ we have $\qquad \dfrac{d(\sin x)}{dx} = \dfrac{CE}{\text{line } AC} \qquad = \quad \cos x$

(ii) $y = \cos x$

In triangle AOB $\qquad \cos x = \dfrac{OB}{OA} \qquad = \quad OB$

In triangle DOC $\quad \cos(x + \delta x) = \dfrac{OD}{OC} \qquad = \quad OD$

$$\cos(x + \delta x) - \cos x = \quad \delta(\cos x) \quad = OD - OB = -EA$$

Divide by δx $\qquad \dfrac{\delta(\cos x)}{\delta x} = \dfrac{\cos(x + \delta x) - \cos x}{\delta x} = \dfrac{-EA}{\delta x}$

$$= \dfrac{-EA}{\text{arc } AC}$$

As $\delta x \to 0$ we have $\qquad \dfrac{d(\cos x)}{dx} = \dfrac{-EA}{\text{line } AC} \qquad = \quad -\sin x$

(d) $y = \tan x$

$\tan x$ is treated as a quotient $\dfrac{\sin x}{\cos x}$

Let $u = \sin x,$ $\dfrac{du}{dx} = \cos x$

$v = \cos x,$ $\dfrac{dv}{dx} = -\sin x$

The rule for a quotient is

$$\frac{dy}{dx} = \frac{v\dfrac{du}{dx} - u\dfrac{dv}{dx}}{v^2}$$

$$= \frac{\cos x \cos x - \sin x(-\sin x)}{\cos^2 x}$$

$$= \frac{\cos^2 x + \sin^2 x}{\cos^2 x}$$

$$= \frac{1}{\cos^2 x}$$

$$= \sec^2 x$$

$$\boxed{\frac{d(\tan x)}{dx} = \sec^2 x}$$

$\qquad\qquad$ (4.11)

(e) $y = \cot x$

$\cot x$ is treated as a quotient $\dfrac{\cos x}{\sin x}$

Let $u = \cos x,$ $\dfrac{du}{dx} = -\sin x$

$v = \sin x,$ $\dfrac{dv}{dx} = \cos x$

$$\frac{dy}{dx} = \frac{\sin x(-\sin x) - \cos x \cos x}{\sin^2 x}$$

$$= \frac{-\sin^2 x - \cos^2 x}{\sin^2 x}$$

$$= -\frac{1}{\sin^2 x}$$

$$= -\text{cosec}^2 x$$

$$\boxed{\frac{d(\cot x)}{dx} = -\text{cosec}^2 x}$$

$\qquad\qquad$ (4.12)

(f) $y = \text{cosec } x$

cosec x is treated as a quotient $\dfrac{1}{\sin x}$

Let $u = 1,$ $\dfrac{du}{dx} = 0$

$v = \sin x,$ $\dfrac{dv}{dx} = \cos x$

therefore $\dfrac{dy}{dx} = \dfrac{(\sin x)0 - 1\cos x}{\sin^2 x}$

$= -\dfrac{\cos x}{\sin x \sin x}$

$$\boxed{\dfrac{d(\text{cosec } x)}{dx} = -\cot x \ \text{cosec } x}$$ (4.13)

(g) $y = \text{sec } x$

sec x is treated as a quotient $\dfrac{1}{\cos x}$

Let $u = 1,$ $\dfrac{du}{dx} = 0$

$v = \cos x,$ $\dfrac{dv}{dx} = -\sin x$

$\dfrac{dy}{dx} = \dfrac{(\cos x)0 - 1(-\sin x)}{\cos^2 x}$

$= \dfrac{\sin x}{\cos^2 x}$

$= \dfrac{\sin x}{\cos x \cos x}$

$= \tan x \ \text{sec } x$

$$\boxed{\dfrac{d(\text{sec } x)}{dx} = \tan x \ \text{sec } x}$$ (4.14)

4.15 Differentiation of more difficult trigonometric functions

(a) Type 1: When the trigonometric function contains a compound angle

For example, $y = \sin(4x - 7)$. This is differentiated as a function of a function.

Let

$$u = 4x - 7$$

therefore

$$\frac{du}{dx} = 4$$

so that

$$y = \sin u, \qquad \frac{dy}{du} = \cos u$$

From rule (4.6)

$$\frac{dy}{dx} = \frac{dy}{du} \times \frac{du}{dx}$$

$$= \cos u \,.\, 4$$

$$= 4\cos(4x - 7)$$

The rule below summarises the procedure.

1. Differentiate the trigonometric function.
2. Differentiate the compound angle.

3. $\dfrac{dy}{dx}$ = product of (1) and (2)

Example 4.7   Differentiate $y = 4\tan(x^2 + 1)$

1. Differentiate the trigonometric function: $4\sec^2(x^2 + 1)$
2. Differentiate the compound angle: $2x$

3. $\dfrac{dy}{dx} = 8x\sec^2(x^2 + 1)$

(b) Type 2: A simple trigonometric function raised to a power

For example, $y = \cos^5 x$. This is differentiated as a function of a function and for convenience written first as

$$y = (\cos x)^5$$

Let $u = \cos x, \qquad \dfrac{du}{dx} = -\sin x$

$\qquad y = u^5, \qquad \dfrac{dy}{du} = 5u^4$

$\dfrac{dy}{dx} = 5u^4 \times (-\sin x)$

$\qquad = -5 \sin x \cos^4 x$

Instead of making this substitution the following rule may be used.

1.   Differentiate the bracket as a simple variable.
2.   Differentiate the trigonometric function inside the bracket.
3. $\dfrac{dy}{dx} =$ product of (1) and (2)

Example 4.8   Differentiate $y = \operatorname{cosec}^5 x$

Rewrite using brackets: $y = (\operatorname{cosec} x)^5$

1.   Differentiate the bracket: $5(\operatorname{cosec} x)^4$
2.   Differentiate the function inside the bracket: $-\cot x \operatorname{cosec} x$

3. $\dfrac{dy}{dx} = 5 \operatorname{cosec}^4 x \times -\cot x \operatorname{cosec} x$

$\qquad = -5 \cot x \operatorname{cosec}^5 x.$

(c) Type 3: Trigonometric functions which are a combination of Types 1 and 2

For example, $y = 4 \sin^3(3x - 2)$.
Rewrite using brackets:

$\qquad y = 4[\sin(3x - 2)]^3$

As in type 2

$\qquad \dfrac{dy}{dx} = 4 \times 3 \sin^2(3x - 2) \times \dfrac{d}{dx} [\sin(3x - 2)]$

$\qquad = 12 \sin^2(3x - 2) \times \cos(3x - 2) \times 3$

$\qquad = 36 \cos(3x - 2)\sin^2(3x - 2)$

EXERCISE 4.10

Differentiate the following with respect to x:

1. $\sin(7x - 4)$ 2. $\cos(4x^2 - 3)$
3. $\sec(3x^2 - 1)$ 4. $\tan(\frac{3}{2}x^2 - 3x + 1)$
5. $\operatorname{cosec} 3x$ 6. $\cot(5x - 3)$
7. $\cos^6 x$ 8. $\cot^{\frac{2}{3}} x$
9. $\sec^2 x$ 10. $5 \operatorname{cosec}^{\frac{1}{2}} x$
11. $\tan^4 x$ 12. $\frac{3}{4} \sin^{\frac{5}{6}} x$
13. $\sin^4(3x^2)$ 14. $4 \cos^{\frac{3}{2}}(3x - 2)$
15. $6 \tan^{\frac{1}{2}}(4x^2 - 1)$

4.16   Differentiation of an implicit function

When an expression can be written in the form $y = f(x)$, such as $y = 3x^3 - 2x + 7$, y is said to be an explicit function of x. Another class of expressions, such as, for example,

$$y^4 + x^2 = x + y^2 \tag{i}$$

cannot be written in the form $y = f(x)$. In such expressions y is said to be an **implicit** function of x.

It is now necessary to differentiate an implicit function of x, with respect to x; that is, to find $\dfrac{dy}{dx}$ for an implicit function such as (i) above.

Differentiating (i) with respect to x:

$$\frac{d(y^4)}{dx} + \frac{d(x^2)}{dx} = \frac{d(x)}{dx} + \frac{d(y^2)}{dx} \tag{ii}$$

But using rule (4.6)

$$\frac{d(y^4)}{dx} = \frac{d(y^4)}{dy}\frac{dy}{dx} = 4y^3 \frac{dy}{dx}$$

$$\frac{d(x^2)}{dx} \qquad = 2x$$

$$\frac{d(x)}{dx} \qquad = 1$$

$$\frac{d(y^2)}{dx} \qquad = \frac{d(y^2)}{dy}\frac{dy}{dx} = 2y\frac{dy}{dx}$$

Using these results in equation (ii) we obtain

$$4y^3 \frac{dy}{dx} + 2x = 1 + 2y \frac{dy}{dx}$$

$$4y^3 \frac{dy}{dx} - 2y \frac{dy}{dx} = 1 - 2x$$

$$2y(2y^2 - 1) \frac{dy}{dx} = 1 - 2x$$

$$\frac{dy}{dx} = \frac{1 - 2x}{2y(2y^2 - 1)}$$

Example 4.9   Differentiate $\cos y = x^2$ with respect to x.

In the equation y is an implicit function of x. Differentiate with respect to x.

$$\frac{d(\cos y)}{dx} = \frac{d(x^2)}{dx}$$

$$\frac{d(\cos y)}{dy} \frac{dy}{dx} = 2x$$

$$-\sin y \frac{dy}{dx} = 2x$$

$$\frac{dy}{dx} = -\frac{2x}{\sin y}$$

EXERCISE 4.11

Differentiate the following implicit functions, with respect to x:

1. $y^2 + y^3 = 4x^3$ 2. $3y^4 + 2 = 3x^{\frac{1}{2}} + 4x$
3. $\sin y = x^3$ 4. $\sin y^2 = 3x^2$
5. $3y^{\frac{1}{2}} + 4y = 3x^{\frac{1}{2}} + 4x$ 6. $y + \tan x = \tan y + x^2$
7. $x \sin y = y \sin x$ 8. $xy = 10$

4.17   To differentiate the exponential function

(a) Simple exponential

In Chapter 10 the exponential function is seen to be

$$e^x = 1 + x + \frac{x^2}{1.2} + \frac{x^3}{1.2.3} + \cdots.$$

If this expression is differentiated term by term, we have

$$\frac{d(e^x)}{dx} = 0 + 1 + \frac{2x}{1.2} + \frac{3x^2}{1.2.3} + \cdots$$

$$= 1 + x + \frac{x^2}{1.2} + \cdots$$

which is the original function; that is,

$$\frac{d(e^x)}{dx} = e^x$$

The result may be verified by differentiating from first principles.

At a point x: $\qquad\qquad y = e^x$

At $x + \delta x$: $\qquad\quad y + \delta y = e^{x+\delta x} = e^x e^{\delta x}$

Subtract: $\qquad\qquad\quad \delta y = e^x e^{\delta x} - e^x$

$$= e^x(e^{\delta x} - 1)$$

$$= e^x\left(1 + \delta x + \frac{(\delta x)^2}{1.2} + \cdots - 1\right)$$

$$= e^x \delta x + e^x\frac{(\delta x)^2}{2} + \cdots$$

Divide by δx: $\qquad\quad \frac{\delta y}{\delta x} = e^x + e^x\frac{\delta x}{2} + \cdots$

As $\delta x \to 0$, $\delta y/\delta x \to dy/dx$, and the second and subsequent terms on the right disappear.

$$\boxed{\frac{d(e^x)}{dx} = e^x} \qquad\qquad (4.5)$$

(b) Exponential with a compound exponent

If the exponent of the exponential is a compound function, as in

$$y = e^{x^2 - 4}$$

the differentiation is carried out as a function of a function.

Let $\qquad\qquad u = x^2 - 4 \qquad \frac{du}{dx} = 2x$

therefore $\qquad y = e^u \qquad\qquad \frac{dy}{du} = e^u$

Using rule (4.6):

$$\frac{dy}{dx} = \frac{dy}{du} \times \frac{du}{dx}$$

$$= e^u 2x$$

$$= 2xe^{(x^2 - 4)}$$

Rule 4.16 can be used to differentiate such a function making the substitution unnecessary; that is,

$$\boxed{\frac{dy}{dx} = \text{original ponential} \times \frac{d}{dx}(\text{exponent})} \tag{4.16}$$

Example 4.10 Differentiate with respect to x:

(i) $y = 4e^{(7x^2 - 3x + 1)}$

(ii) $y = 7e^{\sin x}$

Using rule (4.16),

(i) $\dfrac{dy}{dx} = 4e^{(7x^2 - 3x + 1)} \dfrac{d}{dx}(7x^2 - 3x + 1)$

$\qquad = 4(14x - 3)e^{(7x^2 - 3x + 1)}$

(ii) $\dfrac{dy}{dx} = 7e^{\sin x} \cdot \dfrac{d}{dx}(\sin x)$

$\qquad = 7\cos x\, e^{\sin x}$

EXERCISE 4.12

Differentiate the following, with respect to the appropriate variable

1. $y = e^{5x}$
2. $y = 7e^{x^2}$
3. $y = 2e^{\frac{1}{2}x} - \frac{1}{2}x^2$
4. $y = 3e^{(4x^2 - 6)} + x^3$
5. $y = e^{4t^3}$
6. $y = 7e^{\cos x} + 3e^{4\sin x}$
7. $y - 4e^{\tan x} + e^{-3x^2}$
8. $e^y - 4x^2 - 6$
9. $e^{3y} = 7x^{\frac{1}{2}}$
10. $e^y + x^3 = y^2 - 4$.

4.18   To differentiate the logarithm function

(a) Simple logarithm function

Transforming $y = \ln x$ into the equivalent indicial equation as in Chapter 1 gives $e^y = x$. y is now an implicit function of x. Differentiating with respect to x,

$$\frac{d}{dx}(e^y) = \frac{d(x)}{dx}$$

$$\frac{d}{dy}(e^y)\frac{dy}{dx} = 1$$

$$e^y\frac{dy}{dx} = 1$$

$$\frac{dy}{dx} = \frac{1}{e^y}$$

Substitute $e^y = x$

$$\boxed{\frac{d(\ln x)}{dx} = \frac{1}{x}}$$

(4.17)

(b) Logarithm of a compound function

If the logarithm function contains a compound function such as

$$y = \ln(4x^3 - 1)$$

it can be differentiated as a function of a function in a similar manner to paragraph 4.17(b). Rule (4.18) can be used to avoid making a substitution.

$$\boxed{\frac{dy}{dx} = \frac{1}{\textbf{inner function}} \times \frac{d}{dx}(\textbf{inner function})}$$

(4.18)

Therefore in the above function

$$\frac{dy}{dx} = \frac{1}{(4x^3 - 1)} \cdot \frac{d}{dx}(4x^3 - 1)$$

$$= \frac{12x^2}{(4x^3 - 1)}$$

If the logarithm function contains a product or a quotient the law of logarithms should first be used. In this way the differentiation becomes easier. This is illustrated in Example 4.11.

Example 4.11 Differentiate $y = \ln \dfrac{x^2 + 1}{x^2 - 1}$, with respect to x.

Using the second law of logarithms (and rule (4.18))

$$y = \ln(x^2 + 1) - \ln(x^2 - 1)$$

$$\frac{dy}{dx} = \frac{1}{x^2 + 1} \quad 2x - \frac{1}{(x^2 - 1)} \quad 2x$$

$$= \frac{2x[(x^2 - 1) - (x^2 + 1)]}{(x^2 + 1)(x^2 - 1)}$$

$$= -\frac{4x}{(x^2 + 1)(x^2 - 1)}$$

(c) Logarithm function with a base of 10

In obtaining rule (4.17) for the differential coefficient of $y = \ln x$, the property of the exponential e^y was used. The result given in equation (4.17) will therefore not be true for the base 10. In order to differentiate a log function on base 10 it must first be changed into a form containing the base e, using the change of base equation (1.2), as shown in example 4.12.

Example 4.12 Differentiate $y = \log x$, with respect to x.

Using the change of base formula

$$y = \log x = 0.4343 \ln x$$

$$\frac{dy}{dx} = 0.4343 \frac{d}{dx} (\ln x)$$

$$= 0.4343 \times \frac{1}{x}$$

EXERCISE 4.13

Differentiate the following with respect to the appropriate variable

1. $y = \ln 7x$
2. $y = \ln(8x^2 - 1)$
3. $y = \ln(4x^2 - 7)$
4. $z = \ln(3t^2 - 7t + 1)$
5. $t = \ln(7s^{\frac{1}{2}})$
6. $y = \ln\left(\dfrac{4x}{x - 2}\right)$
7. $y = \ln\left(\dfrac{3x^2 - 2}{3x^2 - 4}\right)$
8. $y = \ln\left(\dfrac{1}{7x^2}\right)$
9. $y = \ln \sin x$
10. $y = \ln \tan x$

11. $y = \log 7x$

13. $y = \log 3x$

12. $\ln(y^2 + 2) = 4x^2 - 6$

14. $\ln(3y - 6) = y^2 + x^2$

4.19 Examples of the differentiation of products and quotients of mixed functions

Example 4.13   Differentiate $y = x^3 \sin 4x$

This is a product of two functions. Let

$$u = x^3, \qquad\qquad \frac{du}{dx} = 3x^2$$

$$v = \sin 4x, \qquad\qquad \frac{dv}{dx} = 4\cos 4x$$

$$\frac{dy}{dx} = v\frac{du}{dx} + u\frac{dv}{dx}$$

$$= \sin 4x\, 3x^2 + x^3\, 4\cos 4x$$

$$= x^2(3\sin 4x + 4x\cos 4x)$$

Example 4.14   Differentiate $y = e^{3x} \cos 2x$

This is a product of two functions. Let

$$u = e^{3x}, \qquad\qquad \frac{du}{dx} = 3e^{3x}$$

$$v = \cos 2x, \qquad\qquad \frac{dv}{dx} = -2\sin 2x$$

$$\frac{dy}{dx} = v\frac{du}{dx} + u\frac{dv}{dx}$$

$$= \cos 2x\, 3e^{3x} + e^{3x}(-2\sin 2x)$$

$$= e^{3x}(3\cos 2x - 2\sin 2x)$$

Example 4.15   Differentiate $y = \dfrac{\ln x}{x^4}$

This is a quotient of two functions. Let

$$u = \ln x, \qquad \frac{du}{dx} = \frac{1}{x}$$

$$v = x^4 \qquad \frac{dv}{dx} = 4x^3$$

$$\frac{dy}{dx} = \frac{v\dfrac{du}{dx} - u\dfrac{dv}{dx}}{v^2}$$

$$= \frac{x^4\dfrac{1}{x} - \ln x\, 4x^3}{x^8}$$

$$= \frac{1 - 4\ln x}{x^5}$$

EXERCISE 4.14

Differentiate the following, with respect to the appropriate variable:

1. $\dfrac{1+x}{x^3-1}$ ⠀⠀⠀ 2. $(x^2+1)e^{3x}$
3. $(x^2-1)\sqrt{x}$ ⠀⠀⠀ 4. $(1+x^2)\ln x$
5. $\ln(1-x)^{\frac{1}{2}}$ ⠀⠀⠀ 6. $e^{2x}\cos x$
7. $x^3 e^{2x}$ ⠀⠀⠀ 8. $\ln(x^3\cos x)$
9. $e^{\frac{2}{3}x}\tan 4x$ ⠀⠀⠀ 10. $\dfrac{\sin 5t}{\cos 10t}$
11. $(9x^2-6x+2)e^{3x}$ ⠀⠀⠀ 12. $2\theta\sin\theta\cos\theta$
13. $\dfrac{\sin 5y}{\ln 5y}$ ⠀⠀⠀ 14. $\ln x^x$
15. $e^{2x}\log 2x$.

4.20 Differential equations

In Section 4.9 second- and higher-order differential coefficients were described and defined. For example, if

$$y = \sin x \tag{i}$$

then ⠀⠀ $$\frac{dy}{dx} = \cos x \tag{ii}$$

$$\frac{d^2y}{dx^2} = -\sin x \tag{iii}$$

It is often possible to obtain relationships between the various differential coefficients and the original function. For example, from equations (i) and (iii) above, it is obvious that

$$\frac{d^2y}{dx^2} = -y$$

Such an equation is called a **differential equation**.

The idea is further illustrated using the function

$$y = x e^x \tag{i}$$

where we obtain

$$\frac{dy}{dx} = x e^x + e^x \tag{ii}$$

and $$\frac{d^2y}{dx^2} = x e^x + 2e^x \tag{iii}$$

From equations (ii) and (iii)

$$\frac{d^2y}{dx^2} - 2\frac{dy}{dx} = -x e^x = -y$$

that is

$$\frac{d^2y}{dx^2} - 2\frac{dy}{dx} + y = 0$$

This equation is a differential equation.

Example 4.16   Given that $y = \ln(x^2 + 1)$ show that

$$x \frac{d^2y}{dx^2} + x\left(\frac{dy}{dx}\right)^2 = \frac{dy}{dx}$$

Since $$y = \ln(x^2 + 1)$$

$$\frac{dy}{dx} = \frac{2x}{x^2 + 1}$$

and $$\frac{d^2y}{dx^2} = \frac{-2x^2 + 2}{(x^2 + 1)^2}$$

Making use of these two expressions on the left-hand side of the differential equation,

$$x \frac{d^2y}{dx^2} + x\left(\frac{dy}{dx}\right)^2 = x \frac{-2x^2 + 2}{(x^2 + 1)^2} + x \frac{4x^2}{(x^2 + 1)^2}$$

$$= \frac{x}{(x^2 + 1)^2} \{-2x^2 + 2 + 4x^2\}$$

$$= \frac{x}{(x^2 + 1)^2} 2(x^2 + 1)$$

$$= \frac{2x}{x^2 + 1}$$

$$= \frac{dy}{dx}$$

EXERCISE 4.15

1.   Given that $x = \sin(\ln t)$ show that

$$t^2 \frac{d^2x}{dt^2} + t\frac{dx}{dt} + x = 0$$

2.   Given that $y = \sin^3 x$, prove that

$$\frac{d^2y}{dx^2} - 2\cot x\frac{dy}{dx} + 3y = 0$$

3.   Given that $y = e^x \sin x$, show that

$$\frac{d^2y}{dx^2} - 2\frac{dy}{dx} + 2y = 0$$

4.   Given that $y = x^2 \sin x$, show that

$$x^2\frac{d^2y}{dx^2} - 4x\frac{dy}{dx} + y(6 + x^2) = 0$$

5.   Given that $y = e^{4Ax}$, where A is a constant, show that

$$\frac{d^2y}{dx^2} + 4A\frac{dy}{dx} - 32A^2y = 0$$

   for all values of A. Find the value of A if $dy/dx = 12$ when $x = 0$.

6.   Given that $y = x e^{2x}$, prove that

$$\frac{d^2y}{dx^2} - 4\frac{dy}{dx} + 4y = 0$$

7.   Given that $y = \tan x$, show that

$$\frac{d^2y}{dx^2} = 2y\frac{dy}{dx}$$

4.21   The differentiation of the inverse trigonometric functions

(a) $y = \sin^{-1}x$

Rewriting:

$$\sin y = x$$

Differentiate as an implicit function with respect to x:

$$\frac{d}{dx}(\sin y) = 1$$

$$\frac{d}{dy}(\sin y) \times \frac{dy}{dx} = 1$$

$$\cos y \times \frac{dy}{dx} = 1$$

$$\frac{dy}{dx} = \frac{1}{\cos y} = \frac{1}{\sqrt{1 - \sin^2 y}}$$

$$= \frac{1}{\sqrt{1 - x^2}}$$

$$\boxed{\frac{d}{dx}(\sin^{-1} x) = \frac{1}{\sqrt{1 - x^2}}}$$

(b) $y = \cos^{-1} x$

Rewriting:

$$\cos y = x$$

Differentiate as an implicit function with respect to x:

$$\frac{d}{dx}(\cos y) = 1$$

$$\frac{d}{dy}(\cos y) \times \frac{dy}{dx} = 1$$

$$-\sin y \times \frac{dy}{dx} = 1$$

$$\frac{dy}{dx} = \frac{1}{-\sin y} = -\frac{1}{\sqrt{1 - \cos^2 y}}$$

$$= -\frac{1}{\sqrt{1 - x^2}}$$

$$\boxed{\frac{d}{dx}(\cos^{-1} x) = -\frac{1}{\sqrt{1 - x^2}}}$$

(c) $y = \tan^{-1} x$

Rewriting:

$$\tan y = x$$

Differentiate as an implicit function with respect to x

$$\frac{d}{dx}(\tan y) = 1$$

$$\frac{d}{dy}(\tan y) \times \frac{dy}{dx} = 1$$

$$\sec^2 y \times \frac{dy}{dx} = 1$$

$$\frac{dy}{dx} = \frac{1}{\sec^2 y} = \frac{1}{1 + \tan^2 y}$$

$$= \frac{1}{1 + x^2}$$

$$\boxed{\frac{d}{dx}(\tan^{-1} x) = \frac{1}{1 + x^2}}$$

(d) $y = \operatorname{cosec}^{-1} x$

Rewriting:

$$\operatorname{cosec} y = x$$

Differentiate as an implicit function with respect to x:

$$\frac{d}{dx}(\operatorname{cosec} y) = 1$$

$$\frac{d}{dy}(\operatorname{cosec} y) \times \frac{dy}{dx} = 1$$

$$-\cot y \operatorname{cosec} y \times \frac{dy}{dx} = 1$$

$$\frac{dy}{dx} = -\frac{1}{\cot y \operatorname{cosec} y} = -\frac{1}{x \cot y}$$

$$= -\frac{1}{x\sqrt{\operatorname{cosec}^2 y - 1}} = -\frac{1}{x\sqrt{x^2 - 1}}$$

$$\boxed{\frac{d}{dx}(\operatorname{cosec}^{-1} x) = -\frac{1}{x\sqrt{x^2 - 1}}}$$

(e) $y = \sec^{-1} x$

Rewriting:

$$\sec y = x$$

Differentiate as an implicit function with respect to x:

$$\frac{d}{dx}(\sec y) = 1$$

$$\frac{d}{dy}(\sec y) \times \frac{dy}{dx} = 1$$

$$\tan y \sec y \times \frac{dy}{dx} = 1$$

$$\frac{dy}{dx} = \frac{1}{\tan y \sec y} = \frac{1}{x \tan y}$$

$$= \frac{1}{x\sqrt{\sec^2 y - 1}} = \frac{1}{x\sqrt{x^2 - 1}}$$

$$\boxed{\frac{d}{dx}(\sec^{-1} x) = \frac{1}{\sqrt{x^2 - 1}}}$$

(f) $y = \cot^{-1} x$

Rewriting:

$$\cot y = x$$

Differentiate as an implicit function with respect to x:

$$\frac{d}{dx}(\cot y) = 1$$

$$\frac{d}{dy}(\cot y) \times \frac{dy}{dx} = 1$$

$$-\mathrm{cosec}^2 y \times \frac{dy}{dx} = 1$$

$$\frac{dy}{dx} = -\frac{1}{-\mathrm{cosec}^2 y} = -\frac{1}{1 + \cot^2 y}$$

$$= -\frac{1}{1 + x^2}$$

$$\boxed{\frac{d}{dx}(\cot^{-1} x) = -\frac{1}{1 + x^2}}$$

Example 4.17   Differentiate $y = x^2 \sin^{-1} 5x$

This is the product of two functions. Let

$$u = x^2, \qquad \frac{du}{dx} = 2x$$

$$v = \sin^{-1} 5x \qquad \frac{dy}{dx} = \frac{1}{\sqrt{1 - 25x^2}} \frac{d(5x)}{dx}$$

$$= \frac{1}{\sqrt{1 - 25x^2}} 5$$

$$\frac{dy}{dx} = v\frac{du}{dx} + u\frac{dv}{dx}$$

$$= \sin^{-1} 5x \times 2x + x^2 \frac{5}{\sqrt{1 - 25x^2}}$$

$$= x\left(2\sin^{-1} 5x + \frac{5x}{\sqrt{1 - 25x^2}}\right)$$

MISCELLANEOUS EXERCISE 4

1. Differentiate the following functions with respect to x, simplifying first where necessary:

   (a) $\dfrac{1}{(4\pi x)^3}$,   (b) $-0.32x^{\frac{1}{2}} - 2.4x^{-\frac{1}{2}} - 9.6$

   (c) $(3x^2 - \frac{1}{3})(\frac{1}{3}x^{-2} + 3)$.

2. Differentiate the following expressions with respect to x and simplify as far as possible:

   (a) $\sqrt{(1 - x)}$ (b) $\dfrac{1}{2x^3}$ (c) $\dfrac{4x}{1 - 3x}$

3. Given that $y = x^3 + \dfrac{7}{x^4}$, find the values of dy/dx and d^2y/dx^2 when $x = 2$.

4. (a) Find $\dfrac{dy}{dx}$ when $y = 3x^2 + 4x - \dfrac{3}{x^2}$

   (b) Given that $\theta = 4t^3 + 30t - 12$, find $d\theta/dt$ when $t = 3$

   (c) Find the gradient of the curve $y = x^3 - 3x^2 + 8x - 3$ at the point $(1, 3)$.

5. Evaluate (a) $\dfrac{d}{dx}\left(3x^2 - \dfrac{1}{x}\right)$ (b) $\dfrac{d}{dt}(t^2 - 2t)^{\frac{3}{2}}$

   (c) $\dfrac{d}{dp}\left(\dfrac{p^3 + p^2 + p + 1}{p^2}\right)$

6. (a) Differentiate with respect to x the quotient $x^3/(x^2+1)$.

   (b) Find the rate of increase of y with respect to x when $x = \frac{1}{3}\pi$ rad, given that $y = \frac{1}{2}\sin x$.

      What is the smallest positive value of x for which the rate of increase is zero?

   (c) For the expression $y = 4 + 7x - 3x^3$ find dy/dx, d^2y/dx^2.

      Hence find the value of dy/dx when $d^2y/dx^2 = 9$.

7. (a) Differentiate the following functions with respect to x:

   (i) $\dfrac{(x+2)^2}{x}$ (ii) $\dfrac{1}{x^3}$ (iii) $4.2x^{1.6} - 2.7x^{4.3}$

   (b) Find the gradient of the function

   $$y = 3x + \frac{2}{x^2}$$

   and show that for very large values of x this gradient has almost the same value as the gradient of the straight line $y = 3x$.

8. (a) Find $\dfrac{dy}{dx}$ when

   (i) $y = 4x^3 + 7$

   (ii) $y^2 = x^2 - 8x + 16$

   (iii) $y^{\frac{2}{3}} = x$

   (b) Evaluate

   $$\frac{d^2}{dx^2}\left(1 + \frac{7}{x} - \frac{2}{x^2}\right)$$

9. (a) Differentiate the following expressions with respect to x, simplifying the results where possible:

   (i) $\ln(x^2 + 5x - 3)$ (ii) $\dfrac{3x}{x^2+5}$ (iii) $\cos^4 3x$

   (b) For $y = 3x\,e^{-2x}$ find the value of x which makes $\dfrac{d^2y}{dx^2} = 0$.

10. (a) Differentiate the following with respect to x:

   (i) $\sqrt{(1-2x)}$ (ii) $2x^2\cos\frac{1}{2}x$

   (b) Find the gradient of the curve $y = \sin 3\theta$ at the point where $\theta = \pi/18$ rad.

11. (a) Find dy/dx if

(i) $y = \dfrac{2}{\sqrt{x}} - \dfrac{1}{x^3} + x^{\frac{3}{2}}$

(ii) $y = \dfrac{\sin 3x}{e^{3x}}$

(b) Given that $y = \ln(x^2 + 1)$, find dy/dx and d^2y/dx^2 and obtain, in its simplest form, an expression for

$$\dfrac{d^2y}{dx^2} + \dfrac{1}{x}\dfrac{dy}{dx}$$

12. (a) Differentiate with respect to x, simplifying your results where possible:

(i) $\dfrac{\cos 2x}{2 + \sin x}$ (ii) $\dfrac{1}{\sqrt{(2 - x^3)}}$ (iii) $\ln \sin 3x$

(b) Given that $x = e^{-3t}\sin 2t$, show that

$$\dfrac{d^2x}{dt^2} + 6\dfrac{dx}{dt} + 13x = 0$$

13. (a) Differentiate the following, with respect to x:

(i) $\sqrt{(1 - x^2)}$ (ii) $e^{-x}\cos x$ (iii) $\ln\left(\dfrac{2}{x}\right)$

(b) Given that

$$y = \dfrac{\sin x}{\sin x + \cos x}$$

show that

$$\dfrac{dy}{dx} = \dfrac{1}{1 + \sin 2x}$$

(c) Find the gradient of the curve $xy + y^2$ at the point $(1, -2)$.

14. Differentiate

(a) $x^3 \cos^{-1}x$ (b) $4x^5 \sin^{-1}2x$

(c) $e^x \tan^{-1}x$ (d) $\dfrac{x^3}{\ln x}$

Chapter 5

Applications of differentiation

5.1   Application to gradients of curves

In the study of gases the variation of pressure p and volume V can take place in two ways:

1. isothermally according to the law $pV = k$
2. adiabatically according to the law $pV^\gamma = k$ where γ and k are constants > 1.

The gradient of each curve, $\mathrm{d}p/\mathrm{d}V$, can be found by differentiating. For the isothermal case $p = kV^{-1}$

$$\frac{\mathrm{d}p}{\mathrm{d}V} = -kV^{-2}$$

Substitute for k,

$$k = pV$$

therefore

$$\frac{\mathrm{d}p}{\mathrm{d}V} = -\frac{pV}{V^2} = -\frac{p}{V} \qquad\qquad (5.1)$$

For the adiabatic case

$$p = kV^{-\gamma}$$

$$\frac{\mathrm{d}p}{\mathrm{d}V} = -\gamma kV^{-\gamma-1}$$

Substitute for k,:

$$k = pV^\gamma$$

therefore

$$\frac{\mathrm{d}p}{\mathrm{d}V} = -\gamma \frac{pV^\gamma}{V^{\gamma+1}} = -\gamma \frac{p}{V} \qquad\qquad (5.2)$$

Since $\gamma > 1$ the gradient of the adiabatic curve is always greater than the isothermal curve, as shown in Fig. 5.1.

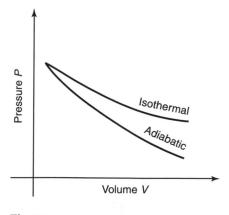

Fig. 5.1

5.2   Velocity of sound in a gas

The velocity of sound U in a gas is given by

$$U = \sqrt{\left(\frac{-V\dfrac{\mathrm{d}p}{\mathrm{d}V}}{\text{Density}}\right)}$$

where $-V\,\mathrm{d}p/\mathrm{d}V$ is called the adiabatic bulk modulus because the change in volume is taking place adiabatically. Using equation (5.2),

$$U = \sqrt{\left\{\frac{\gamma V\dfrac{p}{V}}{\rho}\right\}}$$

$$= \sqrt{\left(\frac{\gamma p}{\rho}\right)}$$

where ρ is the density of the gas.

5.3   Velocity and acceleration

If a body has a constant velocity U, the velocity can be expressed as

$$U = \frac{s}{t}$$

where s is the distance travelled in t seconds. That is, velocity is the gradient of the distance–time graph, as shown in Fig. 5.2a.

If the velocity is not constant, the distance-time graph is a curve, as in Fig. 5.2(b). The average velocity over a short time δt and distance δs is given

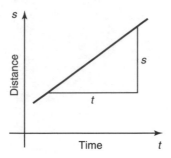

Fig. 5.2(a)

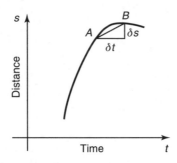

Fig. 5.2(b)

by the gradient of the chord AB, that is, the average velocity over the small interval is

$$\frac{\delta s}{\delta t}$$

As $\delta t \rightarrow 0$, the chord becomes a tangent, so that at the point A, the velocity U is given by

$$\boxed{U = \frac{\mathrm{d}s}{\mathrm{d}t}}$$
(5.3)

which is the gradient of the tangent at the point A.

Acceleration (a) is defined as

$$\frac{\text{change in velocity}}{\text{time interval}}$$

Over a short period of time δt, let the change in velocity be δU. Therefore

$$a = \frac{\delta U}{\delta t}$$

As $\delta t \to 0$

$$\boxed{a = \frac{dU}{dt}} \qquad (5.4)$$

Using equation (5.3) for U

$$a = \frac{d}{dt}\left(\frac{ds}{dt}\right)$$

i.e.

$$\boxed{a = \frac{d^2s}{dt^2}} \qquad (5.5)$$

Example 5.1 A body moves according to the equation

$$s = 25t - t^2$$

where s is the distance in metres from a reference point. Find

(i)   the velocity of the body at $t = 3$ seconds;
(ii) the acceleration;
(iii) the time when the velocity becomes 0.

$$s = 25t - t^2$$

(i)   velocity $u = \dfrac{ds}{dt} = 25 - 2t$

When $t - 3$,

$$u = 25 - 2 \times 3$$
$$= 19 \, \text{m/s}$$

(ii) acceleration $a = \dfrac{d^2s}{dt^2}$

$$= -2 \, \text{m/s}^2$$

(iii) Since $u = 25 - 2t$, then $u = 0$

when $\quad 25 - 2t = 0$

$$t = 12\tfrac{1}{2} \, s.$$

EXERCISE 5.1

1. Given that a particle obeys the equation

    $$s = 4t - 64t^2$$

    find the expression for its velocity and acceleration.
    Find its velocity when $t = 2$ seconds.

2. A car travels along a road so that its distance s in metres from 0 is given by

    $$s = 2t + 6t^2$$

    Find
    (a)   its distance from 0 when $t = 4$ seconds
    (b)   the time at which velocity is 62 m/s

3. When viewed through a microscope a bacterium is seen to move in accordance with the equation

    $$s = (4t + 6t^2) \times 10^{-6}$$

    Find its velocity after 2 seconds.

4. The distance s metres moved by a body in t seconds is given by

    $$s = 2t^3 - 13t^2 + 24t + 10$$

    Find
    (a)   the velocity when $t = 4$ seconds
    (b)   the value of t when the body comes to rest
    (c)   the value of t when the acceleration is $10\,m/s^2$

5. The displacement s metres of a particle from a fixed point at a time t seconds is given by

    $$s = 10 - 5te^{-t/10}$$

    Find expressions for velocity and acceleration. Find the value of t at which velocity is zero.

6. A missile is fired into the air. The height h of the missile after t seconds is given by

    $$h = 19.2t - 4.8t^2$$

    Find
    (a)   the initial velocity of the missile (i.e. at $t = 0$)
    (b)   the height attained when its velocity is one half of its initial velocity.

5.4 Turning points

(a) Maximum and minimum values

There are many examples in engineering and science where the maximum or minimum value of a parameter is required, e.g, the maximum stress in a beam, or the minimum amount of sheet metal to make a can. Such values are examples of turning points as shown in Fig. 5.3a. The so-called **maximum point** at A is not necessarily the maximum value of the graph, but rather a 'hump' about which the gradient of the curve changes sign from $+$ to $-$. Similarly the **minimum point** at B is not necessarily the minimum y value of the curve, but a 'dip' about which the gradient of the curve changes from $-$ to $+$. At the points A and B the gradient is zero, that is,

$$\frac{dy}{dx} = 0$$

The gradient of the curve $y = f(x)$ is given by dy/dx. Fig. 5.3b shows the values of dy/dx about each point A and B. At the actual points A, B,

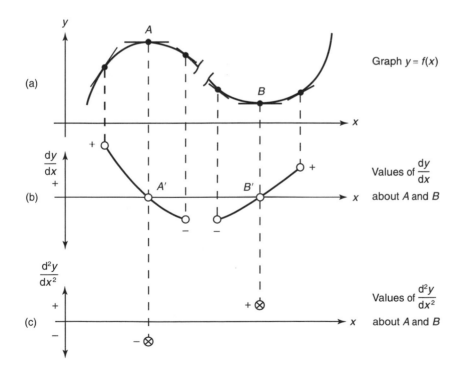

Fig. 5.3

$dy/dx = 0$, which is represented by the points A', B', in Fig. 5.3b. This condition allows the position of the turning points to be determined.

The values of the gradients of the curves in Fig. 5.3b at A' and B' are shown in Fig. 5.3c. The gradient of the dy/dx graph is

$$\frac{d}{dx}\left(\frac{dy}{dx}\right) = \frac{d^2y}{dx^2}$$

As shown they are negative and positive respectively.

In order to determine the turning points and to distinguish between them the following sequence is adopted:

(i)   Find $\dfrac{dy}{dx}$

(ii) Find $\dfrac{d^2y}{dx^2}$

(iii) Set $\dfrac{dy}{dx} = 0$ to find the values of x at the turning points.

(iv) Substitute these x values into $\dfrac{d^2y}{dx^2}$.

if d^2y/dx is $+$ the turning point is a minimum;
if d^2y/dx^2 is $-$ the turning point is a maximum.

Example 5.2   Find the turning points of $y = 2x^3 - \frac{1}{2}x^2 - x + 4$.
Distinguish between them, find the y value at each turning point and hence sketch the curve.

$$y = 2x^3 - \tfrac{1}{2}x^2 - x + 4 \tag{i}$$

$$\frac{dy}{dx} = 6x^2 - x - 1 \tag{ii}$$

For turning points $dy/dx = 0$. At these points let the x value be called x_m

therefore
$$6x_m^2 - x_m - 1 = 0$$
$$(3x_{mi} + 1)(2x_m - 1) = 0$$
$$x_m = -\tfrac{1}{3} \quad \text{and} \quad x_m = +\tfrac{1}{2}$$

The turning points occur at $x = -\frac{1}{3}$ and $x = +\frac{1}{2}$ (see Fig. 5.4). To distinguish between them, we find d^2y/dx^2 from (ii)

$$\frac{d^2y}{dx^2} = 12x - 1$$

When $x = -\frac{1}{3}$, $\dfrac{d^2y}{dx^2} = -4 - 1 = -5$: hence a maximum point.

When $x = +\frac{1}{2}$, $\dfrac{d^2y}{dx^2} = 6 - 1 = +5$: hence a minimum point.

If the x_m values are substituted into equation (i) the y values of the turning points are obtained:

when $x = -\frac{1}{3}$, $y_m = 2(-\frac{1}{3})^3 - \frac{1}{2}(-\frac{1}{3})^2 - (-\frac{1}{3}) + 4 = 4.2$.

when $x = +\frac{1}{2}$, $y_m = 2(+\frac{1}{2})^3 - \frac{1}{2}(\frac{1}{2})^2 - (\frac{1}{2}) + 4 = 3.6$.

The minimum occurs at the point $(+\frac{1}{2}, 3.6)$ and the maximum at the point $(-\frac{1}{3}, 4.2)$. The curve is sketched in Fig. 5.4.

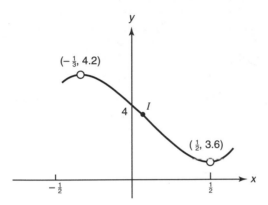

Fig. 5.4

(b) Points of inflexion

In this class of turning points the *gradient* does not change sign but reaches a maximum or minimum value, as shown in Fig. 5.5.

At the point of inflexion I the gradient of the dy/dx curve, i.e., d^2y/dx^2, is zero. The condition for the point of inflexion is

$$\frac{d^2y}{dx^2} = 0$$

Example 5.3.   In the function in Example 5.2 find

(i)   The x value at the inflexion point
(ii)   the gradient of the curve at this point.

The condition for a point of inflexion is

$$\frac{d^2y}{dx^2} = 0$$

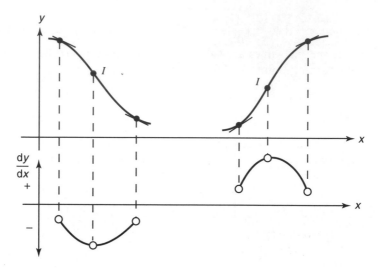

At I gradient reaches a
minimum value

At I gradient reaches a
maximum value

Fig. 5.5

As shown in Example 5.3,

$$\frac{d^2y}{dx^2} = 12x - 1$$

therefore $12x - 1 = 0$

$$x = \tfrac{1}{12}$$

i.e. the point of inflexion occurs at $x = \frac{1}{12}$, which is the point I in
Fig. 5.4.

At I $\dfrac{dy}{dx} = 6x^2 - x - 2,$

when $x = \frac{1}{12}$

$$\frac{dy}{dx} = \tfrac{1}{24} - \tfrac{1}{12} - 1$$

$$= -1\tfrac{1}{24}$$

In some cases the gradient at the point of inflexion is zero as shown at I in
Fig. 5.6, in which case the point is called a stationary point of inflexion, i.e.

$$\frac{dy}{dx} = 0 \qquad \text{and} \qquad \frac{d^2y}{dx^2} = 0$$

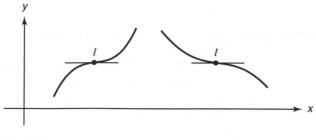

Fig. 5.6

In this case dy/dx is zero at the turning point I, so that the condition $dy/dx = 0$ may represent a maximum, minimum or point of inflexion. The point of inflexion may be distinguished because $d^2y/dx^2 = 0$. The complete sequence for determining the turning points is shown in Fig. 5.7.

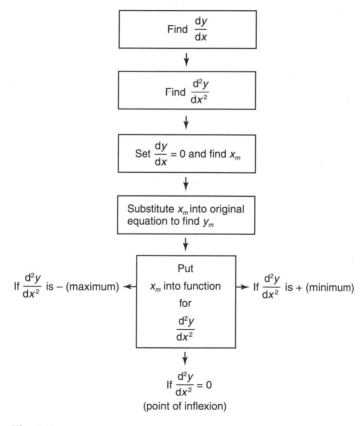

Fig. 5.7

EXERCISE 5.2

1.   Find and distinguish between the maximum and minimum values of
     $2x^3 - 5x^2 - 4x$.
2.   Calculate the maximum and minimum values of the function
     $x^3 - 4x^2 - 3x + 2$, and distinguish between them.
        Sketch the graph for $y = x^3 - 4x^2 - 3x + 2$ from $x = -2$ to $x = +4$.
     Show the maximum and minimum points clearly on the sketch.
3.   Find the turning points on the curve $y = 4x + 1/x$ and distinguish
     between them. Sketch the curve.
4.   Find the coordinates of the points at which the gradient of the curve
     $y = x^3 - 6x^2 + 9x$ is zero and determine, for each point, whether y has
     a maximum or minimum value.
5.   In question 4 find the position of the point of inflexion. Calculate the
     gradient of the function at this point.
6.   The equation of a curve is $y = 2x^4 - 2x^3 + 2x^2 - 3x + 1$. Show that,
     when $x = \frac{3}{4}$, the gradient of the curve is zero. Does this give a
     maximum or minimum value of y?

5.5   Practical examples of turning points

A simple cylinder in the form of a metal can is a good example of the use of
the idea of turning points. For a given volume the metal can have an infinite
number of dimensions. For example, it may be long and thin, or short and
fat as shown in Fig. 5.8.

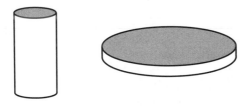

Fig. 5.8

The dimensions which use the minimum amount of sheet metal to make a
can of given volume can be determined.
    First of all the equation for the area of the sheet metal is obtained in terms
of h and r, the height and radius of the cylinder.
    Area A = area of two end faces + curved surface
             $= 2\pi r^2 + 2\pi rh$
    In order to obtain maximum and minimum values, this expression must be
differentiated. But since it contains two independent variables, r and h, one

of them must first be eliminated. This can be done using the volume equation,

$$V = \pi r^2 h$$

with

$$h = \frac{V}{\pi r^2}$$

therefore

$$A = 2\pi r^2 + \frac{2V}{r} \qquad\qquad (i)$$

For a given volume, the only independent variable now is r, so that the coefficient dA/dr may be found. The procedure is summarised as follows:

1. Set up the equation in two variables for the quantity whose maximum or minimum value is required.
2. Substitute for one variable, and differentiate.

For example, in the problem of the metal can, if the volume is required to be $128\pi \times 10^3 \text{ mm}^3$ find the dimensions to make the area a minimum.

$$V = 128\pi \times 10^3 \text{ mm}^3$$

From equation (i):

$$A = 2\pi r^2 + 2\frac{128\pi}{r} \times 10^3$$

$$\frac{dA}{dr} = 4\pi r - 256\pi r^{-2} \times 10^3$$

For maximum or minimum value $dA/dr = 0$
therefore

$$4\pi r = \frac{256\pi}{r^2} \times 10^3$$

therefore

$$r = \sqrt[3]{(64 \times 10^3)} = 40 \text{ mm}$$

Hence

$$h = \frac{128 \times 10^3 \pi}{\pi \times (40)^2}$$

$$= 80 \text{ mm}$$

At the value $r = 40$ mm, it can be shown that d^2A/dr^2 is positive, indicating that the area A has a minimum value.

Example 5.4   An open gutter, made from strip metal 160 mm wide, has a rectangular cross-section. Find the width and depth which give the maximum area of cross-section.

In Fig. 5.9, the cross-section is shown, depth h, width x. Area of cross-section is

$A = xh$

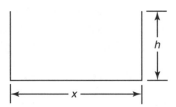

Fig. 5.9

Eliminate one of the variables from the fact that the perimeter P of the actual cross-section is

$P = 2h + x = 160 \text{ mm}$

therefore

$x = 160 - 2h$ (i)

Hence

$A = h(160 - 2h) = 160h - 2h^2$

$\dfrac{dA}{dh} = 160 - 4h$

For maximum or minimum value of A, $dA/dh = 0$

therefore $4h = 160$

$h = 40 \text{ mm}$

Now, $d^2A/dh^2 = -4$, so that the turning point is a maximum; that is, at $h = 40$ mm, A is a maximum.
From equation (i), width x is $160 - 2 \times 40 = 80$ mm.
The width of 80 mm and depth of 40 mm give a maximum area of cross-section.

EXERCISE 5.3

1. A rectangular block, with square base of side x mm, has a total surface area of 150 mm^2. Show that the volume of the block is given by

$$V = \tfrac{1}{2}(75x - x^3)$$

and hence find the maximum volume of the block.

2. The total internal surface area of a closed cement storage silo, which is in the form of an inverted right pyramid with a square top is 32 m^2. Given that the length of a side of the top is x metres, show that the volume of the silo is given by

$$V = \frac{4x}{3}(16 - x^2)^{\frac{1}{2}}$$

Calculate the maximum volume of cement that can be stored in the silo.

3. The area, S square metres, of sheet metal used in the manufacture of a closed, hollow cylindrical container, of a given fixed volume and base radius r metres, is given by

$$S = \frac{32}{9r} + 6r^2$$

Find
(a)   the value r for which the area of sheet metal used is a minimum;
(b)   the resulting minimum area of sheet metal used.

4. Show that the area $A \text{ mm}^2$ of a rectangle having a perimeter of 36 mm and the shorter side of length x mm is given by

$$A = 18x - x^2$$

If this area is to be a maximum, show that the figure must be a square of side 9 mm.

5. A right circular cone of maximum volume is to be cut from a sphere of unit radius. Determine its dimensions.

6. The total area of sheet metal used in the construction of a closed cylindrical can is $24\pi \text{ mm}^2$. Given that the base radius of the can is r mm, show that the volume $V \text{ mm}^3$ is given by

$$V = (12 - r^2)\pi r$$

Determine the maximum volume of the can.

7. A window frame is made in the shape of a rectangle with a semicircle on top. Given that the area is to be 8 m^2, show that the perimeter of the frame is

$$P = \frac{8}{r} + r\left(\frac{\pi}{2} + 2\right)$$

Find the minimum cost of producing the frame if, correct to the nearest penny, 1 metre costs 75p.

5.6   Rates of change with time

There are countless examples in engineering and science of a physical quantity changing with time. The rate of change with time of a physical quantity is the amount of change taking place in a unit of time, as shown below.

(a) Rate of radioactive disintegration

A lump of radioactive material contains, at a time t, N atoms, that are disintegrating. The rate of disintegration is $\delta N/\delta t$, where δN is the actual number of atoms disintegrating in a short time δt. As $\delta t \to 0$, the rate becomes dN/dt. It is found that this rate of disintegration is proportional to the number of radioactive atoms in the lump at any time, i.e.

$$\frac{dN}{dt} \propto -N$$

or

$$\frac{dN}{dt} = -\lambda N \qquad \text{where } \lambda \text{ is a constant}$$

Very often a rate of change of one variable, say V, can be obtained from a knowledge of the rate of change of another variable h, **provided that the equation between V and h is known**. It is possible to write

$$\frac{\delta V}{\delta t} = \frac{\delta V}{\delta h} \times \frac{\delta h}{\delta t}$$

As $\delta t \to 0$, then $\delta h \to 0$ and $\delta V \to 0$, so that

$$\frac{dV}{dt} = \frac{dV}{dh} \times \frac{dh}{dt} \tag{5.6}$$

The procedure for finding dV/dt is shown in Example 5.5.

Example 5.5   Water flows at a constant rate into a tank, which is an inverted cone of depth 3 m and radius 2 m. Given that the rate at which the level of water rises in the tank is 0.03 m/s at a depth of 2 m, find the rate of flow of water into the tank.

Let V be the volume of water in the tank at any time t, and let h be the depth of water in the tank at the same time t.

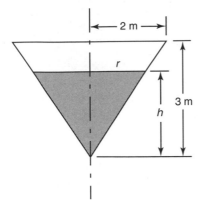

Fig. 5.10

The rate at which water flows into the tank is the rate at which V is changing, that is dV/dt. We find dV/dt from a knowledge of dh/dt. From equation (5.6),

$$\frac{dV}{dt} = \frac{dV}{dh} \times \frac{dh}{dt} \tag{i}$$

In equation (i) dV/dh must first be determined from the formula for the volume of a cone

$$V = \tfrac{1}{3}\pi r^2 h$$

The expression cannot be differentiated because the right-hand side contains two variables. From Fig. 5.10

$$\frac{r}{h} = \frac{2}{3} \qquad \text{so that} \qquad r = \frac{2}{3}h$$

Hence

$$V = \frac{4}{27}\pi h^3,$$

V is now a function of one variable h. Therefore

$$\frac{dV}{dh} = 3 \times \frac{4\pi}{27}h^2.$$

Since dh/dt is known when $h = 2\,\text{m}$, then dV/dt is found at the same value of h.

We have

$$\frac{dV}{dh} = \frac{4}{9}\pi 2^2 = \frac{16}{9}\pi \qquad \text{when } h = 2$$

Also

$$\frac{dh}{dt} = 0.03 \, \text{m/s} \qquad \text{when } h = 2$$

From equation (i)

$$\frac{dV}{dt} = \frac{16}{9} \pi \times 0.03$$

$$= 0.17 \, \text{m}^3/\text{s}$$

$dV/dt = 0.17 \, \text{m}^3/\text{s}$ represents the rate at which the volume of water in the cone is increasing and hence the rate at which it is flowing in.

EXERCISE 5.4

1. A tank has a uniform cross-section in the form of a symmetrical inverted triangle. Grain is poured into the tank at a constant rate. Given that the rate at which the level of grain rises is 5 mm/s at a depth of 7 m, find the rate at which the grain enters the tank. Depth of tank 10 m, width across the top 6 m, length 10 m.

2. The rate of increase of radius of a balloon as it rises in the air is 0.2 mm/s at a radius of 60 mm. Find the rate of increase of volume at this radius.

3. A vessel contains 1000 ml of water in which 300 g of salt is dissolved, giving a concentration of 0.3 g/ml. Water is added at a constant rate of 200 ml/min. Obtain an expression for the concentration. At what rate is the concentration changing after water has been added for two minutes?

4. A conical container whose height is equal to the base diameter is placed vertex downwards. Water is pouring into the container at the rate of $0.08 \, \text{m}^3/\text{s}$. Calculate
   (a) the volume of water in the container when the depth of water is x.
   (b) the rate at which the depth is increasing when the depth is 2 m. (Give the answer in mm/s.)

5. The circumference of a circle is increasing at the rate of $200\pi \, \text{mm/s}$ when the radius is 500 mm. Calculate the rate of increase at this time of (a) radius, (b) area.

6. An electric current through a resistor of $10 \, \Omega$ is varying. The rate of change of current is 0.2 A/s when the current is 4 A. Find the rate of change of power developed at this value of the current.

7. The capacity of a parallel plate capacitor is given by

$$C = \frac{40}{d} \, \mu F$$

where d is the separation in mm. As the plates move apart the rate of separation is 0.01 mm/s when the separation is 2 mm. Calculate the rate of change of capitance at this separation.

5.7 Partial differentiation

(a) Partial derivatives

ABCD in Fig. 5.11 is a section of a right cylinder, of radius R and height H.

If the height only of the cylinder is allowed to increase by δH, keeping the radius constant, the volume increases by δV_1, as shown by the shaded region Y. Now

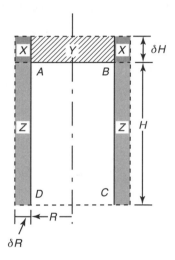

Fig. 5.11

$$\frac{\delta V_1}{\delta H} \to \frac{\mathrm{d} V_1}{\mathrm{d} H} \quad \text{as} \quad \delta H \to 0.$$

Therefore

$$\frac{\delta V_1}{\delta H} \simeq \frac{\mathrm{d} V_1}{\mathrm{d} H}$$

that is

$$\delta V_1 \simeq \frac{\mathrm{d} V_1}{\mathrm{d} H} \delta H \tag{i}$$

Since the volume of the cylinder depends upon R and H, in order to show that (i) is at constant radius, dV/dH is written as $\partial V/\partial H$

$$\delta V_1 \simeq \frac{\partial V}{\partial H} \times \delta H \tag{ii}$$

Similarly, if the radius only is allowed to increase by δR, keeping the height constant, the volume increases by δV_2 shown by the dotted shaded region Z. Again

$$\delta V_2 \simeq \frac{dV}{dR} \times \delta R$$

Since this result is obtained at constant H, dV/dR is written as $\partial V/\partial R$

i.e. $$\delta V_2 \simeq \frac{\partial V}{\partial R} \times \delta R \tag{iii}$$

$\partial V/\partial H$ and $\partial V/\partial R$ are called *partial derivatives*. They may be obtained by differentiating the equation for the volume of a cylinder:

$$V = \pi R^2 H$$

$\partial V/\partial R$ is obtained by differentiating the equation keeping H as constant.

$$V = (\pi H)R^2$$

$$\frac{\partial V}{\partial R} = (\pi H)2R$$

$$= 2\pi RH$$

$\partial V/\partial H$ is obtained by keeping R constant.

$$V = (\pi R^2)H$$

$$\frac{\partial V}{\partial H} = \pi R^2$$

The method developed can be extended to algebraic functions having no direct physical meaning, as shown in Example 5.6.

Example 5.6   Given $z = 3x^2y^3 + 4x^2 + 4y^3$

find $\dfrac{\partial z}{\partial x}$ and $\dfrac{\partial z}{\partial y}$

(i)   In order to find $\partial z/\partial x$, y is regarded as a constant.

$$z = (3y^3)x^2 + 4x^2 + \text{constant.}$$

$$\frac{\partial z}{\partial x} = (3y^3)2x + 8x + 0 \qquad \text{(since the derivative of a}$$
$$\text{constant term} = 0)$$

$$= 6xy^3 + 8x \tag{iv}$$

(ii)   In order to find $\dfrac{\partial z}{\partial y}$, x is regarded as a constant

$$z = (3x^2)y^3 + \text{constant} + 4y^3$$

$$\frac{\partial z}{\partial y} = (3x^2)3y^2 + 0 + 12y^2$$

$$= 9x^2y^2 + 12y^2 \tag{v}$$

The second derivatives may be obtained in the same way. In the above example, from equations (iv) and (v):

$$\frac{\partial^2 z}{\partial x^2} = 6y^3 + 8 \qquad \text{(keeping } y \text{ constant)}$$

and

$$\frac{\partial^2 z}{\partial y^2} = 18x^2y + 24y \qquad \text{(keeping } x \text{ constant)}$$

One other second-order derivative may be obtained, by differentiating (iv) with respect to y keeping x constant. This derivative is written as

$$\frac{\partial^2 z}{\partial x \, \partial y} = 18xy^2 + 0$$

$$= 18xy^2$$

Similarly from (v), by differentiating with respect to x, at constant y

$$\frac{\partial^2 z}{\partial y \, \partial x} = 18xy^2 + 0$$

$$= 18xy^2$$

It can be seen that

$$\frac{\partial^2 z}{\partial x \, \partial y} = \frac{\partial^2 z}{\partial y \, \partial x}$$

This is true for all functions which can be represented graphically by a smooth surface.

(b) Total change

Returning to Fig. 5.11, it is obvious that if R and H are both increasing the total increase in volume δV is given by all the shaded parts. Therefore

$$\delta V = \delta V_1 + \delta V_2 + \text{shaded region } X$$

The shaded region X is very much smaller than δV_1 or δV_2 since it contains the product δH and δR, which are both small quantities. To a first approximation it may be neglected. Therefore

$$\delta V = \delta V_1 + \delta V_2$$

$$= \frac{\partial V}{\partial H} \delta H + \frac{\partial V}{\partial R} \delta R$$

The total change δV to a first approximation is given in terms of changes in two variables H and R. The result may be extended for any number of variables. For example, if $V = f(x, y, z)$, then

$$\boxed{\delta V = \frac{\partial V}{\partial x} \delta x + \frac{\partial V}{\partial y} \delta y + \frac{\partial V}{\partial z} \delta z} \tag{5.7}$$

5.8   Application of partial differentiation to the estimation of error

The result given by equation (5.7) may be readily applied to the estimation of error in values of variables determined experimentally in science and engineering. Consider the experiment of measuring the acceleration due to gravity using a simple pendulum. The equation for the period of oscillation T is

$$T = 2\pi \sqrt{\frac{L}{g}}$$

where L is the length of pendulum and g is the acceleration due to gravity.
   Rearranging the formula,

$$g = \frac{4\pi^2 L}{T^2}$$

Let the errors in measuring L be $+1\%$ and T be $+2\%$, i.e. $\delta L = 0.01L$, $\delta T = 0.02T$.   Let the calculated error in g be δg. Using equation (5.7),

$$\delta g = \frac{\partial g}{\partial L} \delta L + \frac{\partial g}{\partial T} \delta T$$

$$= \frac{\partial g}{\partial L} (0.01L) + \frac{\partial g}{\partial T} (0.02T)$$

From the equation for g,

$$\frac{\partial g}{\partial L} = \frac{4\pi^2}{T^2}$$

$$\frac{\partial g}{\partial T} = 4\pi^2 L(-2)T^{-3}$$

Hence

$$\delta g = \frac{4\pi^2}{T^2} L(0.01) - \frac{8\pi^2 L}{T^3} T(0.02)$$

$$= g(0.01) - g(0.04)$$

$$= g(0.01 - 0.04)$$

$$= -0.03g$$

Therefore percentage error in g is $(\delta g/g) \times 100 = -0.03 \times 100 = -3\%$.

EXERCISE 5.5

1. Given $z = 3x^2 y + 5xy^2$ find

$$\frac{\partial z}{\partial x}, \frac{\partial^2 z}{\partial x^2}, \frac{\partial^2 z}{\partial y^2}$$

2. Given $t = \sin\left(\dfrac{x}{y}\right)$, find $\partial t/\partial x$ and $\partial t/\partial y$. Show that
    $t = -y^2 \, \partial^2 t/\partial x^2$.

3. Given $s = e^{xy}$, determine

$$\frac{\partial s}{\partial x}, \frac{\partial s}{\partial y}, \frac{\partial^2 s}{\partial x \, \partial y}$$

4. If $U = \cos xy$, show that

$$\frac{1}{y}\frac{\partial U}{\partial x} = \frac{1}{x}\frac{\partial U}{\partial y}$$

5. Obtain $\partial^2 V/\partial x \, \partial y$ from $V = x^2 \sin(2x + 3y)$.

6. Given that $U = \ln(5x + 3y)$, show that

$$\frac{\partial U}{\partial x}\frac{\partial U}{\partial y} + \frac{\partial^2 U}{\partial x \partial y} = 0$$

7. In an experiment to measure the heat H generated by a coil, the equation is

$$H = I^2 R t$$

    If the error in measuring the current I is -1.5% and the error in R is 2%, find the error in the calculated value of H.

8. The gas constant is calculated using the equation

$$R = \frac{PV}{T}$$

If the error in V is $+2\%$, in P is -1% and in T is 2%, find the error in the calculated value of R.

MISCELLANEOUS EXERCISE 5

1. Figure 5.12 shows a rectangular strip of metal of length 400 mm and $30x$ mm. The four holes drilled in the strip each have a diameter of $20x$ mm.
   Find the radius of each hole such that the area of the shaded portion is a maximum.

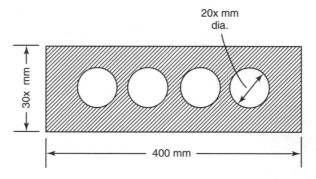

Fig. 5.12

2. (a) Find the values of x at which there is a maximum or minimum of the function

   $$y = x^3 - 3x^2 - 24x + 6$$

   and distinguish between them.
   (b) Calculate the maximum volume of the cylinder which can be cut out of a cone of base radius 1 m and height 2 m.

3. (a) Find the maximum and minimum values of

   $$2x^3 + 3x^2 - 36x + 21$$

   and distinguish between them.
   (b) A vessel is in the form of a hollow cone of semi-vertical angle α, where $\tan \alpha = \dfrac{1}{\sqrt{3}}$. Its axis is vertical and vertex downward.
   Given that water is poured in at the rate of $50 \text{ mm}^3/\text{s}$ find the rate at which the water level is rising at the instant when the depth of water is 20 mm.

4. In Fig. 5.13, *ABCD* is a square of side 200 mm. *AFE* is an isosceles triangle, right-angled at *F*, and on *EB* as diameter a semicircle is drawn. Given $AE = x$ mm, express the shaded area in terms of x and hence find the length of *AE* which makes the shaded area a maximum.

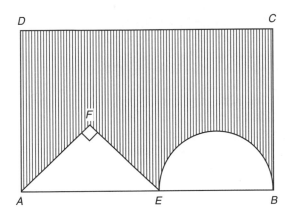

Fig. 5.13

5. (a) A box with sides of length x, y and z metres is expanding along the x and y sides at 0.02 and 0.03 m/s but contracting along the z side at 0.04 m/s. Find, by partial differentiation, the rate of change of volume when $x = y = 0.1$ m and $z = 0.2$ m.

    (b) The length of the hypotenuse of a right-angled triangle is calculated from the length of its sides. Given that these are measured as 12 mm and 5 mm respectively, with an error of 0.2 mm in each, find, by partial differentiation, the possible error in the calculated value of the hypotenuse.

6. Two straight roads intersect at right angles. A car travelling at 40 km/h along one of the roads passes the crossing when the other car, travelling at 30 km/h along the other road towards the crossing, is 10 km from it.

    Find the minimum distance apart of the two cars.

7. Given $y = x^3 + 6x^2 - 63x + 6$ find (a) the values of x for which y is a maximum and a minimum, (b) the maximum and minimum values of y.

8. A rectangular tank of height h metres has a base length $\dfrac{5x}{4}$ metres and width x metres. It has no lid. Write down
    (a) the total volume $V\,\text{m}^3$
    (b) the total surface area $S\,\text{m}^2$.
    Express S in terms of x and V.
       Given that the capacity of such a tank is to be $150\,\text{m}^3$ find the dimension x such that the surface area is a minimum (proving that it is a minimum and not a maximum).

9. The displacement of a moving object from a fixed point is given by $s = 3 - t + 3t^2 - t^3$.
    (a) Find an expression for the velocity at time t and an expression for the acceleration at time t.
    (b) When is the acceleration zero?
    (c) Find the maximum velocity.

10. A particle starts at the origin and moves along the positive x-axis such that its distance from the origin at any time is determined by the equation $x = 2t^3 - 9t^2 + 12t$.
    Find
    (a) the times when the velocity is zero,
    (b) the acceleration when the velocity is zero,
    (c) the value of the minimum velocity.

11. Prove that the function $y = (4/x) + 6x$ has two turning points.
    Determine the nature of these turning points.

12. Given $U = \sin(x/y)$, find

$$\frac{\partial U}{\partial x}, \quad \frac{\partial U}{\partial y}, \quad \frac{\partial^2 U}{\partial x^2}, \quad \frac{\partial^2 U}{\partial y^2} \quad \text{and} \quad \frac{\partial^2 U}{\partial x \, \partial y}$$

13. Given $z = \sin(x + ay)$, prove that

$$\text{(a)} \ \frac{\partial z}{\partial y} = a\frac{\partial z}{\partial x} \qquad \text{(b)} \ \frac{\partial^2 z}{\partial y^2} = a^2 \frac{\partial^2 z}{\partial x^2}$$

and evaluate $\dfrac{\partial^2 x}{\partial x \, \partial y}$ if $x = y = \tfrac{1}{4}\pi$ and $a = 2$.

14. The pressure, volume and absolute temperature of a gas are related by the law $V = RT/P$, where R is a constant. Show that if δT and δP denote small changes in T and P respectively, then the corresponding change δV in V is given approximately by

$$\delta V = \frac{R}{P}\delta T - \frac{RT}{P^2}\delta P$$

15. (a) Given that $u = \ln(e^x + e^{-y})$ show that

$$\frac{\partial u}{\partial x} - \frac{\partial u}{\partial y} = 1$$

(b) Using the method of partial derivatives find the approximate percentage change in the total surface area of a closed cylinder when the radius increases from 40 to 40.1 mm and the height decreases from 100 to 99.8 mm.

Chapter 6

Integration

6.1 The area between a curve and the x-axis

There are many examples in engineering and science where an area beneath a graph represents the magnitude of a physical quantity. One such example is the work done by an object moving a distance s against a constant force F. It is well known that the work done W is given by

$$W = Fs$$

In Fig. 6.1a CD is a graph of this constant force F. The product Fs is the area of the shaded rectangle. Hence it is seen that the area under a force–distance curve represents work done.

When the force is not constant the area under the curve will still represent the work done. In this case, however, the area cannot be determined by geometrical methods (Fig. 6.1b).

In both examples the area required is bounded by two ordinates such as LM and NP.

Another example is that of the area under a velocity–time curve, where the area represents the distance travelled.

The concept of negative area is shown in Fig. 6.1c where it is the area between a curve and the x-axis, below the axis.

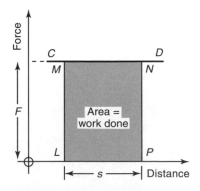

Fig. 6.1(a)

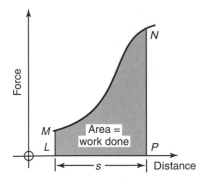

Fig. 6.1(b)

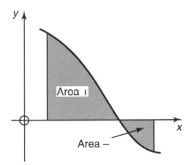

Fig. 6.1(c)

6.2 Integration: the method of obtaining the exact area under a curve

There are several ways of determining the area under a curve. Three of these are

1. graphical method: the area is found by counting squares;
2. mid-ordinate rule;
3. Simpson's rule: see Chapter 11.

These methods produce approximate results. In order to determine an area exactly the method of **integration** must be used.

It was seen in Chapter 4 that the area of a strip such as $ABCD$ (Fig. 6.2) was very nearly the area of the rectangle, when the width was small; that is,

$$\text{area of strip} \approx y\,\delta x$$

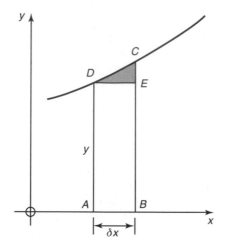

Fig. 6.2

As the width becomes smaller the area of the strip gets closer to the area of the rectangle, so that as $\delta x \to 0$, the above equation becomes exact. At this limit δx is written as dx. Therefore

$$\text{area of strip} = y \cdot dx.$$

In order to find the area under the curve in Fig. 6.3, between the ordinates at $x = a$ and $x = b$, the area is divided into a large number, N, of strips, each of width δx. Since δx is small the shaded area at the top of each strip is small in comparison to the area of the rectangle. The area $ABCD$ is given approximately by the sum of the areas of the rectangles.

$$\text{Area of } ABCD \simeq y_0 \, \delta x + y_1 \, \delta x + y_2 \, \delta x + \cdots + y_{N-1} \, \delta x$$

Such a sum is written in the form

$$\text{Area of } ABCD \simeq \sum_0^{N-1} y \, \delta x$$

where $\sum$ denotes the sum from 0 to $N - 1$.

As $\delta x \to 0$ the shaded areas become negligibly small, and the above sum gives the exact area under the curve. In order to denote that the sum is carried out under the special condition of $\delta x \to 0$, the $\sum$ sign is replaced by $\int$, which is called the integral sign. As before the δx is replaced by dx. The area can now be written

$$\text{Area} = \int y \, dx$$

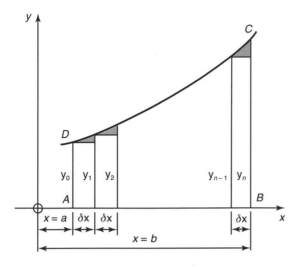

Fig. 6.3

The area is determined between the ordinates $x = a$ and $x = b$, which are called the **limits of integration**. These limits are shown on the integral sign.

$$\text{Area} = \int_a^b y \, dx$$

The operation of finding the area is called integration and the right-hand side of the above equation is called a **definite integral**.

6.3   The areas under simple curves

(a) The area under $y = 1$, between the limits $x = a$ and $x = b$

The required area is $ABCD$ in Fig. 6.4, which can be found by simple geometry.

$$\text{Area} = 1. \, (b - a)$$

By integration

$$\text{Area} = \int_a^b y \, dx = \int_a^b dx, \quad \text{since } y = 1.$$

Therefore

$$\int_a^b 1 \, dx = b - a \tag{6.1}$$

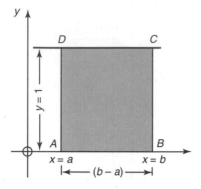

Fig. 6.4

(b) The area under $y = x$, between the limits $x = a$ and $x = b$

The required area is $ABCD$ in Fig. 6.5. The figure is a trapezium so that

$$\text{area} = \tfrac{1}{2}(y_1 + y_2)(b - a)$$

Since $y = x$, then $y_1 = a$ and $y_2 = b$

$$\text{Area} = \tfrac{1}{2}(a + b)(b - a)$$
$$= \tfrac{1}{2}b^2 - \tfrac{1}{2}a^2$$

By integration

$$\text{area} = \int_a^b y\,\mathrm{d}x = \int_a^b x\,\mathrm{d}x, \qquad \text{since } y = x$$

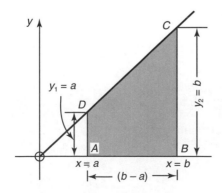

Fig. 6.5

Therefore

$$\int_a^b x\,dx = \tfrac{1}{2}b^2 - \tfrac{1}{2}a^2 \tag{6.2}$$

(c) The area under $y = x^2$, between the limits $x = a$ and $x = b$

Simple geometry can no longer be used to determine the area, and the method of integration is used instead. Consider first the area between the limits $x = 0$ and $x = a$. The area from O to A (Fig. 6.6) is divided into a *large number of strips N*, each of width δx. The area OAD is approximately the sum of the areas of the rectangles. As the widths of the strips tend to zero the area OAD becomes exactly equal to the sum of the rectangles. Let A_r be the sum of the areas of the rectangles. Then

$$A_r = y_0\,\delta x + y_1\,\delta x + y_2\,\delta x \ldots + y_{N-1}\,\delta x.$$

Since $y = x^2$, $y_0 = 0$

$$y_1 = (\delta x)^2 \qquad \text{at } x = \delta x$$
$$y_2 = (2\,\delta x)^2 \qquad \text{at } x = 2\,\delta x$$
$$\vdots$$
$$y_{N-1} = [(N-1)\delta x]^2 \quad \text{at } x = (N-1)\delta x$$

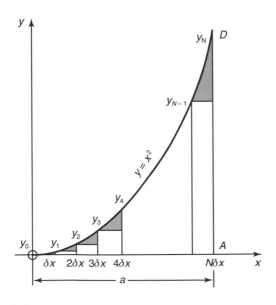

Fig. 6.6

Using these values for the ordinates then A_r becomes

$$A_r = 0\,\delta x + (\delta x)^2\delta x + (2\,\delta x)^2\,\delta x + \cdots + [(N-1)\delta x]^2\,\delta x$$
$$= (\delta x)^3[1^2 + 2^2 + 3^2 + \cdots + (N-1)^2]$$

In Appendix 2 the sum of the series in the square brackets is shown to be $\frac{1}{6}N(N-1)(2N-1)$. Hence

$$A_r = (\delta x)^3\,\tfrac{1}{6}\,N(N-1)(2N-1)$$
$$= (\delta x)^3\,\tfrac{1}{6}\,[2N^3 - 3N^2 + N]$$
$$= \tfrac{1}{3}(N\,\delta x)^3 - \tfrac{1}{2}(N\,\delta x)^2\,\delta x + \tfrac{1}{6}(N\,\delta x)(\delta x)^2$$

From Fig. 6.6, $N\,\delta x = a$
Therefore

$$A_r = \tfrac{1}{3}a^3 - \tfrac{1}{2}a^2\,\delta x + \tfrac{1}{6}a(\delta x)^2 \qquad\qquad\text{(i)}$$

As $\delta x \to 0$ the shaded areas become negligible in comparison to the rectangles, and in the limit the sum of the areas of the rectangles becomes identical with the area under the curve. Therefore

$$\text{Area under the curve} = \lim A_r \quad\text{as}\quad \delta x \to 0$$
$$= \tfrac{1}{3}a^3 \text{ on making use of equation (i)}$$

From this result the area under the curve between the limits $x = a$ and $x = b$ can be determined (Fig. 6.7):

$$\text{area } OAD = \tfrac{1}{3}a^3$$
$$\text{area } OBC = \tfrac{1}{3}b^3$$

The shaded area $ABCD = \tfrac{1}{3}b^3 - \tfrac{1}{3}a^3$

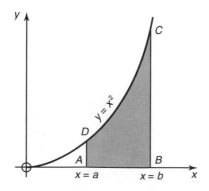

Fig. 6.7

Using the integral notation the area under the curve is

$$\int_a^b y \, dx = \int_a^b x^2 \, dx, \qquad \text{since } y = x^2.$$

Therefore

$$\int_a^b x^2 \, dx = \tfrac{1}{3} b^3 - \tfrac{1}{3} a^3 \tag{6.3}$$

The results given by (6.1), (6.2) and (6.3) are listed in Table 6.1.

Table 6.1

Function	Integral	Area between $x = a$ and $x = b$
$y = 1$	$\displaystyle\int_a^b 1 \, dx$	$b - a$
$y = x$	$\displaystyle\int_a^b x \, dx$	$\tfrac{1}{2} b^2 - \tfrac{1}{2} a^2$
$y = x^2$	$\displaystyle\int_a^b x^2 \, dx$	$\tfrac{1}{3} b^3 - \tfrac{1}{3} a^3$

In each of these examples shown above the result can be put in terms of a function $F(x)$ and the area obtained as $F(b) - F(a)$, as shown in Table 6.2.

A pattern is seen to exist in the integrals of algebraic functions, as listed in this latter table. The results obtained so far can be extended for any function such as

$$y = x^n.$$

Table 6.2

Function	Integral	$F(x)$	Area $= F(b) - F(a)$
$y = 1$	$\displaystyle\int_a^b 1 \, dx$	x	$b - a$
$y = x$	$\displaystyle\int_a^b x \, dx$	$\tfrac{1}{2} x^2$	$\tfrac{1}{2} b^2 - \tfrac{1}{2} a^2$
$y = x^2$	$\displaystyle\int_a^b x^2 \, dx$	$\tfrac{1}{3} x^3$	$\tfrac{1}{3} b^3 - \tfrac{1}{3} a^3$

The area under such a curve between the limits $x = a$ and $x = b$ is

$$\int_a^b y \, dx = \int_a^b x^n \, dx = \left[\frac{x^{n+1}}{n+1} \right]_a^b$$

that is $$\boxed{\int_a^b x^n \, dx = \frac{b^{n+1}}{n+1} - \frac{a^{n+1}}{n+1}}$$ (6.4)

Example 6.1   Find the area beneath the curve $y = x^3$, between the ordinates $x = 1$ and $x = 2$.

Using rule (6.4), the area beneath the curve is

$$\int_1^2 y \, dx = \int_1^2 x^3 \, dx$$

$$= \left[\tfrac{1}{4} x^4 \right]_1^2$$

$$= \frac{2^4}{4} - \frac{1^4}{4}$$

$$= \frac{16}{4} - \frac{1}{4}$$

$$= \frac{15}{4} \quad \text{square units}$$

6.4   Integration of algebraic functions containing coefficients

The discussion in Section 6.3(c) is now extended to a function containing a coefficient, for example

$$y = 5x^2$$

In this case the ordinates will be five times as large as those for $y = x^2$. Therefore the area under the curve will be five times the area under $y = x^2$, between the same limits, that is:

$$\int_a^b y \, dx = \int_a^b 5x^2 \, dx = 5 \int_a^b x^2 \, dx$$

The coefficient can be taken outside the integral sign. This conclusion can be applied to any function containing a coefficient p.

If $y = px^n$ then provided $n \neq -1$ (see section 6.13)

$$\int_a^b px^n \, dx = p \int_a^b x^2 \, dx = p \left[\frac{x^{n+1}}{n+1} \right]_a^b = p \left[\frac{b^{n+1}}{n+1} - \frac{a^{n+1}}{n+1} \right] \qquad (6.5)$$

Example 6.2 The following definite integrals are evaluated in accordance with the discussion so far.

(a) $\int_a^b dx \qquad = [x]_a^b \qquad = b - a$

(b) $\int_a^b 5 \, dx \qquad = 5 \int_a^b dx \qquad = 5[x]_a^b \qquad = 5(b - a)$

(c) $\int_a^b 8x \, dx \qquad = 8 \int_a^b x \, dx \qquad = 8 \left[\tfrac{1}{2} x^2 \right]_a^b \qquad = 8(\tfrac{1}{2} b^2 - \tfrac{1}{2} a^2)$

(d) $\int_a^b - 7x^5 \, dx = -7 \int_a^b x^5 \, dx = -7 \left[\frac{x^6}{6} \right]_a^b = -7 \left(\frac{b^6}{6} - \frac{a^6}{6} \right).$

EXERCISE 6.1

In the following obtain the area under the curve of the function between the limits shown:

1. $\int_4^5 3x \, dx$

2. $\int_3^6 4 \, dx$

3. $\int_{-2}^1 - 9x^2 \, dx$

4. $\int_{-2}^2 \tfrac{1}{2} x^3 \, dx$

5. $\int_0^4 - 3x^4 \, dx.$

Evaluate the following definite integrals:

6. $\int_{-1}^1 \tfrac{3}{4} x^2 \, dx$

7. $\int_{-7}^9 dx$

8. $\int_3^{+2} - 5x^2 \, dx$

9. $\int_a^b 10x^{29} \, dx$

10. $\int_a^{2a} 8x^3 \, dx.$

6.5   Area between a curve and the y-axis

The definite integral has been presented as the area under a curve, and expressed as

$$\int_a^b y \, dx$$

where y is the letter denoting the length of the ordinate of the curve.

Using rule 6.3, it is seen that the integration is evaluated in terms of one variable only. Therefore, before any integral can be worked out y must be expressed in terms of x. This conclusion can be stated as follows, **a variable cannot be integrated with respect to another variable.**

It is possible to illustrate this by considering the area between the y-axis and the curve (Fig. 6.8). By a similar treatment, it is obvious that the area is

$$\int_c^d x \, dy$$

In order to evaluate this integral for a particular curve x must be expressed in terms of y.

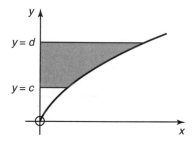

Fig. 6.8

Example 6.3   Find the area between the y axis and the curve $y^2 = x$, between the limits $y = 1$ and $y = 3$.

$$\text{Area} = \int_1^3 x \, dy$$

$$= \int_1^3 y^2 \, dy$$

$$= \left[\frac{y^3}{3} \right]_1^3 , \text{ using equation 6.4}$$

$$= 9 - \tfrac{1}{3}$$

$$= 8 \tfrac{2}{3} \text{ square units}$$

6.6   Integration as the reverse of differentiation – the indefinite integral

Using the result obtained in rule (6.5) we see that

$$\int x^n \, dx = \frac{x^{n+1}}{(n+1)}$$

If we now look at the right-hand side and differentiate the function

$$\frac{x^{n+1}}{(n+1)}$$

we obtain the function x^n; that is, the function on the left-hand side of the above rule. *Therefore, it is clear that integration is the reverse of differentiation.*

Consider the differential coefficient of the two functions:

    (1) $\tfrac{1}{3} x^3 + 7$ and   (2) $\tfrac{1}{3} x^3 + 9.$

The result in each case is x^2. Reversing the process, and integrating x^2 produces the result $\tfrac{1}{3} x^3$. However, no information is obtained about the constant in the original functions. Therefore, treating integration as the reverse of differentiation produces an indefinite result as far as the constant is concerned. Consequently a constant of integration C is always added to the result. This process is called **indefinite integration**. The indefinite integral will still have the same sign $\int$, but without the limits such as a and b in the previous paragraph. For example,

$$\int x^2 \, dx = \tfrac{1}{3} x^3 + C$$

$$\int 5x^6 \, dx = 5 \int x^6 \, dx = \tfrac{5}{7} x^7 + C$$

The constant C cannot be determined unless some further information is known about the resulting function, as shown in Example 6.4.

Example 6.4   Evaluate $\int x^4 \, dx$. Determine the constant of integration if the graph of the resulting function passes through the point (1, 3).

Using the results of the previous section

$$\int x^4 \, dx = \frac{x^5}{5} + C$$

Now the curve $y = (x^5/5) + C$ passes through the point (1, 3). We substitute $x = 1$, $y = 3$ into this equation and obtain

$$3 = \frac{1^5}{5} + C$$

$$= \frac{1}{5} + C$$

Therefore

$$C = 2\tfrac{4}{5} = \frac{14}{5}$$

Hence

$$y = \frac{x^5}{5} + \frac{14}{5}$$

EXERCISE 6.2

Evaluate the following:

1. $\int x^3 \, dx$

2. $\int x^{10} \, dx$

3. $\int 7x \, dx$

4. $\int -3x^6 \, dx$

5. $\int -\tfrac{1}{3} x^5 \, dx$

6. $\int \tfrac{1}{4} x^4 \, dx$

7. $\int \frac{x^5}{6} \, dx$

8. $\int \frac{x^3}{10} \, dx$

9. Integrate $\int 3x^6 \, dx$ and determine the constant C of integration if the graph of the resulting function passes through the point (1, −1).

6.7   Integration of more difficult algebraic functions

(a) Type 1; Negative and fractional indices

The general indefinite integral is given by

$$\int px^n \, dx = p \int x^n \, dx = \frac{pn^{n+1}}{(n+1)} + C \qquad (6.6)$$

and can apply to any value of n, including fractional and negative indices, as shown in Example 6.5. The exception to this result is when $n = -1$. When $n = -1$, the above rule gives

$$\int \frac{1}{x} \, dx = \int x^{-1} \, dx = \frac{x^0}{0} + C$$

which gives an infinite result. Such an integral is evaluated in section 6.13.

Example 6.5   Determine the following integrals:

(a) $\int x^{-3} \, dx$ (b) $\int x^{\frac{3}{4}} \, dx$ (c) $\int \frac{1}{x^5} \, dx$ (d) $\int \frac{1}{\sqrt{x}} \, dx.$

(a) $\int x^{-3} \, dx = \dfrac{x^{-3+1}}{(-3+1)} + C$

$\qquad\qquad = \dfrac{x^{-2}}{-2} + C$

(b) $\int x^{\frac{3}{4}} \, dx \quad = \dfrac{x^{\frac{3}{4}+1}}{(\frac{3}{4}+1)} + C$

$\qquad\qquad = \dfrac{x^{\frac{7}{4}}}{\frac{7}{4}} + C$

$\qquad\qquad = \frac{4}{7} x^{\frac{7}{4}} + C$

(c) $\int \frac{1}{x^5} \, dx.$

In order to evaluate the integral, the function inside the integral sign must be put in the form x^n in the numerator. Therefore, the integral becomes

$$\int x^{-5}\, dx = \frac{x^{-5+1}}{(-5+1)} + C$$

$$= \frac{-x^{-4}}{4} + C$$

(d) $\displaystyle\int \frac{1}{\sqrt{x}}\, dx = \int \frac{dx}{x^{\frac{1}{2}}}$

$$= \int x^{-\frac{1}{2}}\, dx$$

$$= \frac{x^{\frac{1}{2}}}{\frac{1}{2}} + C$$

$$= 2x^{\frac{1}{2}} + C$$

Type 2: Integration of polynomials

In this type each term can be integrated separately as shown in Example 6.6.

Example 6.6   Determine the following integral:

$$\int (x^2 + 3x + 4)\, dx.$$

Each term is considered separately:

$$\int (x^2 + 3x + 4)\, dx = \int x^2\, dx + \int 3x\, dx + \int 4\, dx$$

$$= \tfrac{1}{3}x^3 + \tfrac{3}{2}x^2 + 4x + C$$

Type 3: Integration of functions reducible to polynomials

Some functions can be integrated by converting them into the form shown in Example 6.6. Two functions which can be integrated in this way are shown in Example 6.7 below.

Example 6.7 Determine

(a) $\displaystyle\int_1^2 (x+3)^2 \, dx$

(b) $\displaystyle\int \frac{x^2 + 3x + 4}{x^4} \, dx.$

(a) The first step is to expand the bracket, and then integrate term by term.

$$\int_1^2 (x+3)^2 \, dx = \int_1^2 (x^2 + 6x + 9) \, dx$$

$$= \left[\tfrac{1}{3} x^3 + 3x^2 + 9x \right]_1^2$$

$$= \left[\tfrac{8}{3} + 12 + 18 \right] - \left[\tfrac{1}{3} + 3 + 9 \right]$$

$$= 20 \tfrac{1}{3}$$

(b) The first step is to divide x^4 into each term in the numerator, that is

$$\int (x^{-2} + 3x^{-3} + 4x^{-4}) \, dx = \frac{x^{-1}}{-1} + \frac{3x^{-2}}{-2} + \frac{4x^{-3}}{-3} + C$$

$$= -x^{-1} - \frac{3}{2} x^{-2} - \frac{4}{3} x^{-3} + C$$

EXERCISE 6.3

Determine the following integrals,

1. $\displaystyle\int x^{-6} \, dx$

2. $\displaystyle\int_1^2 3x^{-9} \, dx$

3. $\displaystyle\int t^{\frac{3}{4}} \, dt$

4. $\displaystyle\int \tfrac{1}{4} x^{-\frac{1}{4}} \, dx$

5. $\displaystyle\int 6x^{-\frac{2}{3}} \, dx$

6. $\displaystyle\int \tfrac{2}{3} t^{-\frac{1}{3}} \, dt$

7. $\displaystyle\int_0^1 (x+1)^2 \, dx$

8. $\displaystyle\int_{-1}^{+1} (2x-1)^2 \, dx$

9. $\displaystyle\int (x-2)^3 \, dx$

10. $\displaystyle\int_{-2}^3 (2x+3)(x-4) \, dx$

11. $\displaystyle\int (x-3)(x+3) \, dx$

12. $\displaystyle\int (-x^2 + 1)^2 \, dx$

13. $\displaystyle\int \sqrt{x}\,dx$

14. $\displaystyle\int \left(\sqrt{x}+\frac{1}{\sqrt{x}}\right) dx$

15. $\displaystyle\int \frac{x+4}{x^3}\,dx$

16. $\displaystyle\int \frac{2x^2-3x+6}{x^4}\,dx$

17. $\displaystyle\int \frac{x^2-1}{-3x^5}\,dx$

18. $\displaystyle\int \frac{x^3-3x^2-2x}{x^{\frac{1}{2}}}\,dx$

19. $\displaystyle\int \frac{x^{\frac{1}{4}}-x^{\frac{3}{4}}}{x^{\frac{1}{2}}}\,dx$

20. $\displaystyle\int x^{-\frac{1}{3}}(2x^{\frac{2}{3}}+\tfrac{3}{4}x^{\frac{1}{3}})\,dx$

6.8   Further examples of areas under curves

(a) The area enclosed between a curve and the x-axis

In Section 6.1 the area under a curve, between two limits, was shown to be the sum of strips $y\,\delta x$ which becomes

$$\int_a^b y\,dx \quad \text{as} \quad x \to 0.$$

In some cases a curve cuts the axis in two places as shown in Fig. 6.9. The enclosed area is still given by

$$\int_a^b y\,dx \quad ,$$

the limits of integration now being the x values of the points of intersection of curve and axis.

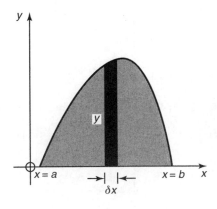

Fig. 6.9

Example 6.8 Find the area enclosed between the curve

$$y = 2x^2 + x - 3 \text{ and the } x\text{-axis.}$$

First of all the positions of the points of intersection of the curve with the x-axis are required. At these points $y = 0$ so that the x values are given by the solution of the equation

$$2x^2 + x - 3 = 0$$

that is $(2x + 3)(x - 1) = 0$

$$x = -\tfrac{3}{2} \quad \text{or} \quad x = 1$$

These values are the limits of the integral so that the area under the curve is

$$\int_{-\frac{3}{2}}^{1} (2x^2 + x - 3)\,\mathrm{d}x = \left[\frac{2x^3}{3} + \frac{x^2}{2} - 3x\right]_{-\frac{3}{2}}^{1}$$

$$= [\tfrac{2}{3} + \tfrac{1}{2} - 3] - [\tfrac{2}{3}(-\tfrac{3}{2})^3 + \tfrac{1}{2}(-\tfrac{3}{2})^2 - 3(-\tfrac{3}{2})]$$

$$= -5\tfrac{5}{24} \quad \text{square units.}$$

(b) The area of a circle

A circle of radius r cuts the x-axis at $x = -r$ and $x = +r$, as shown in Fig. 6.10(a). It is possible to find the area of the upper semi-circle by integrating $\int_{-r}^{+r} y\,\mathrm{d}x$. The integral turns out to be difficult to evaluate.

An easier method is possible making use of the symmetry of a circle.

Consider a thin strip, of thickness δx and radius x about the centre of the circle. This strip is the dark shaded portion in Fig. 6.10. Such a strip is called

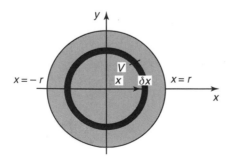

Fig. 6.10(a)

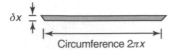

δx

Circumference $2\pi x$

Fig. 6.10(b)

an annulus. Because the annulus is thin the inner and outer circumferences are practically the same, so that if it were cut at V, and straightened out, a thin rectangle would be obtained, of area

$$2\pi x \delta x$$

As $\delta x \to 0$ the area of the annulus becomes $2\pi x\,dx$. The area of the circle is now the sum of all such annuli from the centre to the outer edge at $x = r$; that is,

$$\text{area of circle} = \int_0^r 2\pi x\,dx$$

$$= \left[2\pi\frac{x^2}{2}\right]_0^r$$

$$= \pi r^2$$

(c) Area enclosed between a curve and a straight line

In Fig. 6.11 a straight line intersects the curve at the points $x = a$ and $x = b$. The enclosed shaded area $CDEF$ is the difference between the areas $ABCDE$ and $ABCFE$, which are, respectively, the area under the curve and the area under the straight line. As indicated already, these areas can be determined by integration.

The procedure is shown in Example 6.9.

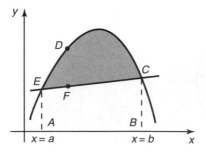

Fig. 6.11

Example 6.9 Find the area between the curve $y = -x^2 + 4x$ and the line $y = x + 2$.

At the points of intersection the y values of both functions are the same. The x values of the points of intersection are found by solving the equation

$$-x^2 + 4x = x + 2$$

that is

$$x^2 - 3x + 2 = 0$$
$$(x-1)(x-2) = 0$$
$$x = 1 \quad \text{or} \quad x = 2$$

These values are now the limits of the two integrals.

$$\text{Area } ABCDE = \int_1^2 y \, dx$$

$$= \int_1^2 (-x^2 + 4x) \, dx$$

$$= \left[-\frac{x^3}{3} + 2x^2 \right]_1^2$$

$$= [-\tfrac{8}{3} + 8] - [-\tfrac{1}{3} + 2]$$

$$= 3\tfrac{2}{3}$$

$$\text{Area } ABCFE = \int_1^2 y \, dx$$

$$= \int_1^2 (x + 2) \, dx$$

$$= \left[\frac{x^2}{2} + 2x \right]_1^2$$

$$= [2 + 4] - [\tfrac{1}{2} + 2]$$

$$= 3\tfrac{1}{2}$$

Enclosed area $= 3\tfrac{2}{3} - 3\tfrac{1}{2} = \tfrac{1}{6}$ square units

EXERCISE 6.4

1. Find the area between the curve $y = x^2 - 3x - 4$ and the x-axis.
2. Find the area under the curve $y = x^3$ enclosed by the x-axis and the ordinates $x = 0$ and $x = 5$.
3. Find the area between the curve $y = 3x^2 + 4$ and the x-axis, between the limits $x = -1$ and $x = +2$.
4. Determine the area enclosed between the curve $y^2 = 4 - x$ and the y-axis.
5. What is the area enclosed by the two curves $y = -x^2 + 6$ and $y = x^2 - 2$?
6. Sketch, using the same axes, the straight line $y = x + 1$ and the curve $y = x^2 + 1$. Find the area enclosed between the curve and the straight line.
7. Find the area bounded by the curve $y = 3x^2 - 10x - 2$, the x-axis and the ordinates at $x = 1$ and $x = 3$. Account for the sign of the answer.
8. Evaluate the area enclosed between the curve $y = -x^2 + 2x$ and the line $y = x$.
9. Sketch the curve $y = -x^2 + 6$ and the lines $y = x$ and $y = 5x$ on the same axes. Find the area enclosed by the three graphs in the first quadrant.

6.9   Volumes of revolution

Fig. 6.12 shows what is meant by a solid of revolution. If triangle OAC is rotated through an angle of 360° about the x-axis it will trace out a right circular cone, which is the solid of revolution.

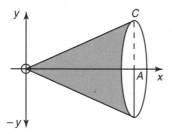

Fig. 6.12

The volume of revolution can be determined in the same way as the area under the curve. The area is divided into a large number of strips. If this area is made to rotate about the x-axis these strips will rotate and produce figures which are very nearly discs with uniform areas of cross section (Fig. 6.13). The summation of the volumes of the discs gives the volume of revolution.

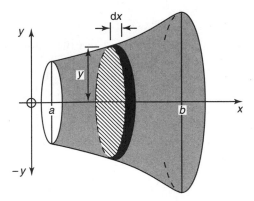

Fig. 6.13

As the thickness of the discs $\delta x \to 0$ they can be regarded as discs with uniform area of cross-section.

Volume of disc = area of cross-section × thickness

$$= \pi y^2 \, dx$$

The volume of revolution between the limits $x = a$ and $x = b$ is the sum of the volumes of these discs. When $\delta x \to 0$, this summation is an integral; that is,

$$\text{Volume of revolution} = \int_a^b \pi y^2 \, dx \tag{6.7}$$

Example 6.10   Find the volume of revolution between the limits $x = 2$ and $x = 5$, when the curve $y = 2x^2$ is rotated about the x-axis.

$$
\begin{aligned}
\text{Volume of revolution} &= \int_2^5 \pi y^2 \, dx \\[2mm]
&= \int_2^5 \pi (2x^2)^2 \, dx \\[2mm]
&= \int_2^5 \pi 4x^4 \, dx \\[2mm]
&= \left[\frac{4\pi x^5}{5} \right]_2^5 \\[2mm]
&= 4\pi 625 - \frac{4\pi 32}{5} \\[2mm]
&= 2474\pi \text{ cubic units}
\end{aligned}
$$

EXERCISE 6.5

1. The line $y = 2x$ is rotated about the x-axis. Find the volume of revolution between the limits $x = 0$ and $x = 4$.
2. The curve $y = 3x^2 + 2x - 4$ is rotated about the x-axis. Find the volume of revolution between $x = 0$ and $x = 1$.
3. The parabola $y^2 = 8x + 2$ is rotated about the x-axis. Find the volume of revolution between $x = +3$ and $x = +4$.
4. The areas in Questions 3, 4, Exercise 6.4, are rotated about the x-axis. Find the volume of revolution between the limits stated in the questions.

6.10   Volume of a cone

In Fig. 6.12 a triangle is rotated about the x-axis to form a cone. Let OA, the height of the cone be h, and AC, the base radius, be r. Therefore the equation of the line OC is

$$y = \frac{r}{h}x$$

since r/h is the gradient of the line.

From rule (6.7),

$$\text{volume of revolution} = \int_0^h \pi y^2 \, dx$$

$$= \pi \int_0^h \frac{r^2 x^2}{h^2} \, dx$$

$$= \pi \frac{r^2}{h^2} \left[\frac{x^3}{3} \right]_0^h$$

$$= \tfrac{1}{3} \pi r^2 h$$

6.11   Volume of a sphere

If a circle, $x^2 + y^2 = r^2$, is rotated about the x-axis it will produce the outline of a sphere, as shown in Fig. 6.14. The limits are seen to be $x = -r$ and $x = +r$.

From Fig. 6.14 it is seen that the volume of the disc is

$$\pi y^2 \, dx$$

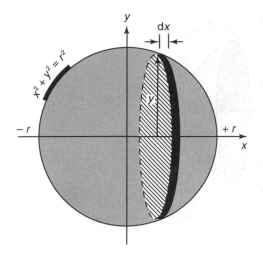

Fig. 6.14

The volume of revolution is the sum of these discs; that is,

$$\text{Volume of revolution} = \pi \int_{-r}^{+r} y^2 \, dx$$

But from the equation of a circle $y^2 = r^2 - x^2$, so that

$$\text{Volume of revolution} = \pi \int_{-r}^{+r} (r^2 - x^2) \, dx$$

$$= \pi \left[r^2 x - \frac{x^3}{3} \right]_{-r}^{=r}$$

$$= \pi [r^3 - \tfrac{1}{3} r^3] - \pi [-r^3 + \tfrac{1}{3} r^3]$$

$$\text{Volume of sphere} = \tfrac{4}{3} \pi r^3$$

The volume of a cap of a sphere of height h can be found in the same way by integrating from $(r - h)$ to r, as in Fig. 6.15.

$$\text{Volume of cap} = \pi \int_{(r-h)}^{r} y^2 \, dx$$

$$= \pi \int_{(r-h)}^{r} (r^2 - x^2) \, dx$$

$$= \pi \left[r^2 x - \frac{x^3}{3} \right]_{r-h}^{r}$$

$$= \pi h^2 (r - \tfrac{1}{3} h)$$

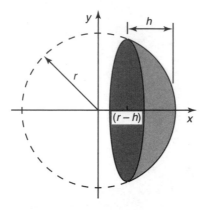

Fig. 6.15

6.12   Integration by substitution

Earlier in the chapter two interpretations of integration were given,

1.   indefinite integration as the reverse of differentiation;
2.   definite integration as an area beneath the curve.

In each case the integration is carried out as the reverse of differentiation, but in the second case limits are inserted.

In this section more complex integrals are worked out by making use of a substitution to reduce the integrals to *standard – forms*. An integral such as

$$\int (2x - 4)^4 \, dx$$

is carried out in this way. A substitution is made for the inner function; that is,

$$u = 2x - 4 \tag{i}$$

The integral becomes

$$\int u^4 \, dx$$

But in Section 6.5 we saw that a function u cannot be integrated with respect to another variable x in dx. From (i)

$$\frac{du}{dx} = 2 \tag{ii}$$

Now

$$\delta u = \frac{\delta u}{\delta x} \delta x$$

As $\delta x \to 0$

$$du = \frac{du}{dx} dx \qquad\qquad (6.8)$$

In this example using (ii)

$$du = 2\,dx$$

or $dx = \frac{1}{2}\,du.$

The integral becomes

$$\int u^4 \tfrac{1}{2}\,du = \tfrac{1}{2}\frac{u^5}{5} + C$$

$$= \tfrac{1}{10}(2x - 4)^5 + C$$

The substitution method can be used when the function outside the bracket is a differential coefficient of the function inside the bracket, apart from a multiplying constant.

Example 6.11   Determine the integrals

(a) $\displaystyle\int (x^2 - 6)^7 x \, dx$

(b) $\displaystyle\int \frac{(x - 1)\,dx}{\sqrt{(x^2 - 2x + 1)}}.$

(a)   The function outside the bracket is x. The differential coefficient of the inner function is $2x$. The above rule is applicable, the multiplying constant being 2.
Let

$$u = x^2 - 6, \qquad \text{then} \quad \frac{du}{dx} = 2x$$

From equation (6.8)

$$du = 2x\,dx$$

$$x\,dx = \tfrac{1}{2}\,du$$

The integral becomes

$$\int u^7 \tfrac{1}{2}\, du = \tfrac{1}{2}\frac{u^8}{8} + C$$

$$= \tfrac{1}{16}(x^2 - 6)^8 + C$$

(b)　The integral can be written as

$$\int \frac{(x-1)\, dx}{(x^2 - 2x + 1)^{\frac{1}{2}}} = \int (x^2 - 2x + 1)^{-\frac{1}{2}}(x-1)\, dx$$

The inner function, when differentiated, gives $2x - 2$, which is the same as the outer function apart from a multiplying constant 2. The substitution method is again applicable.

Let　$u = (x^2 - 2x + 1)$,　　then　$\dfrac{du}{dx} = 2x - 2 = 2(x - 1)$

From equation (6.8)

$$du = 2(x - 1)\, dx$$
$$(x - 1)\, dx = \tfrac{1}{2}\, du$$

The integral becomes

$$\int u^{-\frac{1}{2}}\tfrac{1}{2}\, du = \tfrac{1}{2}\frac{u^{\frac{1}{2}}}{\frac{1}{2}} + C$$

$$= u^{\frac{1}{2}} + C$$

$$= (x^2 - 2x + 1)^{\frac{1}{2}} + C$$

EXERCISE 6.6

Determine the following integrals.

1.　$\displaystyle\int (2x - 5)^3 2\, dx$

2.　$\displaystyle\int (2x^2 - 3)^{\frac{1}{3}}x\, dx$

3.　$\displaystyle\int (2x^2 - 2x + 1)^3(4x - 2)\, dx$

4.　$\displaystyle\int (3x^3 - 2)^{\frac{1}{2}}x^2\, dx$

5.　$\displaystyle\int (6t^2 - 4t + 2)^{-3}(3t - 1)\, dt$

6.　$\displaystyle\int (4x^2 - 1)^6 x\, dx$

7.　$\displaystyle\int \frac{(x + 1)\, dx}{(x^2 + 2x + 2)^{\frac{1}{4}}}$

8.　$\displaystyle\int_0^1 \frac{x^2\, dx}{(2x^3 - 3)^2}$

9.　$\displaystyle\int \frac{dx}{(2x - 4)^{\frac{1}{3}}}$

10.　$\displaystyle\int \frac{x^2\, dx}{(\frac{1}{3}x^3 + 1)^{\frac{1}{2}}}$

6.13   Functions which integrate to give the logarithm function

In Section 4.18(a) it was shown that

$$\frac{d}{dx}(\ln x) = \frac{1}{x}$$

Reversing the procedure

$$\int \frac{1}{x}dx = \ln x + C$$

If we look at the integral it is obvious that the numerator, 1, is an exact derivative of the denominator. Whenever this characteristic is present in an integral the result is the logarithm of the denominator, even when the denominator is complex.

In an integral, if the numerator is the differential coefficient of the denominator apart from a multiplying constant, the integration may be carried out by substituting u for the denominator

---

Example 6.12   Determine

$$\int \frac{(2x + 1)\,dx}{2x^2 + 2x + 7}$$

The differential coefficient of the denominator is $4x + 2$ which is the same as the numerator, apart from a multiplying constant.

Let $u = 2x^2 + 2x + 7;$ then $\dfrac{du}{dx} = 4x + 2 = 2(2x + 1)$

From equation (6.8)

$$du = 2(2x + 1)\,dx$$

$$(2x + 1)\,dx = \tfrac{1}{2}\,du$$

The integral becomes

$$\int \frac{\tfrac{1}{2}\,du}{u} = \tfrac{1}{2}\int \frac{du}{u}$$

$$= \tfrac{1}{2}\ln u + C$$

$$= \tfrac{1}{2}\ln(2x^2 + 2x + 7) + C$$

It is possible to write down the integral without having to go through the process of substitution each time. In the above example the numerator is half the differential coefficient of the denominator (which is $4x + 2$). If we multiply the numerator by two then it becomes the exact differential of the denominator. However, in order to keep the complete integral unchanged, if we multiply the inside by 2, we must multiply the integral by $\frac{1}{2}$. Thus the integral becomes

$$\frac{1}{2} \int \frac{2(2x + 1)\, dx}{2x^2 + 2x + 7}$$

The actual integral is the logarithm of the denominator, with the multiplying factor $\frac{1}{2}$; that is,

$$\frac{1}{2} \ln(2x^2 + 2x + 7) + C$$

This method is used in Example 6.13.

Example 6.13   Find the value of

$$\int_2^3 \frac{x\, dx}{(x^2 + 1)}$$

The numerator is a differential coefficient of the denominator apart from a multiplying constant. The multiplying constant required to make the top an exact differential of the bottom is two. If the top is multiplied by 2, then the integral must be multiplied by $\frac{1}{2}$. The integral becomes

$$\frac{1}{2} \int_2^3 \frac{2x\, dx}{(x^2 + 1)}$$

The actual integral is now the logarithm of the denominator; that is, the result is

$$\left[\frac{1}{2} \ln(x^2 + 1) \right]_2^3 = \frac{1}{2} \ln 10 - \frac{1}{2} \ln 5$$

$$= \frac{1}{2} \ln \frac{10}{5}$$

$$= \frac{1}{2} \ln 2.$$

EXERCISE 6.7

Determine the following integrals either by making a substitution, or by following Example 6.13.

1. $\displaystyle \int \frac{2x\, dx}{(x^2 - 3)}$

2. $\displaystyle \int \frac{(3x + 2)\, dx}{(3x + 4x + 7)}$

3. $\displaystyle\int_0^1 \frac{x\,dx}{(x^2+2)}$

4. $\displaystyle\int \frac{(2x+2)\,dx}{x^2+2x+6}$

5. $\displaystyle\int \frac{x^{\frac{1}{2}}\,dx}{(x^{\frac{3}{2}}+4)}$

6. $\displaystyle\int_1^3 \frac{7t\,dt}{(4t^2-3)}$

7. $\displaystyle\int \frac{2s^{-2}\,ds}{\left(\dfrac{1}{s}+2\right)}$

8. $\displaystyle\int \frac{y^3\,dy}{y^4-1}$

6.14   Integration of the trigonometric functions

Treating integration as the reverse of differentiation, from Section 4.14, we have

(a) Integration of sin x

$$\int \sin x \,dx = -\cos x + C \qquad\qquad (6.9)$$

(b) Integration of cos x

$$\int \cos x\,dx = \sin x + C \qquad\qquad (6.10)$$

(c) Integration of tan x

$\tan x$ is written as $\dfrac{\sin x}{\cos x}$ so that the integral is

$$\int \frac{\sin x\,dx}{\cos x}$$

The differential coefficient of $\cos x$ is $-\sin x$, which is equal to the numerator, apart from a multiplying constant of -1. Let

$$u = \cos x,$$

then

$$\frac{du}{dx} = -\sin x$$

From equation (6.8),

$$du = -\sin x\,dx$$

that is

$$\sin x\,dx = -du$$

The integral is

$$\int \frac{-du}{u} = -\ln u + C$$

$$= -\ln(\cos x) + C$$

$$\boxed{\int \tan x\,dx = -\ln(\cos x) + C} \qquad (6.11)$$

(d) Integration of cot x

cot $x\,dx$ is obtained in a similar manner to give

$$\boxed{\int \cot x\,dx = \ln(\sin x) + C} \qquad (6.12)$$

(e) Integration of cosec x

In order to carry out the integration multiply top and bottom by
($\cosec x - \cot x$); that is,

$$\int \cosec x\,dx = \int \frac{\cosec x\,(\cosec x - \cot x)}{(\cosec x - \cot x)}\,dx$$

$$= \int \frac{-\cosec x\,\cot x + \cosec^2 x\,dx}{\cosec x - \cot x}$$

From Chapter 4 it is seen that

$$\frac{d(\cosec x)}{dx} = -\cosec x\,\cot x$$

and $\qquad \dfrac{d(\cot x)}{dx} = -\cosec^2 x$

Therefore in the integral the top is an exact differential of the bottom, so
that the result is the log of the denominator; that is,

$$\boxed{\int \cosec x\,dx = \ln(\cosec x - \cot x) + C} \qquad (6.13)$$

(f) Integrating sec x

The integral is multiplied, top and bottom, by the factor (sec x + tan x); that is,

$$\int \sec x \, dx = \int \frac{\sec x (\sec x + \tan x) \, dx}{(\sec x + \tan x)}$$

$$= \int \frac{(\sec x \tan x + \sec^2 x)}{\sec x + \tan x} \, dx$$

From equations (4.11) and (4.14) it is seen that the top is an exact derivative of the bottom. Therefore the integral evaluates to the logarithm of the denominator; that is,

$$\boxed{\int \sec x \, dx = \ln(\sec x + \tan x) + C} \qquad (6.14)$$

The integrals covered in (a)–(f) above are known as the **standard forms** of trigonometric functions.

6.15   Integration of the trigonometric functions containing compound angles of the form $(bx + c)$

Consider the integral $\int \sin(4x + 3) \, dx$
Let $u = 4x + 3$. Then $du/dx = 4$
From equation (6.8)

$$du = \frac{du}{dx} dx = 4 \, dx$$

or $dx = \frac{1}{4} du$

Substituting, the integral becomes

$$\int \sin(4x + 3) dx = \int \sin u \cdot \frac{1}{4} \, du$$

$$= -\frac{1}{4} \cos u + C$$

$$= -\frac{1}{4} \cos(4x + 3) + C$$

The result is seen to be the same as for the standard form in equation (6.9), divided by the coefficient of x of the angle, with the compound angle replacing x. This conclusion can be used as a working rule to avoid making a substitution each time.

The integrals of trigonometric functions containing the compound angle $(bx + c)$ give the same results as the standard forms above, with $(bx + c)$ replacing x, and the result divided by the coefficient b.

Example 6.14   Find $\int \tan(\frac{5}{2}x - 3)\,dx$.

The standard form for the integral of tan is $(-\log \cos)$ Therefore using the above rule,

$$\int \tan(\tfrac{5}{2}x - 3)\,dx = [-\ln \cos(\tfrac{5}{2}x - 3) \div \tfrac{5}{2}] + C$$

$$= -\tfrac{2}{5} \ln \cos(\tfrac{5}{2}x - 3) + C$$

Example 6.15   Determine (i) $\int_0^{\pi/4} \cos 2x\,dx$, (ii) $\int \sec(3x + \frac{1}{4})\,dx$

Using the rule and the standard forms,

(i) $$\int_0^{\pi/4} \cos 2x\,dx = \left[\tfrac{1}{2} \sin 2x \right]_0^{\pi/4}$$

$$= \tfrac{1}{2} \sin \frac{\pi}{2} - \tfrac{1}{2} \sin 0$$

$$= \tfrac{1}{2}$$

(ii) $\int \sec(3x + \frac{1}{4})\,dx = \frac{1}{3} \ln[\sec(3x + \frac{1}{4}) + \tan(3x + \frac{1}{4})] + C$

EXERCISE 6.8

Determine

1. $\int \sin 4x\,dx$

2. $\int_0^{\pi/6} \sin 3x\,dx$

3. $\int \cos(\frac{7}{2}x - 3)\,dx$

4. $\int \csc(2x + 4)\,dx$

5. $\int_0^{0.1} \sec(5\theta + 1)\,d\theta$

6. $\int \tan(-\frac{3}{4}t + \frac{1}{2})\,dt$

7. $\int_{-0.5}^{-2} \cot(-3s + 1)\,ds$

8. $\int \csc(-\frac{1}{2}x)\,dx$.

6.16   Integration of the squared trigonometric functions

As the reverse of differentiation we have the following results from Section 4.14(d) and (e).

(a) Integration of $\sec^2 x$

$$\int \sec^2 x \, dx = \tan x + C$$

(b) Integration of $\operatorname{cosec}^2 x$

$$\int \operatorname{cosec}^2 x \, dx = -\cot x + C$$

Using these results it is possible to integrate $\tan^2 x$ and $\cot^2 x$.

(c) Integration of $\tan^2 x$

$$\int \tan^2 x \, dx = \int (\sec^2 x - 1) \, dx, \text{ using the equation in Section 3.10}$$

Using result (a) above we have

$$\int \tan^2 x \, dx = \tan x - x + C$$

(d) Integration of $\cot^2 x$

$$\int \cot^2 x \, dx = \int (\operatorname{cosec}^2 x - 1) \, dx, \text{ using the equation in Section 3.10.}$$

Using result (b) above we have

$$\int \cot^2 x \, dx = -\cot x - x + C$$

(e) Integration of $\sin^2 x$

In order to carry out the integration $\sin^2 x$ is expressed in terms of $\cos 2x$, using the equation in Section 3.12

$$\cos 2x = 1 - 2 \sin^2 x$$

$$\sin^2 x = \tfrac{1}{2} - \tfrac{1}{2} \cos 2x$$

$$\int \sin^2 x \, dx = \int \left(\tfrac{1}{2} - \tfrac{1}{2} \cos 2x\right) dx$$

which gives

$$\int \sin^2 x \, dx = \tfrac{1}{2} x - \tfrac{1}{4} \sin 2x + C \qquad\qquad (6.15)$$

(f) Integration of cos² x

$\cos^2 x$ is now expressed in terms of $\cos 2x$ using the equation in Section 3.12

$$\cos 2x = 2 \cos^2 x - 1$$

so that

$$\cos^2 x = \tfrac{1}{2} + \tfrac{1}{2} \cos 2x$$

$$\int \cos^2 x \, dx = \int \left(\tfrac{1}{2} + \tfrac{1}{2} \cos 2x\right) dx$$

which gives

$$\int \cos^2 x \, dx = \tfrac{1}{2} x + \tfrac{1}{4} \sin 2x + C \qquad\qquad (6.16)$$

(g) Products of sines and cosines

In Chapter 3, Example 3.24 it was shown how to express a product of two cosines as a sum. In order to integrate products of sines and cosines they can first be written as sums or differences in this way. The process is shown in Example 6.16.

Example 6.16   Determine $\int \sin 4x \sin 2x \, dx$

Expressing the product as sums or differences, following Example 3.24:

$$\frac{C+D}{2} = 4x, \qquad \frac{C-D}{2} = 2x$$

that is

$$C + D = 8x, \qquad C - D = 4x$$

Solving the two equations in terms of x,

$$C = 6x$$

$$D = 2x$$

From Section 3.13,

$$-2 \sin \frac{C+D}{2} \sin \frac{C-D}{2} = \cos C - \cos D$$

so that

$$\sin 4x \sin 2x = \tfrac{1}{2} [\cos 2x - \cos 6x]$$

Therefore

$$\int \sin 4x \sin 2x \, dx = \int \tfrac{1}{2} [\cos 2x - \cos 6x] \, dx$$

$$= \frac{\sin 2x}{4} - \frac{\sin 6x}{12} + C$$

EXERCISE 6.9

Determine

1. $\int \sin 5x \sin 4x \, dx$

2. $\int \cos 5x \cos 3x \, dx$

3. $\int \cos 3x \sin 7x \, dx$

4. $\int_{\pi/4}^{\pi/2} \sin \tfrac{1}{2} x \cos \tfrac{3}{2} x \, dx$

5. $\int_{0}^{\pi} \sin \tfrac{1}{4} x \sin \tfrac{3}{4} x \, dx$

6. $\int \sin -4x \sin -6x \, dx$

7. $\int_{0}^{\pi/2} \cos 3x \cos 7x \, dx$

8. $\int \cos -3x \sin 5x \, dx$

(h) Powers of trigonometric functions

Certain integrals of powers of trigonometric functions are extremely easy to determine. This is the class of functions where the trigonometric function raised to a power is multiplied by its differential coefficient. For example,

$$\int \sin^5 x \cos x \, dx$$

where the $\cos x$ is the differential coefficient of the $\sin x$. The integral can easily be recognised and evaluated if it is written in the form

$$\int (\sin x)^5 \cos x \, dx$$

The outside function is a differential coefficient of the inside function so that the integral can be obtained by the substitution method.

$$\text{Let} \quad u = \sin x$$

$$\frac{du}{dx} = \cos x$$

From equation (6.8)

$$du = \cos x \, dx$$

The integral becomes

$$\int u^5 \, du = \frac{u^6}{6} + C$$

$$= \tfrac{1}{6} \sin^6 x + C$$

Example 6.17   Determine $\displaystyle\int \tan^{-1\frac{1}{2}} x \sec^2 x \, dx$.

The integral is written as

$$\int (\tan x)^{-\frac{1}{2}} \sec^2 x \, dx$$

where it can be seen that the outside function is a differential coefficient of the inner function. Using the substitution

$$u = \tan x$$

the integral becomes

$$\int u^{-\frac{1}{2}} \, du = \frac{u^{\frac{1}{2}}}{\frac{1}{2}} + C$$

$$= 2 \tan^{\frac{1}{2}} x + C$$

EXERCISE 6.10

Integrate the following:

1. $\displaystyle\int \sin^{-4} x \cos x \, dx$

2. $\displaystyle\int \cos^{\frac{3}{4}} x \sin x \, dx$

3. $\displaystyle\int \tan^{-\frac{1}{3}} x \sec^2 x \, dx$

4. $\displaystyle\int \cot^5 x \operatorname{cosec}^2 x \, dx$

5. $\displaystyle\int \sin^7 2x \cos 2x \, dx$

6. $\displaystyle\int \tan^{\frac{1}{4}}(-\tfrac{2}{3}x) \sec^2(-\tfrac{2}{3}x) \, dx$

7. $\displaystyle\int \cot^{-\frac{7}{4}} 5x \operatorname{cosec}^2 5x \, dx$

8. $\displaystyle\int \cos^{\frac{3}{2}}(2x - 1) \sin(2x - 1) \, dx.$

6.17 Integration of the exponential function

In Section 4.17(a), equation (4.15), it was shown that

$$\frac{d}{dx}(e^x) = e^x$$

Reversing the process we have

$$\boxed{\int e^x \, dx = e^x + C}$$

When the exponential contains a compound exponent, such as $\int e^{-3x+1} \, dx$, the integral may be evaluated by a substitution method.

Let $u = -3x + 1$, so that

$$\frac{du}{dx} = -3$$

In equation (6.8)

$$du = \frac{du}{dx} dx$$

so that in this case

$$du = -3 \, dx$$

Therefore

$$dx = -\tfrac{1}{3} du$$

Substituting, in terms of u, the integral becomes

$$\int e^{-3x+1} \, dx = \int -\tfrac{1}{3} e^u \, du$$
$$= -\tfrac{1}{3} e^u + C$$
$$= -\tfrac{1}{3} e^{-3x+1} + C$$

Such an integral can be obtained by using the following rule.

The integral of exponential function is the original exponential divided by the coefficient of the variable in the exponent.

Note: The rule only applies when the exponent is of degree 1.

Example 6.18   Evaluate $\displaystyle\int_1^2 e^{2x+1}\,dx$

Using the above rule the integral becomes

$$\left[\frac{e^{2x+1}}{2}\right]_1^2 = \tfrac{1}{2}e^5 - \tfrac{1}{2}e^3$$

$$= \tfrac{1}{2}(148.4) - \tfrac{1}{2}(20.1)$$

$$= 64.15$$

EXERCISE 6.11

Determine:

1. $\displaystyle\int e^{3x}\,dx$

2. $\displaystyle\int e^{-2x-1}\,dx$

3. $\displaystyle\int e^{-\frac{1}{2}x}\,dx$

4. $\displaystyle\int e^{-5\theta+\frac{1}{2}}\,d\theta$

5. $\displaystyle\int_1^2 e^{-t+1}\,dt$

6. $\displaystyle\int_{-1}^1 e^{\frac{1}{2}x-2}\,dx$

7. $\displaystyle\int_{0.1}^{0.2} e^{-5t+1}\,dt$

8. $\displaystyle\int (e^{-3x} + e^{3x})\,dx$

9. $\displaystyle\int (e^{-\frac{3}{4}v} - e^{2v})\,dv$

6.18   Integration by partial fractions

If the denominator of a fraction can be expressed in simple factors, the fraction may be converted into partial fractions, and integrated easily. In

Chapter 1 it was shown that

$$\frac{2x+5}{(x+2)(x+3)} = \frac{1}{(x+2)} + \frac{1}{(x+3)}$$

Therefore

$$\int \frac{2x+5}{(x+2)(x+3)} \, dx = \int \left[\frac{1}{(x+2)} + \frac{1}{(x+3)} \right] dx$$

$$= \ln(x+2) + \ln(x+3) + C$$

$$= \ln(x+2)(x+3) + C$$

Example 6.19   Evaluate $\int_4^5 \frac{dx}{x^2 - 9}$.

The fraction $1/(x^2 - 9)$ can be written as

$1/(x-3)(x+3)$

and expressed as partial fractions:

$$\frac{1}{x^2 - 9} = \frac{A}{x-3} + \frac{B}{x+3}$$

$$= \frac{A(x+3) + B(x-3)}{(x-3)(x+3)}$$

$$= \frac{x(A+B) + (3A - 3B)}{(x-3)(x+3)}$$

Comparing the coefficients of the numerators, since they are identical on both sides:

$x:$ $\qquad\qquad A + B = 0$

constant: $\qquad 3A - 3B = 1$

Solving the two equations we have

$A = \frac{1}{6}$

$B = -\frac{1}{6}$

Therefore

$$\frac{1}{x^2 - 9} = \frac{1}{6(x-3)} - \frac{1}{6(x+3)}$$

so that the integral becomes

$$\int_4^5 \frac{dx}{6(x-3)} - \int_4^5 \frac{dx}{6(x+3)} = \left[\tfrac{1}{6}\ln(x-3) - \tfrac{1}{6}\ln(x+3) \right]_4^5$$

$$= \left[\tfrac{1}{6}\ln\frac{x-3}{x+3} \right]_4^5$$

$$= \tfrac{1}{6}\ln\tfrac{2}{8} - \tfrac{1}{6}\ln\tfrac{1}{7}$$

$$= \tfrac{1}{6}\ln\tfrac{7}{4}$$

EXERCISE 6.12

Using the method of partial fractions carry out the following.

1. $\displaystyle \int \frac{4x}{(x-1)(x-3)}\,dx$

2. $\displaystyle \int \frac{2x-3}{x^2-16}\,dx$

3. $\displaystyle \int \frac{4}{(x-1)(x-4)}\,dx$

4. $\displaystyle \int_1^2 \frac{3x-2}{(2x+1)(3x-1)}\,dx$

5. $\displaystyle \int_0^2 \frac{1}{x^2+5x+6}\,dx$

6. $\displaystyle \int \frac{3}{x^2-4x-5}\,dx$

7. $\displaystyle \int \frac{x^2+x+1}{(x+1)(x+2)(x+3)}\,dx$

8. $\displaystyle \int \frac{u}{(2u-1)^2}\,du$

9. $\displaystyle \int \frac{3t}{(3t-4)^2}\,dt$

10. $\displaystyle \int_5^7 \frac{6}{(x-4)(x-2)}\,dx$

11. $\displaystyle \int \frac{x}{(x-1)(x+1)(x+2)}\,dx$

6.19 Integration of a product of two functions (integration by parts)

When an integral of the product of two functions, such as $\int x \sin x \, dx$, is required, it is often possible to carry out the integral by the method of parts, as described below.

In Chapter 4, the differential coefficient of the product of two functions was shown to be

$$d(uv) = v\,du + u\,dv$$

so that $u\,dv = d(uv) - v\,du$

Integrating both sides

$$\int u \, dv = \int d(uv) - \int v \, du$$

giving

$$\boxed{\int u \, du = uv - \int v \, du}$$ (6.17)

As a result of using this equation it is often easier to evaluate $\int v \, du$ than $\int u \, dv$. The method is applied to

$$\int x \sin x \, dx$$

Let $u = x,$ $dv = \sin x \, dx$

Then $du = dx,$ $v = \int \sin x \, dx$

$$= -\cos x$$

Using these results in rule (6.17)

$$\int x \sin x \, dx = -x \cos x - \int -\cos x \, dx$$

$$= -x \cos x + \int \cos x \, dx$$

$$= -x \cos x + \sin x + C$$

Note: u and dv must always be chosen in such a way that the integral on the right of rule (6.17) has a lower power of x than the integral on the left. This is achieved, as follows.

Choose power of x as u, and the other function as dv   (618)

Example 6.20   Determine $\int x^2 \cos x \, dx$.

Bearing in mind the rule above, let

Let $u = x^2,$ $dv = \cos x \, dx$

Then $du = 2x \, dx,$ $v = \int \cos x \, dx$

$$= \sin x$$

Substituting these results into rule (6.17)

$$\int x^2 \cos x \, dx = x^2 \sin x - \int 2x \sin x \, dx$$

It is now necessary to integrate $\int 2x \sin x \, dx$. By choosing $u = x$ and $dv = \sin x \, \delta x$, and integrating again by parts, the result shown earlier in the section is obtained:

$$\int x \sin x \, dx = -x \cos x + \sin x$$

Hence

$$\int x^2 \cos x \, dx = x^2 \sin x - 2[-x \cos x + \sin x] + C$$

$$= x^2 \sin x + 2x \cos x - 2 \sin x + C$$

An exception to rule (6.18) occurs when one function is the log function, and the other a power of x. In this case the log function is chosen as u, and the power of x as dv.

Example 6.21.   Evaluate $\int x^3 \ln x \, dx$.

Let $u = \ln x, \qquad dv = x^3 \, dx$

Then $du = \dfrac{1}{x} dx, \qquad v = \int x^3 \, dx$

$$= \frac{x^4}{4}$$

Therefore

$$\int x^3 \ln x \, dx = \tfrac{1}{4} x^4 \ln x - \int \tfrac{1}{4} x^4 \frac{1}{x} dx$$

$$= \tfrac{1}{4} x^4 \ln x - \int \tfrac{1}{4} x^3 \, dx$$

$$= \tfrac{1}{4} x^4 \ln x - \frac{x^4}{16} + C$$

6.20   To determine $\int \ln x \, dx$

This integral is a special case of Example 6.21, in which case the product of the two functions is $1 \cdot \ln x$.

Let $\quad u = \ln x, \quad dv = dx$

Then $\quad du = \dfrac{1}{x} dx, \quad v = \displaystyle\int dx = x$

Using equation (6.17)

$$\int \ln x \, dx = x \ln x - \int x \frac{1}{x} dx$$

that is

$$\boxed{\int \ln x \, dx = x \ln x - x + C}$$

EXERCISE 6.13

Integrating by parts, determine the following.

1. $\displaystyle\int x e^x \, dx$ 2. $\displaystyle\int x^4 \ln x \, dx$

3. $\displaystyle\int x^2 \sin 4x \, dx$ 4. $\displaystyle\int_0^2 x e^{-x} \, dx$

5. $\displaystyle\int \frac{\ln x}{x^2} \, dx$ 6. $\displaystyle\int x \tan^2 x \, dx$

7. $\displaystyle\int_0^{\pi/2} e^{3x} \sin 4x \, dx$ 8. $\displaystyle\int e^{5x} \sin 2x \, dx$

9. $\displaystyle\int_3^4 (x^2 - x) \ln x \, dx$ 10. $\displaystyle\int e^{4x} (x - 3)^2 \, dx$

6.21   Integrals involving square roots

Two types of integrals are considered:

(i) $\displaystyle\int \frac{1}{\sqrt{a^2 - x^2}} \, dx$ and   (ii) $\displaystyle\int \sqrt{a^2 - x^2} \, dx$

In both cases we remove the root by substituting

$x = a \sin \theta$, from which $dx = a \cos \theta$

(i) $\displaystyle\int \frac{1}{\sqrt{a^2 - x^2}} \, dx = \int \frac{1}{\sqrt{a^2 - a^2 \sin^2 \theta}} a \cos \theta \, d\theta$

$\displaystyle = \int \frac{a \cos \theta}{a\sqrt{1 - \sin^2 \theta}} \, d\theta$

$\displaystyle = \int 1 \, d\theta$

$= \theta + C$

But $\sin \theta = \dfrac{x}{a}$

$$\theta = \sin^{-1} \dfrac{x}{a}$$

Therefore

$$\boxed{\int \frac{1}{\sqrt{a^2 - x^2}}\, dx = \sin^{-1} \frac{x}{a} + C}$$

(ii) $\displaystyle \int \sqrt{a^2 - x^2}\, dx = \int \sqrt{a^2 - a^2 \sin^2 \theta}\, a \cos \theta\, d\theta$

$$= \int a^2 \cos^2 \theta\, d\theta$$

$$= a^2 \int \cos^2 \theta\, d\theta$$

$$= a^2(\tfrac{1}{2}\theta + \tfrac{1}{4} \sin 2\theta) + C \text{ from equ (6.16)}$$

Again $\sin \theta = \dfrac{x}{a}$ and $\theta = \sin^{-1} \dfrac{x}{a}$

Also $\sin 2\theta = 2 \sin \theta \cos \theta = 2 \sin \theta \sqrt{1 - \sin^2 \theta}$

$$= 2\frac{x}{a} \sqrt{1 - \frac{x^2}{a^2}}$$

$$= 2\frac{x}{a^2} \sqrt{a^2 - x^2}$$

Substituting

$$\boxed{\int \sqrt{a^2 - x^2}\, dx = \tfrac{1}{2} a^2 \sin^{-1} \frac{x}{a} + \tfrac{1}{2}x \sqrt{a^2 - x^2} + C}$$

6.22   The integral $\displaystyle \int \frac{1}{x^2 + a^2}\, dx$

In a similar manner to the previous section we substitute $x = a \tan \theta$, and $dx = a \sec^2 \theta\, d\theta$. Therefore

$$\int \frac{1}{x^2 + a^2}\, dx = \int \frac{1}{a^2 \tan^2 \theta + a^2}\, a \sec^2 \theta\, d\theta$$

But $\tan^2 \theta + 1 = \sec^2 \theta$, so that

$$\int \frac{1}{x^2 + a^2}\, dx = \int \frac{1}{a}\, d\theta$$

$$= \frac{1}{a}\theta + C$$

Since

$$\tan\theta = \frac{x}{a}, \qquad \theta = \tan^{-1}\frac{x}{a}$$

Therefore

$$\boxed{\int \frac{1}{x^2+a^2}\,dx = \frac{1}{a}\tan^{-1}\frac{x}{a} + C}$$

MISCELLANEOUS EXERCISE 6

Obtain the following integrals:

1. $\displaystyle\int x\left\{\frac{1}{\sqrt{x}} - 1\right\}^2 dx$

2. $\displaystyle\int_0^1 \sqrt{x}(1 - x^3)^2\,dx$

3. $\displaystyle\int_1^2 (x^3 - 3x^2 + 5)\,dx$

4. $\displaystyle\int \left(x + \frac{1}{x}\right)^2 dx$

5. $\displaystyle\int_4^9 \left(\sqrt{x} + \frac{1}{\sqrt{x}} + 2\right) dx$

6. $\displaystyle\int (\sqrt{x^5} - \sqrt[5]{x})\,dx$

7. $\displaystyle\int_1^4 \sqrt{x}(x^2 - 1)\,dx$

8. (a) Simplify

   (i) $\displaystyle\int (r^7 - r^6)\,dr$

   (ii) $\displaystyle\int \frac{y^5 - y^3}{y^2}\,dy$.

   (b) Find the area between the curve $y = x^2 - 1$ and the x-axis.

9. Evaluate the following integrals:

   (a) $\displaystyle\int \frac{dx}{6x^2}$,   (b) $\displaystyle\int (3x + 3\sqrt{x})\,dx$,   (c) $\displaystyle\int (x + a)^2\,dx$,

   (d) $\displaystyle\int_1^2 6x^6\,dx$,   (e) $\displaystyle\int_0^{2p} q(\tfrac{1}{3}p + x)\,dx$.

10. Calculate the area in the first quadrant bounded by the curve $y = 10 - 3x - x^2$ and the coordinate axes.

11. Find the coordinates of the points of intersection of the curves $x = 3y^2$ and $x^2 = 9y$. Sketch roughly the curves on the same axes from $x = 0$ to $x = 3$.
   Find
   (a) the area A completely enclosed by these two curves;
   (b) the volume obtained by rotating this area A about the x-axis through $360°$.

12. Sketch approximately to scale the curve $y^2 = 4x$ and the line $y = 2x - 4$ on the same axes and indicate the values at salient points. Calculate the area in the first quadrant between the curve, the line and the x-axis. Calculate the volume swept out if this area is rotated through $180°$ about the x-axis.

13. Plot roughly the curve $xy = 4$ in the first quadrant. On the same axes draw the line $y = 3x + 1$, which cuts the y-axis at A and the curve at P. The line $x = 2$ cuts the x axis at B, and the curve at Q. Given that O is the origin determine the area $APQBO$ correct to two decimal places.

14. The frustum of a cone is generated by rotating about the x-axis the area bounded by the axis of x, the line $y = ax + R$ and the ordinates $x = 0$ and $x = h$. Obtain an expression for the volume of the frustum in terms of a, R and h. Use the result to calculate the volume of a frustum of a cone of height 2 m, whose end radii are 1 m and 3 m respectively.

15. Determine the following integrals:

   (a) $\displaystyle\int_0^{\pi/4} \tan^2 x\, dx,$ (b) $\displaystyle\int_0^{\pi/3} \sin 2x\, dx,$

   (c) $\displaystyle\int_1^2 \frac{(x-1)(x+2)}{x}\, dx.$

16. By means of the substitution $\sin\theta = x$, show that

   $$\int_0^{\pi/2} e^{\sin\theta} \cos\theta\, d\theta = 1.718$$

17. Obtain

   (a) $\displaystyle\int \frac{x^2 + 2}{x}\, dx,$ (b) $\displaystyle\int_0^{\pi/4} \sin x \cos x\, dx,$

   (c) $\displaystyle\int_0^1 e^{-2x}\, dx$

18. Find

   (a) $\displaystyle\int \left(\frac{1}{\sqrt{(2x)}} + \sqrt{x^3} \right) dx,$ (b) $\displaystyle\int 3x \cos 2x\, dx$

(c) $\displaystyle\int_0^1 \left(\frac{x+2}{x^2+4x+5}\right) dx,$ (d) $\displaystyle\int_0^1 2xe^{-x^2} dx.$

19. Find

(a) $\displaystyle\int \frac{x+2}{\sqrt{x}} dx,$ (b) $\displaystyle\int x^2 \ln x \, dx$

(c) $\displaystyle\int_0^2 \frac{1}{(5x-2)^3} dx,$ (d) $\displaystyle\int x^2\sqrt{(x^3+1)} \, dx.$

20. Find (a) $\displaystyle\int \frac{12x-1}{6x^2-x-1} dx,$ (b) $\displaystyle\int_0^{\pi/2} \cos(3x+\tfrac{1}{4}\pi) \, dx.$

21. Evaluate, in each case correct to three significant figures

(a) $\displaystyle\int_1^4 \frac{x^2+2}{\sqrt{x^3}} dx,$ (b) $\displaystyle\int_1^2 \frac{dx}{2x-1},$ (c) $\displaystyle\int_0^4 \frac{x}{\sqrt{(2x+1)}} dx.$

22. Obtain

(a) $\displaystyle\int (e^{3x}+e^{-3x}) \, dx,$ (b) $\displaystyle\int_1^2 x\left(x+\frac{1}{x}\right)^2 dx$

(c) $\displaystyle\int_0^{\pi/4} (\tfrac{1}{2}-\sin^2 x) \, dx.$

23. Show that

(a) $\displaystyle\int_0^1 \frac{dx}{\sqrt{(3-2x)}} = \sqrt{3}-1$

(b) $\displaystyle\int_0^{\pi/4} \sin 3x \sin 2x \, dx = \frac{3\sqrt{2}}{10}.$

(c) $\displaystyle\int_0^{\frac{1}{\sqrt{2}}} \frac{x^2}{\sqrt{(1-x^2)}} dx = \frac{\pi}{8}-\frac{1}{4}.$

Use the substitution $x = \sin\theta$

24. Show that

(a) $\displaystyle\int_0^4 \sqrt{(1+2x)} \, dx = \frac{26}{3}$

(b) $\displaystyle\int_\pi^{2\pi} (\sin x + \cos x)^2 \, dx = \pi$

(c) $\displaystyle\int_0^1 xe^{-2x} \, dx = \tfrac{1}{4}(1-3e^{-2}).$

25. Integrate the following by parts:

(i) $\displaystyle\int xe^2 x \, dx,$ (ii) $\displaystyle\int e^{2x} \cos x \, dx.$

26. Evaluate the following integrals. Definite integrals should be
calculated correct to three significant figures.

(a) $\displaystyle\int_0^3 \frac{dx}{x+5}$, (b) $\displaystyle\int \frac{dv}{v^{1.2}}$ (c) $\displaystyle\int_0^{1/300} \sin(100\pi t + \tfrac{1}{3}\pi)\,dt$

(d) $\displaystyle\int \tan^2 x\,dx$, (e) $\displaystyle\int_1^2 \left(\frac{x+1}{x}\right)^2 dx$

Chapter 7

Differential equations

7.1   Introduction

In Section 4.20 equations were obtained which related differential coefficients dy/dx, d^2y/dx^2, with functions of x and y. Such equations are called **differential equations**. Solving a differential equation means finding the original equation relating y and x, that is, of the form of $y = f(x)$. A differential equation is called a **first-order equation** if the highest differential coefficient is dy/dx and a **second-order equation** if the highest differential coefficient is d^2y/dx^2, etc.

A first-order differential equation is integrated once to obtain a solution, and a second order equation is integrated twice, and so on. Many situations occur in engineering, science and economics which can be modelled as differential equations, thereby providing solutions to complex problems.

7.2   Differential equations involving direct integration (variables separable)

(i) Equations of the form $dy/dx = f(x)$

This form of the equation can be rearranged as $dy = f(x)\,dx$ and solved by direct integration as

$$y = \int f(x)\,dx$$

Consider the equation

$$\frac{dy}{dx} = x^2,$$

which can be rearranged as

$$dy = x^2\,dx$$

Integrating both sides:

$$y = \int x^2 \, dx$$

$$y = \frac{x^3}{3} + C \tag{i}$$

Equation (i) is the **general solution** of the differential equation because C is unknown. If this equation were plotted with different values of C it would produce a family of curves as shown in Fig. 7.1.

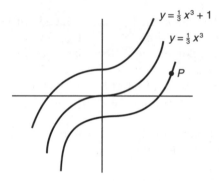

$y = \frac{1}{3} x^3 + 1$

$y = \frac{1}{3} x^3$

P

Fig. 7.1

Each of the curves has a specific value of C. To find C of a particular curve it is necessary to know the x and y values of any point P through which it passes, in other words, to fix an x and y value. These particular x, y values are called **boundary conditions**. Consider the case in which $y = 5$ when $x = 3$, where one of the family of curves passes through the point (3, 5). Substituting $x = 3$, $y = 5$ into (i) gives

$$5 = \frac{1}{3} \times 3^3 + C$$

$$C = -4$$

so that the **particular solution** is

$$y = \frac{x^3}{3} - 4$$

The same procedure is used if the differential equation contains a second-order differential coefficient, as shown in Example 7.1, with the added requirement that two boundary conditions are necessary to obtain the particular solution.

Example 7.1 Solve the equation

$$\frac{d^2y}{dx^2} = 3x$$

given that   (i) $dy/dx = \frac{7}{2}$ when $x = 1$

(ii) $y = 16$ when $x = 2.$

Integrating both sides:

$$\frac{dy}{dx} = \int 3x\,dx$$

$$= \tfrac{3}{2}x^2 + C$$

Using the first boundary condition in this result:

$$\tfrac{7}{2} = \tfrac{3}{2} + C$$
$$C = 2$$
$$\frac{dy}{dx} = \tfrac{3}{2}x^2 + 2$$

Integrating again:

$$y = \int (\tfrac{3}{2}x^2 + 2)dx$$

$$= \tfrac{1}{2}x^3 + 2x + D$$

Using the second boundary condition:

$$16 = \tfrac{1}{2}(2)^3 + 2(2) + D$$
$$= 4 + 4 + D$$
$$D = 8$$

Hence $y = \tfrac{1}{2}x^3 + 2x + 8$

(b) Equations of the form $dy/dx = f(y)$

This form of the equation can be rearranged as $dy/f(y) = dx$ and solved by direct integration as

$$\int \frac{1}{f(y)}\,dy = \int dx$$

Example 7.2 Solve the equation $dy/dx = 2y^2$ given that $y = 0.5$ when $x = 2.$

Rearranging:

$$\frac{dy}{2y^2} = dx$$

Integrating:

$$\int \frac{1}{2} y^{-2} \, dy = x + C$$

$$\frac{y^{-1}}{-2} = x + C$$

$$\frac{1}{y} = -2(x + C)$$

When $x = 2$, $y = 0.5$ so that $\qquad \dfrac{1}{0.5} = -2(2 + C)$

$$C = -3$$

The solution is $\qquad \dfrac{1}{y} = -2(x - 3)$

(c) Equations of the form $dy/dx = f_1(x) \times f_2(y)$

This form of the equation can be rearranged as

$$dy/f_2(y) = f_1(x) \, dx$$

and solved by direct integration as

$$\int \frac{1}{f_2(y)} \, dy = \int f_1(x) \, dx$$

An example of this type of differential equation is given below.

$$\frac{dy}{dx} = \frac{3x^2}{2y}$$

The variables can be separated to each side of the equation; that is,

$$2y \, dy = 3x^2 \, dx$$

Integrating both sides:

$$\int 2y \, dy = \int 3x^2 \, dx$$

$$\frac{2y^2}{2} = \frac{3x^3}{3} + C$$

$$y^2 = x^3 + C$$

The boundary condition must be known before C can be determined.

Example 7.3 Find the solution of

$$\frac{dy}{dx} = e^{x+2y}$$

given that $y = 0$, when $x = 0$.

Separating the variables

$$e^{-2y} \, dy = e^x \, dx$$

Integrating both sides

$$\int e^{-2y} \, dy = \int e^x \, dx$$

$$-\tfrac{1}{2} e^{-2y} = e^x + C$$

Using the boundary condition, $y = 0$, when $x = 0$

$$-\tfrac{1}{2} = 1 + C$$
$$C = -\tfrac{3}{2}$$

Hence $\qquad -\tfrac{1}{2} e^{-2y} = e^x - \tfrac{3}{2}$

EXERCISE 7.1

Find the solutions of the following equations:

1. $\dfrac{dy}{dx} = \cos(5x - 3)$

2. $\dfrac{dy}{dx} = 5e^{3x}$

3. $\dfrac{d^2y}{dx^2} = 4x^2$, given

4. $\dfrac{d^2y}{dx^2} = \sin 2x$

    (i) $\dfrac{dy}{dx} = 37$ when $x = 3$

    (ii) $y = 33$ when $x = 3$.

5. $\dfrac{d^2y}{dx^2} = \dfrac{1}{x}$, given

    (i) $\dfrac{dy}{dx} = 3$ when $x = 3$

    (ii) $y = 4$ when $x = 1$.

6. $\dfrac{dy}{dx} = 3yx^3$, given $y = e^2$ when $x = 2$.

7. $\dfrac{dy}{dx} = 4e^{3x-4y}$

8. $\dfrac{dy}{dx} = y \sin x$

9. $y^2 \dfrac{dy}{dx} = x + 3$

10. $\dfrac{dy}{dx} = \dfrac{2y - 5}{x - 2}$

11. Given that

$$L\frac{di}{dt} + Ri = E$$

where L, R, E are constants, find the equation for i, given that $i = 0$ when $t = 0$.

12. Given that

$$\frac{dy}{dx} = y(1 - \cos 2x)$$

find the solution given that $y = $ e when $x = 0$.

13. Given that

$$\frac{dy}{dx} = x(1 + 2y)$$

find the solution for y, given that $y = 1$ when $x = 0$.

7.3   First-order linear differential equations

Linear differential equations are of the form

$$\frac{dy}{dx} + Py = Q \tag{7.1a}$$

and

$$\frac{d^2y}{dx^2} + R\frac{dy}{dx} + Py = Q \tag{7.1b}$$

The equations are linear because the differential coefficients, and the term in y, all have powers of 1, and do not contain products of differential coefficients, or products with functions of y, The coefficient P, Q or R of each differential coefficient is a function of x only, or a constant.

The order of the equation is the number of the highest differential coefficient. Equation (7.1a), containing dy/dx, is of order 1 and is therefore a first-order linear equation. Equation (7.1b), containing d^2y/dx^2, is of order 2, and is therefore a second-order linear differential equation.

In order to solve an equation such as (7.1a) the procedure is to make the **coefficient of dy/dx equal to** 1 and to multiply the equation on both sides by an integrating factor. The integrating factor is

$$e^{\int P\,dx} \tag{7.2}$$

The equation now becomes

$$\frac{dy}{dx}e^{\int P\,dx} + Pye^{\int P\,dx} = Qe^{\int P\,dx}$$

which reduces to

$$\frac{d}{dx}\left(ye^{\int P\,dx}\right) = Qe^{\int P\,dx} \tag{i}$$

which can be verified by differentiating the product inside the bracket, remembering that

$$\frac{d}{dx}\int P\,dx = P$$

The equation (i) can now be integrated as follows:

$$\int d(ye^{\int P\,dx}) = \int(Qe^{\int P\,dx})dx$$

$$ye^{\int P\,dx} = \int(Qe^{\int P\,dx})dx$$

Example 7.4   Solve the differential equation

$$\frac{dy}{dx} + 4y = 5x + 2$$

Given that the curve of the resulting equation passes through the point (0, 1) find the particular solution.

Comparing this equation with (7.1a) $P = 4$, $Q = 5x + 2$.

The integrating factor is

$$e^{\int P\,dx} = e^{\int 4\,dx} = e^{4x}$$

Multiply throughout by this integrating factor:

$$\frac{dy}{dx}e^{4x} + 4ye^{4x} = e^{4x}(5x + 2)$$

$$\frac{d}{dx}(ye^{4x}) = (5x + 2)e^{4x}$$

Integrating:

$$ye^{4x} = \int(5x + 2)e^{4x}\,dx$$

$$= \int 5xe^{4x}\,dx + \int 2e^{4x}\,dx$$

The first integral can be determined using the method of parts, as follows

Let

$$u = 5x, \qquad \text{then} \qquad du = 5\,dx$$

$$dv = e^{4x}\,dx, \qquad \text{then} \qquad v = \int e^{4x}\,dx = \frac{e^{4x}}{4}$$

so that

$$\int 5xe^{4x}\,dx = uv - \int v\,du$$

$$= 5x\frac{e^{4x}}{4} - \int 5\frac{e^{4x}}{4}\,dx$$

$$= \tfrac{5}{4}xe^{4x} - \tfrac{5}{16}e^{4x} \qquad\qquad\qquad\text{(i)}$$

The second integral is

$$\int 2e^{4x}\,dx = \tfrac{2}{4}e^{4x} \qquad\qquad\qquad\qquad\text{(ii)}$$

Therefore

$$ye^{4x} = \text{(i)} + \text{(ii)}$$

$$= \tfrac{2}{4}e^{4x} + \tfrac{5}{4}xe^{4x} - \tfrac{5}{16}e^{4x} + C$$

reducing to

$$y = \tfrac{3}{16} + \tfrac{5}{4}x + Ce^{-4x} \qquad\qquad\qquad\text{(iii)}$$

Since the curve of this equation passes through the point (0, 1) then $y = 1$ when $x = 0$

$$1 = \tfrac{3}{16} + 0 + Ce^{0}$$

$$C = \tfrac{13}{16}$$

so that $\qquad y = \tfrac{13}{16}e^{-4x} + \tfrac{5}{4}x + \tfrac{3}{16}$

Example 7.5 Solve the differential equation $x\dfrac{dy}{dx} - y = x \ln x$.

Rewrite the equation so that the coefficient of dy/dx is 1 by dividing throughout by x

$$\frac{dy}{dx} - \frac{1}{x}y = \ln x$$

Comparing with equation (7.1) $P = -1/x$.

The integrating factor is

$$e^{\int -\frac{1}{x}dx} = e^{-\ln x} = \frac{1}{e^{\ln x}} = \frac{1}{x}$$

since $e^{\ln x} = x$ as seen in Section 1.6(d).
Multiply throughout by $1/x$:

$$\frac{1}{x}\frac{dy}{dx} - \frac{1}{x^2}y = \frac{1}{x}\ln x$$

$$\frac{d}{dx}\left(\frac{y}{x}\right) = \frac{1}{x}\ln x$$

Integrating both sides $\dfrac{y}{x} = \displaystyle\int \frac{1}{x}\ln x\,dx$

The right-hand side may be integrated by the method of parts.

Let $\qquad u = \ln x \qquad du = \dfrac{1}{x}dx$

$$dv = \frac{1}{x}dx \qquad v = \int \frac{1}{x}dx = \ln x$$

Therefore $\qquad \displaystyle\int \frac{1}{x}\ln x\,dx = (\ln x)^2 - \int \frac{1}{x}\ln x\,dx$

$$2\int \frac{1}{x}\ln x\,dx = (\ln x)^2$$

$$\frac{y}{x} = \tfrac{1}{2}(\ln x)^2 + C$$

EXERCISE 7.2

Solve the following linear differential equations:

1. $\dfrac{dy}{dx} + xy = x$ when $x = \sqrt{\ln 4}$ and $y = 2$.

2. $x\dfrac{dy}{dx} - y = x$ when $y = 10$ and $x = 1$.

3. $(x+2)\dfrac{dy}{dx} + y = x$ when $y = 3$ and $x = 2$.

4. $\dfrac{dy}{dx} + 2y \tan x = 1$ when $y = 0$ and $x = 0$.

5. $x\dfrac{dy}{dx} - y = x \ln x$ given that $y = 2$ and $x = 1$.

7.4   Examples of differential equations in science and engineering

(a) Radioactive decay

A lump of material such as radium or uranium is called a radioactive material because the atoms split into simpler atoms at random. The number of atoms δN disintegrating in this way in a time δt is proportional to the number of the original atoms N left in the material. Therefore

$$\frac{\delta N}{\delta t} \propto -N$$

(The negative sign occurs because N is decreasing.)
As $\delta t \to 0$,

$$\frac{\delta N}{\delta t} \to \frac{\mathrm{d}N}{\mathrm{d}t} = -\lambda N$$

This latter equation is a differential equation of the variable separable type. Rearranging

$$\frac{\mathrm{d}N}{N} = -\lambda \, \mathrm{d}t$$

Integrating both sides

$$\ln N = -\lambda t + C$$

The boundary condition is: at a time $t = 0$ the number of atoms is the original number N_0.

$$\ln N_0 = C$$

Therefore $\ln N = -\lambda t + \ln N_0$

$$\ln N - \ln N_0 = -\lambda t$$

$$\ln \left(\frac{N}{N_0}\right) = -\lambda t$$

Changing to the exponential form $\dfrac{N}{N_0} = \mathrm{e}^{-\lambda t}$

$$N = N_0 \mathrm{e}^{-\lambda t}$$

This equation shows that the number of atoms is decreasing exponentially. The importance of this equation is that the equation can be used from any time $t = 0$ provided that the number of atoms is known at that time.

(b) The current flowing in a capacitor/resistor circuit

The circuit is shown in Fig. 7.2(a).

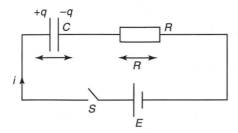

Fig. 7.2(a)

The capacitor C is in series with a resistor R connected to a switch S and a battery of voltage E. At a time $t = 0$ the switch S is closed and a current flows in the circuit. This causes a charge to be built up on the capacitor which sets up an opposing voltage so that the current decreases. At a time t later let the current be i, and let the charge on C be q.

The voltage across C is $\dfrac{q}{C}$ and the voltage across R is Ri.

Therefore

$$E = \frac{q}{C} + Ri \tag{i}$$

But the current i is the charge flowing per second, that is, dq/dt
Equation (i) becomes

$$E = \frac{q}{C} + R\frac{dq}{dt}$$

where E, C and R are constants. Rearranging:

$$R\frac{dq}{dt} = \frac{EC - q}{C}$$

This is a differential equation of the variables separable type.

$$\frac{dq}{EC - q} = \frac{1}{RC}dt$$

Integrating both sides

$$-\ln(EC - q) = \frac{1}{RC}t + K \tag{ii}$$

Applying the boundary condition $t = 0$, $q = 0$ equation (ii) gives

$$-\ln EC = K$$

$$\ln(EC - q) - \ln EC = -\frac{t}{RC}$$

$$\ln\left(\frac{EC - q}{EC}\right) = -\frac{t}{RC}$$

Changing into the exponential form

$$\frac{EC - q}{EC} = e^{-t/(RC)}$$

$$q = EC(1 - e^{-t/RC}) \qquad\qquad\qquad\text{(iii)}$$

As t increases $e^{-t/RC} \to 0$ and equation (iii) becomes $q = EC$.
This means that no further charge will flow, that is, $i = 0$, because the charge
built up on the capacitor produces a voltage equal and opposite to E.
  From equation (iii)

$$i = \frac{dq}{dt} = \frac{E}{R}e^{-t/RC}$$

The variation of i with time t is shown in Fig. 7.2(b).

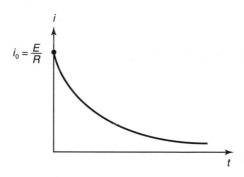

Fig. 7.2(b)

(c) The motion of a ball-bearing falling vertically in a liquid

When a ball-bearing falls under gravity in a liquid two forces will act on it,
the force of gravity acting downwards and the viscous force F acting
upwards because the viscosity of the liquid impedes its motion (Fig. 7.3).

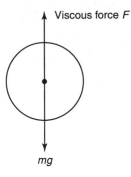

Viscous force F

mg

Fig. 7.3

The resultant force acting on the ball-bearing causes an acceleration downwards so that the equation of motion is

$$m\frac{dv}{dt} = mg - F \qquad (i)$$

where dv/dt is the acceleration, and F is given by Stokes' law: $F = kv$, v being the velocity of the ball-bearing at the time t, and k being a constant containing the coefficient of viscosity of the liquid and the radius of the ball-bearing. Equation (i) reduces to a differential equation of the variable separable type

$$m\frac{dv}{dt} = mg - kv$$

$$\frac{dv}{dt} = g - \frac{k}{m}v$$

$$\frac{dv}{g - \frac{k}{m}v} = dt$$

Integrating both sides:

$$-\frac{m}{k}\ln\left(g - \frac{k}{m}v\right) = t + C \qquad (ii)$$

The boundary condition is that at $t = 0$, $v = 0$. Hence

$$-\frac{m}{k}\ln(g) = C$$

Equation (ii) becomes

$$-\frac{m}{k}\ln\left(g - \frac{k}{m}v\right) + \frac{m}{k}\ln(g) = t$$

$$-\frac{m}{k}\ln\left(\frac{g - \frac{k}{m}v}{g}\right) = t$$

$$\ln\left(1 - \frac{k}{mg}v\right) = -\frac{kt}{m}$$

Changing to the exponential form

$$1 - \frac{k}{mg}v = e^{-kt/m}$$

$$v = \frac{mg}{k}(1 - e^{kt/m}) \tag{iii}$$

As t increases the velocity of the ball-bearing increases, which in turn increases the viscous retarding force F. After a time this viscous force balances the gravitational force and thereafter the velocity remains constant. In equation (iii) as t increases

$$e^{-kt/m} \to 0 \qquad \text{and} \qquad v \to \frac{mg}{k}$$

At this point v is called the terminal velocity.

7.5 Linear differential equations with constant coefficients

We consider equations of the type

$$a\frac{d^2y}{dx^2} + b\frac{dy}{dx} + cy = 0 \tag{7.3}$$

where a, b and c are constants.

Linear differential equations of this type are used for modelling systems such as vibrating mass damper systems, RLC circuits, etc. When $a = 0$ it reduces to a first-order linear differential equation. Equations of the form (7.3) can be solved using the D operator method, as shown below.

(a) The D operator

We write dy/dx as Dy and d^2y/dx^2 as D^2y. D is called an **operator**.

(b) The properties of the D operator

1. $\dfrac{d}{dx}(u+v) = \dfrac{du}{dx} + \dfrac{dv}{dx}$, therefore

$$D(u+v) = Du + Dv \tag{7.4}$$

2. $\dfrac{d}{dx}\left(\dfrac{dy}{dx}\right) = \dfrac{d^2y}{dx^2}$, therefore

$$D(Dy) = D^2y \tag{7.5}$$

3. Consider the operation $(D-2)(D-3)y = 0$. The $(D-3)$ must operate on y first

   that is $\quad (D-2)(Dy - 3y) = 0$

   Then $(D-2)$ operates on $(Dy - 3y)$, as follows:

   $$D(Dy - 3y) - 2(Dy - 3y) = 0$$
   $$D^2y - 3Dy - 2Dy + 6y = 0$$
   $$D^2y - 5Dy + 6y = 0$$
   $$(D^2 - 5D + 6)y = 0$$

   It is seen that although D is an operator it can be treated as a quadratic because

   $$(D-2)(D-3) = D^2 - 5D + 6$$

(c) First-order differential equations with constant coefficients

Consider the solution of the following equation using the integrating factor

$$\frac{dy}{dx} - 4y = 0 \tag{i}$$

The integrating factor is

$$e^{\int P\,dx} = e^{\int -4\,dx} = e^{-4x}$$

Multiplying throughout by e^{-4x}

$$\frac{dy}{dx}e^{-4x} - 4ye^{-4x} = 0$$

that is

$$\frac{d}{dx}(ye^{-4x}) = 0$$

$$ye^{-4x} = A \qquad \text{where } A \text{ is a constant}$$

Therefore

$$y = Ae^{4x} \tag{ii}$$

is the solution

In operator form equation (i) is $Dy - 4y = 0$

$$(D - 4)y = 0$$

giving

$$y = A\,e^{4x} \tag{iii}$$

as the solution

Comparing equations (ii) and (iii) it can be seen that the general solution can be obtained by inserting the solution of $(D - 4) = 0$, that is 4, in the exponential function.

Example 7.6   Find the general solution of

$$5\frac{dy}{dx} + 3y = 0$$

Divide by 5:

$$\frac{dy}{dx} + 0.6y = 0$$

Writing in operator form this is $(D + 0.6)y = 0$ and using the above analysis the constant to be inserted into the exponential is obtained from $D + 0.6 = 0$, that is -0.6. The general solution is

$$y = A\,e^{-0.6x}$$

(d) Second-order differential equations with constant coefficients

Consider the equation

$$\frac{d^2y}{dx^2} - 5\frac{dy}{dx} + 6y = 0$$

In D-operator notation

$$(D^2 - 5D + 6)y = 0 \tag{i}$$

From 7.5(b) above

$$(D^2 - 5D + 6) = (D - 3)(D - 2)$$

Therefore equation (i) becomes $(D-3)(D-2)y = 0$ so that

$$(D-3)y = 0 \quad \text{or} \quad (D-2)y = 0$$

Each of these produces a solution:

$$y_1 = Ae^{3x} \quad \text{and} \quad y_2 = Be^{2x}$$

Substituting these solutions into equation (i)

$$(D^2 - 5D + 6)y_1 = 0 \quad \text{and} \quad (D^2 - 5D + 6)y_2 = 0 \qquad \text{(ii)}$$

Using property (7.4) for the D operator

$$(D^2 - 5D + 6)(y_1 + y_2) = (D^2 - 5D + 6)y_1 + (D^2 - 5D + 6)y_2$$

Therefore from equation (ii)

$$(D^2 - 5D + 6)(y_1 + y_2) = 0$$

so that $y = y_1 + y_2$ is a solution, that is

$$y = Ae^{3x} + Be^{2x} \qquad \text{(iii)}$$

To determine a particular solution two boundary conditions are required so that both A and B can be determined. For example when $x = 0$, $y = 1$ and $dy/dx = 0$.

Substituting $x = 0$ and $y = 1$ in equation (iii):

$$1 = Ae^0 + Be^0 = A + B$$

Differentiating (iii):

$$\frac{dy}{dx} = 3Ae^{3x} + 2Be^{2x} \qquad \text{(iv)}$$

When $x = 0$, $dy/dx = 0$, equation (iv) becomes

$$0 = 3Ae^0 + 2Be^0 = 3A + 2B$$

Solving the two simultaneous equations

$$A + B = 1$$

and

$$3A + 2B = 0$$

gives

$$A = -2 \quad B = 3$$

The particular solution is $y = -2e^{3x} + 3e^{2x}$.

Example 7.7   Find the solution of the equation

$$\frac{d^2y}{dx^2} - 4\frac{dy}{dx} + 2y = 0$$

given that $y = 4$ when $x = 0$ and $y = 10$ when $x = 0.2$.

Writing in D-operator form:

$$(D^2 - 4D + 2)y = 0$$

To solve the equation the quadratic formula is applied to

$$D^2 - 4D + 2 = 0$$

$$D = \frac{4 + \sqrt{16 - 8}}{2} \qquad \text{or} \qquad D = \frac{4 - \sqrt{16 - 8}}{2}$$

$$= 2 + \sqrt{2} \qquad \text{or} \qquad = 2 - \sqrt{2}$$

$$= 3.4 \qquad \text{or} \qquad 0.6$$

These two results produce solutions

$$y_1 = Ae^{3.4x} \qquad \text{and} \qquad y_2 = Be^{0.6x}$$

The general solution is $y = Ae^{3.4x} + Be^{0.6x}$

When $x = 0$

$$4 = A + B$$

When $x = 0.2$

$10 = 2A + B$ (where coefficients are to nearest whole number)

$$A = 6, \qquad B = -2$$

The particular solution is $y = 6e^{3.4x} - 2e^{0.6x}$

The solution to the quadratic D operator can produce the square root of a negative number, as shown in Example 7.8.

Example 7.8   The current flowing in a circuit is given by

$$\frac{d^2i}{dt^2} + 4\frac{di}{dt} \, 5i = 0$$

Find the general solution for the current in terms of time.

Writing the equation in D-operator form

$$(D^2 + 4D + 5)i = 0$$

To solve the differential equation the roots of the quadratic $D^2 + 4D + 5 = 0$ are first determined. Using the quadratic formula

$$D = \frac{-4 + \sqrt{16 - 20}}{2} \quad \text{or} \quad D = \frac{-4 - \sqrt{16 - 20}}{2} .$$

$$= -\frac{4 + 2\sqrt{4 - 5}}{2} \quad \text{or} \quad = \frac{-4 - 2\sqrt{4 - 5}}{2}$$

$$= -2 + j \quad \text{or} \quad = -2 - j$$

where $j = \sqrt{-1}$

The two solutions are

$$i_1 = A e^{(-2 + j)t} \quad \text{and} \quad i_2 = B e^{(-2 - j)t}$$

The general solution is $i = i_1 + i_2 = e^{-2t}(A e^{jt} + B e^{-jt})$

In Chapter 14 it is shown that

$$e^{jt} = \cos(t) + j \, \sin(t)$$

$$e^{-jt} = \cos(t) - j \, \sin(t)$$

so that equation (i) becomes

$$i = e^{-2t}[(A + B) \cos t + j(A - B) \sin t]$$

Let $\quad A + B = C \sin \alpha$

and $\quad j(A - B) = C \cos \alpha$

so that

$$i = e^{-2t}[C \sin \alpha \cos t + C \cos \alpha \sin t]$$
$$= C e^{-2t} \sin(t + \alpha) \text{ using formula (3.8)}$$

To find the value of the constants C and α two boundary conditions are required. For example, if

$$i = 10 \quad \text{when} \quad t = 0$$

and

$$i = 10.2 \quad \text{when} \quad t = 1$$

we have, when

$$t = 0 \; : \qquad 10 = C \sin \alpha \qquad\qquad\qquad \text{(i)}$$

$$t = 1 \; : \qquad 10.2 = C e^{-2} \sin(1 + \alpha) \qquad\quad \text{(ii)}$$

Rewriting (ii) by expanding $\sin(1 + \alpha)$ and remembering that the angles are in radians:

$$10.2 = Ce^{-2}[\sin 1 \cos \alpha + \cos 1 \sin \alpha]$$

$$= C\,0.14[0.84 \cos \alpha + 0.54 \sin \alpha]$$

$$= C\,0.114 \cos \alpha + C\,0.073 \sin \alpha$$

Substituting from equation (i)

$$10.2 = C\,0.114 \cos \alpha + 0.073 \times 10$$

$$83.0 = C \cos \alpha \qquad\qquad\qquad\qquad\qquad\qquad\text{(iii)}$$

Dividing equation (i) by equation (iii):

$$\frac{10}{83} = \tan \alpha$$

$$\alpha = 0.12 \text{ rad}$$

From equation (i)

$$10 = C \sin 0.12$$

$$C = 83$$

Therefore $i = 83e^{-2t} \sin(t + 0.12)$

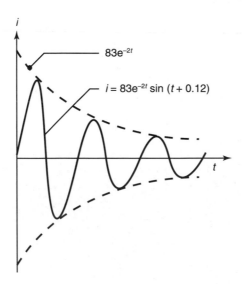

Fig. 7.4

This equation represents a vibration $83e^{-2t} \sin(t + 0.12)$ with $\omega = 1$ so that its frequency is $f = 1/2\pi$, and phase angle $\alpha = 0.12$ rad, and an amplitude $83A$.

The exponential indicates a decay in amplitude so that the equation represents a decaying sinusoidal current. This is illustrated in Fig. 7.4.

EXERCISE 7.3

Solve the following differential equations:

1. $\dfrac{d^2y}{dx^2} + 6\dfrac{dy}{dx} + 5y = 0$ with $y = 0$ and $\dfrac{dy}{dx} = 1$ when $x = 0$

2. $\dfrac{d^2y}{dx^2} + 2\dfrac{dy}{dx} + 5y = 0$

3. $\dfrac{d^2y}{dx^2} + 3\dfrac{dy}{dx} + 2y = 0$ with $y = 0$ and $\dfrac{dy}{dx} = 3$ when $x = 0$

4. $\dfrac{d^2y}{dx^2} - \dfrac{dy}{dx} - 6y = 0$

5. $\dfrac{d^2y}{dx^2} + 2\dfrac{dy}{dx} = 0$ given that $y = 2$ and $\dfrac{dy}{dx} = 1$ when $x = 0$

6. $\dfrac{d^2y}{dt^2} + 5\dfrac{dy}{dt} + 8.5t = 0$ with $y = 3$ when $t = 0$ and $y = 2$ when $t = 1$

7.6   Simple harmonic motion

It was seen in Example 7.3 that when the roots of the operator quadratic equation are complex the solution of the differential equation represents a decaying or damped oscillation, the rate of decay being controlled by the exponential part of the solution. Consider equation (7.3) when the coefficient $b = 0$, and the equation is a second-order linear equation. There will be no exponential component to the solution; in which case the amplitude of the oscillations will be constant (that is, undamped). Such a motion is called **simple harmonic motion** (SHM) of a particle.

Consider the equation

$$a\dfrac{d^2y}{dt^2} + cy = 0$$

where y is the displacement from a fixed point, and t is the time.

Therefore

$$\frac{d^2y}{dt^2} + \frac{c}{a}y = 0$$

In operator form

$$(D^2 + \omega^2)y = 0$$

where $\omega = \sqrt{\dfrac{c}{a}}$

The solution is given by

$$D^2 + \omega^2 = 0$$

$$D^2 = -\omega^2$$

$$D = \pm\sqrt{-\omega^2} = \pm\omega\sqrt{-1} = \pm j\omega \qquad \text{(i)}$$

The general solution is $y = Ae^{j\omega t} + Be^{-j\omega t}$

$$= (A + B)\cos \omega t + j(A - B)\sin \omega t$$

$$= C\sin(\omega t + \alpha) \qquad \text{(ii)}$$

as in Example 7.3.

This is an undamped oscillation of constant amplitude C with a frequency $\omega/2\pi$ hertz and a phase angle α. It is seen that there is no exponential term in the solution because there is no real part to the complex root in (i), contrary to the situation in Example 7.3.

Let $y = 0$, when $t = 0$. In this case the object passes through a fixed point $y = 0$ at the starting time of the observation. Therefore equation (ii) reduces to

$$C \sin \alpha = 0$$

$$\sin \alpha = 0$$

$$\alpha = 0$$

Equation (ii) describing the motion of the particle is

$$y = C \sin \omega t$$

The velocity of the particle is given by $v = dy/dt = \omega C \cos \omega t$

$$= \omega C\sqrt{1 - \sin^2 \omega t}$$

$$= \omega\sqrt{C^2 - C^2 \sin^2 \omega t}$$

$$= \omega\sqrt{C^2 - y^2} \qquad \text{(iii)}$$

The acceleration of the particle is given by

$$a = dv/dt = -\omega^2 C \sin \omega t$$

$$a = -\omega^2 y \tag{iv}$$

It is seen that the acceleration is proportional to the displacement from the fixed point O, which is the definition of simple harmonic motion. The graph of SHM with zero phase angle is shown in Fig. 7.5.

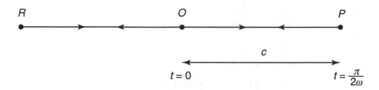

Fig. 7.5(a)

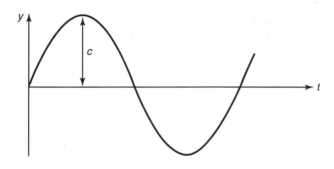

Fig. 7.5(b)

(a) Characteristics of simple harmonic motion

(i) From equation (iv) the acceleration is always in the direction towards the fixed point O (Fig. 7.5a). When the displacement is increasing the acceleration is negative which means that the object is slowing down. When the displacement is decreasing the object is accelerating towards O.

(ii) From equation (iii) the velocity is zero when the displacement $y =$ amplitude C at P and R.

(iii) From equation (iii) at the point O, where $y = 0$, $v = \omega C$.

(iv) The period of oscillation is $T = 2\pi/\omega$. It moves from O to P to O to R to O in a time T.

MISCELLANEOUS EXERCISE 7

1. (a) Given that

   $$\frac{d^2y}{dx^2} = x - \frac{1}{x^2}$$

   and that when $x = 1$

   $$\frac{dy}{dx} = \tfrac{5}{2} \quad \text{and} \quad y = \tfrac{4}{3}$$

   find y when $x = 2$.

   (b) Given that

   $$\frac{dy}{dx} = \frac{y}{1+x}$$

   and that $y = 1$ when $x = 0.5$, find y when $x = 5$.

2. For a certain beam

   $$160\,\frac{d^2y}{dx^2} = x - 1$$

   while it is known that when $x = 2$, $y = dy/dx = 0$.
   Express y in terms of x and calculate y when $x = 0$.

3. Solve the differential equations given below, expressing y in terms of x:

   (a) $\dfrac{d^2y}{dx^2} - 2x = 5$, given that $y = 3$ when $x = 0$ and $y = 17$ when $x = 2$.

   (b) $\dfrac{dy}{dx} = xe^{-2y}$, given that $y = 0$ when $x = 0$.

4. (a) Find y given that

   $$\frac{d^2y}{dx^2} = \frac{4}{\sqrt{x}}, \qquad \frac{dy}{dx} = 1 \qquad \text{when } x = 0$$

   and $y = 160$ when $x = 9$.

   (b) A calorimeter containing hot water at $40\,°C$ is placed on a table to cool in an ambient temperature of $10\,°C$. After t minutes the temperature of the water is $\theta\,°C$, the change in temperature being governed by the differential equation

   $$\frac{d\theta}{dt} = -0.05(\theta - 10)$$

   Solve the equation for θ as a function of t and hence calculate how long it takes for the water temperature to fall to $25\,°C$.

5. The equation

$$EI\frac{d^2y}{dx^2} = 5x - \frac{x^3}{45}$$

represents the bending moment at a point x along the beam, EI being a constant. Find an expression for the deflection y in terms of E, I and x, if $y = 0$ when $x = 0$ and when $x = 15$.

6. Bacterial rate of growth with respect to time is proportional to the amount present. If y is the number of bacteria at any time t, write down the differential equation for the rate of growth. Solve the equation, expressing y in terms of t, if $y = y_0$ at $t = 0$.

   If the population of bacteria doubles in one hour, by how much will it have increased from noon to midnight?

7. A capacitor of 20×10^{-6} F is connected in series with a resistor of $1000\,\Omega$ across a battery of voltage 20 V. Find the equation for the decay in the current flowing in the circuit. From the equation determine the time for the current to decrease to half its original value.

8. The induced voltage in a coil is given by $L\,di/dt$, where L is the self-inductance of the coil. Given that a resistor R is placed in series with the coil and connected to a battery of e.m.f. E via a switch, find the equation for the current i in terms of the time t, given that $i = 0$ when $t = 0$.

9. The rate of cooling of a boiler (that is the rate at which the temperature is falling) is $d\theta/dt$. This rate is proportional to the difference in its temperature above the surroundings:

$$\frac{d\theta}{dt} = -0.02(\theta - \theta_0)$$

where θ_0 is the temperature of the surroundings. Find the expression for θ in terms of the time t.

   Given that the boiler cools from $100\,^\circ$C to $90\,^\circ$C in 5 minutes, find how long it will take to cool from $90\,^\circ$C to $80\,^\circ$C.

10. A mass oscillating on a spring obeys the differential equation

$$\frac{d^2x}{dt^2} + 6\frac{dx}{dy} + 10x = 0$$

Find the general equation for the distance x of the mass from a fixed point in terms of the time t. What is the frequency of the oscillation?

Chapter 8

Applications of integration

8.1   Mean values and root mean square values

(a) Mean values

The arithmetic mean or average of a finite set of N numbers $y_1, y_2, y_3, ..., y_N$ is given by

$$\bar{y} = \frac{y_1 + y_2 + y_3 + \cdots + y_N}{N}$$

For example, the mean of 100, 102 and 104 is

$$\bar{y} = \frac{100 + 102 + 104}{3}$$

$$= 102$$

Consider, now, the function $y = f(x)$ for which the mean value of y is required over a given range of x values, from $x = a$ to $x = b$, as in Fig 8.1. There is an infinite number of y values corresponding to all the x values in this range. Therefore the mean value $\bar{y}$ is

$$\bar{y} = \frac{y_1 + y_2 + y_3 + y_4 + \cdots + y_\infty}{\infty} = \frac{\infty}{\infty}$$

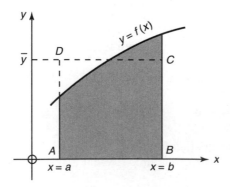

Fig. 8.1

Since this fraction is indeterminate $\bar{y}$ must be defined in a different way. It is defined as the value which makes the area $ABCD$ equal to the area beneath the curve $y = f(x)$ between the ordinates $x = a$ and $x = b$; that is,

$$AB.AD = \int_a^b y\,\mathrm{d}x$$

$$(b - a)\bar{y} = \int_a^b y\,\mathrm{d}x$$

$$\boxed{\bar{y} = \frac{1}{(b - a)} \int_a^b y\,\mathrm{d}x}$$

Example 8.1 Find the mean value of $y = 4x^3$ between $x = 2$ and $x = 4$.

From equation (8.1)

$$\bar{y} = \frac{1}{(4 - 2)} \int_2^4 y\,\mathrm{d}x$$

$$= \tfrac{1}{2} \int_2^4 4x^3\,\mathrm{d}x$$

$$= \tfrac{1}{2} \left[x^4\right]_2^4$$

$$= 120$$

(b) Root mean square (r.m.s) values

In some situations the mean value does not have much significance. such as, for example, when the set contains negative as well as positive values. In the sine curve (Fig. 8.2) the ordinates in the first half-cycle are $(+)$ and in the second half cycle are $(-)$. Otherwise the two half cycles are numerically identical, so that the mean value is zero. Therefore the mean value over a complete cycle gives no information about the magnitude of the amplitude of the curve.

In order to overcome this defect the square values are used, since on squaring a negative quantity the result is positive. Fig. 8.3 shows the squared sine curve; that is, $y = p^2 \sin^2 x$. The curve is symmetrical about the line $y_m = \tfrac{1}{2} p^2$. Because of symmetry the shaded areas are identical so that the area under the curve $y = (p \sin x)^2$ is equal to the area under the line $y_m = \tfrac{1}{2} p^2$. This means that

$\tfrac{1}{2}p^2$ is the mean value of $(p \sin x)^2$

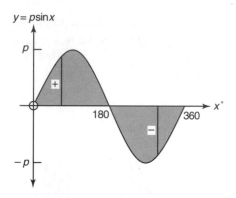

$y = p\sin x$

Fig. 8.2

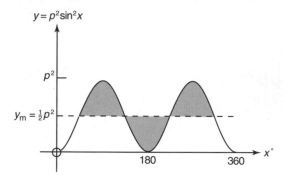

$y = p^2\sin^2 x$

Fig. 8.3

or

$\tfrac{1}{2}p^2$ is the mean square value of $p \sin x$.

Therefore $p/\sqrt{2}$ is the root mean square value of $p \sin x$.

A practical example of such a situation is the passage of an alternating current through a resistor. The mean value is zero, which gives no indication of the effect of the current, such as the power generated in the resistor. The power generated by a current is

$$i^2 R$$

where i is the current flowing. If the current is alternating i^2 must be replaced by the mean square value. The r.m.s. current is then seen to be the effective steady current which produces the same power as the alternating one.

The mean square value of $p \sin x$ can be obtained from equation (8.1). If $y = p \sin x$, using the range $x = 0$ and $x = \pi$,

$$\overline{(y^2)} = \frac{1}{(\pi - 0)} \int_0^\pi p^2 \sin^2 x \, dx$$

$$= \frac{1}{\pi} p^2 \int_0^\pi \sin^2 x \, dx$$

$$= \frac{1}{\pi} p^2 \left[\tfrac{1}{2} x - \tfrac{1}{4} \sin 2x \right]_0^\pi \qquad \text{using equation (16.15)}$$

$$= \tfrac{1}{2} p^2$$

Therefore, r.m.s. value $= \dfrac{p}{\sqrt{2}}$

Example 8.2 Find the r.m.s. current passing through a resistor, if the current is sinusoidal and obeying the equation

$i = 5 \sin 100\pi t.$

Consider any time interval equivalent to a half cycle, which takes the angle from 0 to π radians; that is, from $t = 0$ to $t = \frac{1}{100}$ s. Using equation (8.1) the mean square current is

$$\frac{1}{(0.01 - 0)} \int_0^{0.01} i^2 \, dt = 100 \int_0^{0.01} 25 \sin^2 100\pi t \, dt$$

$$= 2500 \left[\tfrac{1}{2} t - \tfrac{1}{2} \frac{\sin 200\pi t}{100\pi} \right]_0^{0.01}$$

$$= \frac{25}{2}$$

$$\text{R.m.s. current} = \frac{5}{\sqrt{2}} A$$

EXERCISE 8.1

1. Find the mean value of $y = x^{\frac{1}{2}}$ between $x = 0$ and $x = 4$.
2. Find the mean value of $y = 3 \cos 2x$ between $x = 0$ and $x = \pi$.
3. Find the mean value of $y = e^{-x}$ between $x = 1$ and $x = 3$.
4. Find the r.m.s. value of $y = 3x^3$ between $x = 2$ and $x = 4$.
5. Find the r.m.s. value of $y = x^{-1}$ between $x = 1$ and $x = 10$.
6. Find the mean value of $y = \ln 2x$ between $x = \frac{1}{2}$ and $x = \frac{1}{2} e^2$.
7. The voltage $V = V_0 \cos 2\pi t + 2V_0 \sin 4\pi t$ is applied across a circuit. Find its r.m.s. value.

8. The curve $y = -2x^2 + 8$ cuts the x-axis at two points. Find the mean value of y between these two points.

9. Find the r.m.s. value of $y = x + 1/x$ between the limits $x = \frac{1}{2}$ and $x = 2$.

10. In radioactive decay the number of atoms N left in a lump of material after a time t is given by

$$N = N_0 e^{-\lambda t}$$

Find the mean number of atoms in the lump between $t = 0$ and $t = \dfrac{2}{\lambda}$.

8.2   Centre of gravity and centre of mass

From a study of engineering and science it is well known that an object can be made to balance about a specific point. A dinner knife will balance about a point on the handle, or a billiard cue about a point near the thick end (Fig. 8.4). This specific point is called the **centre of gravity** G. Any body suspended by a single string will only balance when the centre of gravity is vertically in line with the string, as in Fig. 8.4. As there are only two forces acting on the object – the weight and the tension in the string – these two forces must be equal and opposite. Hence the weight can be considered to act at the centre of gravity.

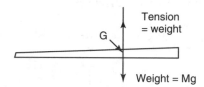

Fig. 8.4

Any object may be considered to be made up of a large number of elementary masses $m_1, m_2, m_3, ..., m_N$, as in Fig. 8.5. The force of gravity on each element is $m_1 g, m_2 g, m_3 g, ..., m_N g$. Let these forces be at perpendicular distances $x_1, x_2, x_3, ..., x_N$ from an axis YY. The moment of a force about an axis is the product of the force and perpendicular distance, so that the total turning moment of the object about YY is

$$m_1 g x_1 + m_2 g x_2 + m_3 g x_3 + \cdots + m_N g x_N$$

But the total weight Mg of the object is regarded as acting at G, at a distance $\bar{x}$ from YY. The total turning moment is $Mg\bar{x}$. Therefore

$$Mg\bar{x} = m_1 g x_1 + m_2 g x_2 + m_3 g x_3 + \cdots + m_N g x_N$$

$$\bar{x} = \frac{m_1 g x_1 + m_2 g x_2 + \cdots + m_N g x_N}{Mg} = \frac{\sum mgx}{Mg}$$

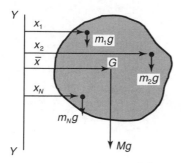

Fig. 8.5

giving

$$\bar{x} = \frac{m_1 x_1 + m_2 x_2 + \cdots + m_N x_N}{M} = \frac{\sum mx}{M} \tag{8.2a}$$

In this last equation $\bar{x}$ is expressed in terms of the masses. When used in this way G is called the centre of mass. Note that $M = m_1 + m_2 + m_3 + \cdots + m_N$.

It is also possible to define G in terms of the distance $\bar{y}$ from some horizontal axis XX, where

$$\bar{y} = \frac{\sum my}{M} \tag{8.2b}$$

$(\bar{x}, \bar{y})$ are the coordinates of the centre of mass from two perpendicular axes YY and XX.

Example 8.3 A rod of circular cross-section increases in radius from zero at one end to 20 mm at the other. The rod is 1000 mm long. Find the position of the centre of mass from the pointed end.

Let the density of the rod be ρ. Consider a small element of rod (hatched in the figure) of radius r and thickness δx, and distance x from YY.

Mass of element = volume $\times \rho$

$$= \pi r^2 \delta x \rho$$

Now for similar triangles

$$\frac{r}{x} = \frac{20}{1000}$$

Substituting for r, we obtain

$$\text{mass of element} = \pi \left(\frac{1}{50}\right)^2 x^2 \rho \, \delta x$$

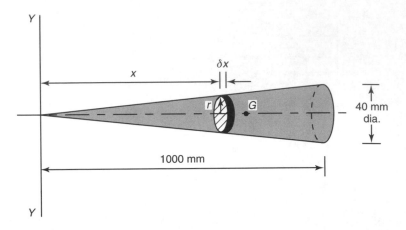

Fig. 8.6

Using this expression for m in equation (8.2a):

$$\bar{x} = \frac{\sum \pi \left(\frac{1}{50}\right)^2 x^3 \rho \, \delta x}{\sum \pi \left(\frac{1}{50}\right)^2 x^2 \rho \, \delta x} = \frac{\sum x^3 \, \delta x}{\sum x^2 \delta x}$$

As $\delta x \to 0$ the summations become integrals:

$$\bar{x} = \frac{\displaystyle\int_0^{1000} x^3 \, \mathrm{d}x}{\displaystyle\int_0^{1000} x^2 \, \mathrm{d}x} = \frac{\left[\frac{1}{4}x^4\right]_0^{1000}}{\left[\frac{1}{3}x^3\right]_0^{1000}}$$

$$= \tfrac{3}{4} \times 1000$$

$$750 \, \text{mm}$$

8.3   Centroids

If the equation (8.2a) is now used for a lamina of uniform thickness, then the mass $m_1, m_2, m_3, \ldots$ of each element is proportional to the area of each element, that is

$$m_1 \propto a_1, \qquad m_2 \propto a_2, \qquad \text{etc.}$$

so that

$$\bar{x} = \frac{\sum ax}{\sum a} \tag{8.3}$$

The point G defined by this value $\bar{x}$ is called the **centroid**. Centroids are used in conjunction with flat areas, where the $x_1, x_2, x_3, ...$ are the distances from the centres of the area elements to the axis YY. Equation (8.3) is applied to simple areas in the following examples.

Example 8.4 Find the position of the centroid in the area shown in Fig. 8.7.

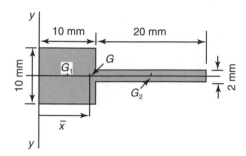

Fig. 8.7

The figure is composed of a rectangle and a square, G_1 and G_2 being their centroids. Equation (8.3) is now applied to this problem in which there are two elements.

Note: The centroid of a symmetrical figure such as a circle or square is at the centre of symmetry.

Let $\bar{x}$ be the distance of the centroid G from YY.

$$\bar{x} = \frac{a_1 x_1 + a_2 x_2}{a_1 + a_2}$$

where a_1 is the area of the square $= 100 \text{ mm}^2$

a_2 is the area of the rectangle $= 40 \text{ mm}^2$

x_1 is the distance of G_1 from $YY = \ \ 5 \text{ mm}$

x_2 is the distance of G_2 from $YY = \ 20 \text{ mm}$

Therefore

$$\bar{x} = \frac{100 \times 5 + 40 \times 20}{140}$$

$$= \frac{1300}{140}$$

$$= 9.3 \text{ mm}$$

The position of G lies on the axis of symmetry so that there is no need to calculate $\bar{y}$. Where there is no axis of symmetry G will lie on the line joining the centroids of the two elements.

Example 8.5   Find the position of the centroid of the lamina shown in Fig. 8.8.

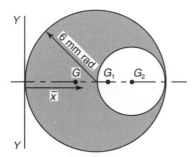

Fig. 8.8

Equation (8.3) is again used with two elements. Now, however, the smaller circle (unshaded) is removed, so that the moment must be subtracted. Using the same symbols as in Example 8.4,

$$\bar{x} = \frac{a_1 x_1 - a_2 x_2}{a_1 - a_2}$$

$$= \frac{\pi 6^2 \times 6 - \pi 3^2 \times 9}{\pi 6^2 - \pi 3^2}$$

$$= 5\,\text{mm}$$

EXERCISE 8.2

1.   Find the position of the centroid from the axis YY in each of the laminae shown in Fig. 8.9.
2.   Find the centre of mass of a right circular cone height 200 mm and base radius 100 mm.
3.   A billiard cue is 2 m long. Its radius at the thin end is 5 mm and at the thick end is 15 mm. Find the position of the centre of mass from the thin end.
4.   The line $y = 3x + 2$ is rotated about the x-axis. Find the centre of mass of the volume generated, between the limits $x = 0$ and $x = 5$.
5.   The curve $y = x^2$ is rotated about the x-axis between the limits $x = 1$ and $x = 2$. Find the centre of mass of the volume so generated.

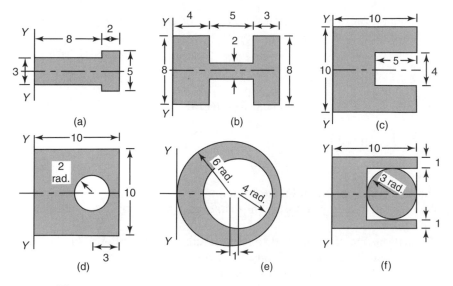

Fig. 8.9

6. The curve $y = 4 \sin 2x$ is rotated about the x-axis, between the limits $x = 0$ and $x = \frac{1}{2}\pi$. Find the centre of mass of the volume produced.

8.4   Centroids of more complicated areas and the centroid of a triangle

For more complicated areas as shown in Fig. 8.10, both the $\bar{x}$ and $\bar{y}$ of the centroid are required.

Small elements δA_1, δA_2, are considered at coordinates (x_1, y_1), (x_2, y_2), ...

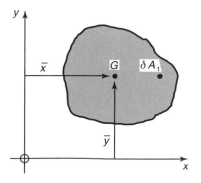

Fig. 8.10

Using equation (8.3), in which the masses of elements are proportional to the small areas, we have

$$\bar{x} = \frac{\sum \delta A\, x}{\sum \delta A}, \qquad \bar{y} = \frac{\sum \delta A\, y}{\sum \delta A}$$

As $\delta A \to 0$, the summations tend to integrals,

$$\boxed{\bar{x} = \frac{\int x\, \mathrm{d}A}{A}, \qquad \bar{y} = \frac{\int y\, \mathrm{d}A}{A}} \tag{8.4}$$

where A is the total area of the lamina.

Note: The distances x and y are the distances of the centroid of each element from the axes.

Equations (8.4) are applied to the triangle shown in Fig. 8.11, having a height h and base length b. Consider a thin strip, width $\mathrm{d}y$ at a distance y from apex 0. Let the length of the strip be x. Then its area is

$$\mathrm{d}A = x\, \mathrm{d}y$$

Because of similar triangles,

$$\frac{x}{y} = \frac{b}{h}$$

$$x = \frac{b}{h}y$$

Therefore, substituting for x

$$\mathrm{d}A = \frac{b}{h}y\, \mathrm{d}y$$

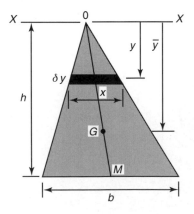

Fig. 8.11

From equation (8.4)

$$\bar{y} = \dfrac{\displaystyle\int_0^h y\,\dfrac{b}{h}\,y\,dy}{A}$$

$$= \dfrac{\dfrac{b}{h}\displaystyle\int_0^h y^2\,dy}{\frac{1}{2}bh} \qquad \text{since the area of the}$$
$$\text{triangle is } \tfrac{1}{2}\text{ base} \times \text{height}$$

$$= \dfrac{2}{h^2}\left[\dfrac{y^3}{3}\right]_0^h = \dfrac{2h}{3}$$

The centroid G is $\frac{1}{3}h$ from the base. Since the centroid of each element G_1 lies at its mid-point then by similar triangles the actual centroid G lies on the median OM, $\frac{1}{3}$ of its length from M.

8.5   The centroid of an area beneath a curve $y = f(x)$

In Fig. 8.12 a curve $y = f(x)$ is shown, with the area between the limits $x = a$ and $x = b$ shaded. The centroid of this shaded area is required. The area is divided into a large number of strips each of width dx. Area of each strip $dA = y\,dx$.

The centroid of each strip, for example G_1, has coordinates $(x, \frac{1}{2}y)$; that is, it is in the middle of the strip.

Using equations (8.4), and the above expression for dA,

$$\bar{x} = \dfrac{\displaystyle\int_a^b xy\,dx}{\displaystyle\int_a^b y\,dx}, \qquad \bar{y} = \dfrac{\displaystyle\int_a^b \tfrac{1}{2}y \cdot y\,dx}{\displaystyle\int_a^b y\,dx} = \dfrac{\tfrac{1}{2}\displaystyle\int_a^b y^2\,dx}{\displaystyle\int_a^b y\,dx} \qquad (8.5)$$

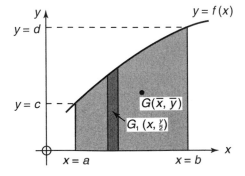

Fig. 8.12

If the position of the centroid of the area between the curve and the y-axis is required expressions (8.5) can be used after making the interchange

$$y \leftrightarrow x$$

that is

$$\bar{y} = \frac{\displaystyle\int_c^d xy \, dy}{\displaystyle\int_c^d x \, dy}, \qquad \bar{x} = \frac{\frac{1}{2}\displaystyle\int_c^d x^2 \, dy}{\displaystyle\int_c^d x \, dy} \tag{8.6}$$

Example 8.6 Find the centroid of the area between the curve $y = 2x^2 + 3$, the x-axis, and the ordinates $x = 0$ and $x = 3$.

Using equation (8.5)

$$\bar{x} = \frac{\displaystyle\int_0^3 x(2x^2 + 3)dx}{\displaystyle\int_0^3 (2x^2 + 3)dx}, \qquad \bar{y} = \frac{\frac{1}{2}\displaystyle\int_0^3 (2x^2 + 3)^2 dx}{\displaystyle\int_0^3 (2x^2 + 3)dx}$$

$$= \frac{\left[\dfrac{2x^4}{4} + \dfrac{3x^2}{2}\right]_0^3}{\left[\frac{2}{3}x^3 + 3x\right]_0^3} \qquad = \frac{\frac{1}{2}\left[\dfrac{4x^5}{5} + \dfrac{12x^3}{3} + 9x\right]_0^3}{\left[\frac{2}{3}x^3 + 3x\right]_0^3}$$

$$= \frac{54}{27} \qquad\qquad\qquad = \frac{164.7}{27}$$

$$= 2 \qquad\qquad\qquad\quad = 6.1$$

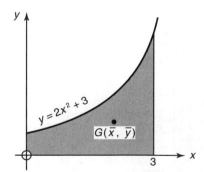

Fig. 8.13

EXERCISE 8.3

1. Find the centroid of the area under the curve $y = 4x^2$, between $x = 1$ and $x = 3$.
2. Find the centroid of the area between $y = 4x^2$, the y-axis, and the limits $y = 1$ and $y = 4$.
3. The curve $y = -x^2 - 2x$ cuts the x-axis in two places. Find the centroid of the enclosed area.
4. Find the centroid of the area enclosed between $y = \sin x$, and the x-axis, between the limits $x = 0$ and $x = \pi/2$.
5. Find the centroid of the area enclosed by $y = x^3$, the x-axis and the ordinates $x = 0$ and $x = 5$.
6. The parabola $y^2 = 4x + 9$ cuts the y-axis at two points. Find the centroid of the area enclosed by the curve and the y-axis.
7. Find the centroid of the area between the curve $y = 2x^2$ and $y = 2x$.
8. Find the centroid of the area enclosed between the curves $y = x^3$ and $y = x^2$.

8.6   Centroid of an arc and a semicircle

Using the theory developed so far it is possible to obtain the centroid of an arc such as BC (Fig. 8.14). The arc is divided into elementary lengths ds_1, ds_2, ... whose centroids G_1, G_2 ... have coordinates (x_1, y_1), (x_2, y_2), ... In equations (8.4) dA is replaced by ds.

$$\bar{x} = \frac{\displaystyle\int_B^C x \, ds}{\displaystyle\int_B^C ds} = \frac{\displaystyle\int_B^C x \, ds}{L} \qquad\qquad (8.7a)$$

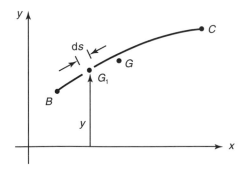

Fig. 8.14

$$\bar{y} = \frac{\int_B^C y \, ds}{\int_B^C ds} = \frac{\int_B^C y \, ds}{L} \tag{8.7b}$$

where L is the length of arc between B and C.

It is now possible to find the centroid of an arc of a semicircle (Fig. 8.15). The radius of the arc is r. Because of symmetry the centroid lies on the y axis, so that its coordinates are $(0, \bar{y})$.

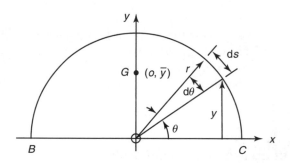

Fig. 8.15

Equation (8.7b) cannot be integrated directly. Both y and ds must be put in terms of the same variable. Let the small element of arc ds subtend an angle $d\theta$ at the centre. Let its distance from the x-axis be y. Referring to Fig. 8.15,

$$y = r \sin \theta$$

$$ds = r \, d\theta \quad \text{(for an arc of a circle)}$$

where θ is measured in radians.

The integrals extend from C to B as θ goes from 0 to π.

$$\bar{y} = \frac{\int_0^\pi r \sin \theta \, r \, d\theta}{\pi r}$$

$$= \frac{r^2 \int_0^\pi \sin \theta \, d\theta}{\pi r}$$

$$= \frac{r}{\pi} \big[- \cos \theta \big]_0^\pi$$

$$\bar{y} = \frac{2r}{\pi} \tag{8.7c}$$

Coordinates of the centroid are $\left(0, \dfrac{2r}{\pi} \right)$.

8.7   The theorems of Pappus (or Guldinus)

Theorem 1   If any arc which does *not* cut the x-axis (Fig. 8.16), is rotated about the x-axis, then

surface area generated = length of arc × distance travelled by its centroid.

The theorem applies to a complete revolution or part of a revolution.

The arc BC is rotated through an angle θ radians about the x-axis, thus forming a surface $BCDE$. G is the centroid of the arc BC. The distance moved by the centroid is GG which is $\bar{y}\theta$.

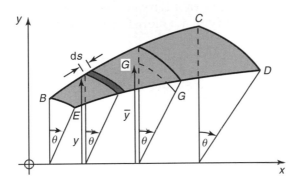

Fig. 8.16

If we consider a small length of arc ds, which rotates to produce a thin strip (heavy shading),

$$\text{area of strip so formed} = ds \times \text{distance moved}$$

$$= ds\, y\theta$$

$$\text{Total surface area } BCDE = \int_{B}^{C} ds\, y\theta$$

$$= \theta \int_{B}^{C} y\, ds, \quad \text{since } \theta \text{ is constant.} \qquad \text{(i)}$$

But from Section 8.6, equation (8.7b),

$$\int_{B}^{C} y\, ds = L\bar{y}$$

Substituting for the integral in (i), gives

total surface area $= L\bar{y}\theta$.

$\bar{y}\theta$ is GG, which is the distance moved by the centroid, hence proving the theorem.

When the arc is rotated through a complete revolution $\theta = 2\pi$ so that

surface area generated $= 2\pi L\bar{y}$ (8.9)

Theorem 2   If any area A, lying in the x–y plane (Fig. 8.17), and *not* cutting the x-axis, is rotated about the x-axis, then

volume generated $=$ area $\times$ distance travelled by its centroid.

Again the theorem applies to a complete revolution or part of a revolution. The centroid of the area is G. The area A is rotated through an angle θ radians to form the shaded volume. The distance moved by the centroid is GG, is

$\bar{y}\theta$.

An elementary area dA, rotated through θ, generates an elementary volume. The distance moved by the elementary area dA is

$y\theta$

so that the elementary volume is

$dA \times \text{distance moved} = dA\, y\theta$

The total volume generated is the sum of all these elementary volumes; that is,

$$\text{volume generated} = \int y\theta\, dA$$

$$= \theta \int y\, dA$$

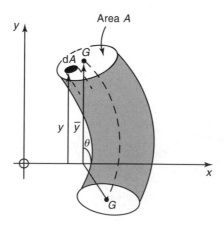

Fig. 8.17

From equation (8.4)

$$\int y \, dA = A\bar{y}$$

Substituting for the integral in (i)

$$\text{volume generated} = A\bar{y}\theta \qquad\qquad (8.10)$$

As seen previously $\bar{y}\theta$ is the length GG, which is the distance moved by the centroid, hence proving the theorem.

When the area is rotated through a complete revolution $\theta = 2\pi$ so that

$$\text{volume generated} = 2\pi A\bar{y} \qquad\qquad (8.11)$$

8.8   Application of Pappus' theorems

(a) To determine the surface area of a sphere

If an arc of a semicircle is rotated through one complete revolution about the diameter (Fig. 8.15) the surface of a sphere is produced. From equation (8.9)

$$\text{surface area generated} = 2\pi L\bar{y}$$

But from equation (8.7c)

$$\bar{y} = \frac{2r}{\pi}$$

Also L is half the circumference.

$$L = \pi r$$

Therefore

$$\text{surface area of a sphere} = 2\pi \times \pi r \frac{2r}{\pi}$$

$$= 4\pi r^2$$

(b) The centroid of a semicircle (area not arc)

Fig. 8.18 shows a semicircle with its diameter lying on the x-axis, with its centre at the origin. If the area rotates through 1 complete revolution about the x-axis it generates a sphere. From equation (8.11),

$$\text{Volume of sphere generated} = 2\pi\bar{y}A$$

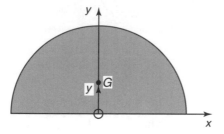

Fig. 8.18

In this equation we wish to determine $\bar{y}$. The volume of a sphere is $\frac{4}{3}\pi r^3$, and $A = \frac{1}{2}\pi r^2$, since A is a semicircle.
Therefore

$$\bar{y} = \frac{\frac{4}{3}\pi r^3}{2\pi \cdot \frac{1}{2}\pi r^2}$$

$$= \frac{4r}{3\pi}$$

Example 8.7   A rubber ring has a cross-section as shown in Fig. 8.19. Find the volume of rubber in the ring.

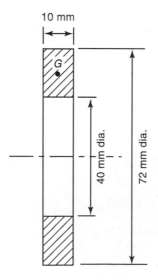

Fig. 8.19

G is the centroid of the upper area of cross-section of the ring, which is 28 mm from the axis of the ring:

$$\bar{y} = 28 \text{ mm}, \qquad A = 160 \text{ mm}^2$$

If the shaded area is rotated about the axis it generates the ring. The volume of the ring is given by equation (8.11):

$$\text{Volume generated} = 2\pi\bar{y}A$$
$$= 2\pi 28 \times 160$$
$$= 8960\pi \text{ mm}^3$$

Example 8.8 A triangular cut is made in a circular shaft, as shown in Fig. 8.20. Find the volume of metal removed.

The triangle ABC, rotated through 1 revolution about the axis of the shaft, generates the volume of metal removed. The centroid G of ABC is $\frac{2}{3}$ the height of the triangle from C, that is, $\frac{2}{3}$ of $15 = 10$ mm from C. Therefore

$$\bar{y} = 20 \text{ mm}$$

Also

$$A = 150 \text{ mm}^2$$

From equation (8.11)

$$\text{volume removed} = 2\pi\bar{y}A$$
$$= 2\pi 20 \times 150$$
$$= 6000\pi \text{ mm}^3$$

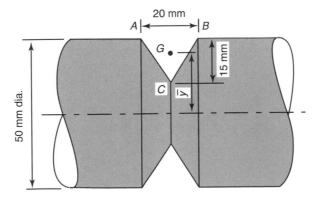

Fig. 8.20

EXERCISE 8.4

1. Determine the volume of the solid formed when the semicircle shown in Fig. 8.21 is rotated through 360° about *XX*.

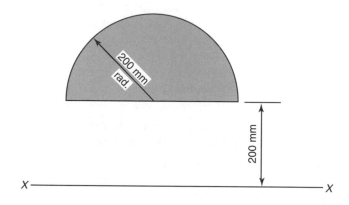

Fig. 8.21

2. An equilateral triangle *ABC*, of side 60 mm, is rotated through 360° about the side *AB*. Find (i) the total surface area generated, (ii) the volume generated.
3. A pulley is of the form of a steel disc of diameter 1 m and thickness 200 mm, from which a v-groove has been machined around the rim. The depth of the groove is 150 mm and its greatest width is the same thickness as the disc. Calculate the volume of the pulley.
4. By rotating a quadrant of a circle about one edge find the coordinates of its centroid.
5. A chemical flask has an outline shown in Fig. 8.22. Find
   (a) the surface area of the flask
   (b) the volume of the flask.

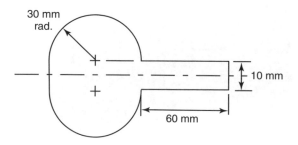

Fig. 8.22

6. (a) Find the centroid of the area shown in Fig. 8.23. Hence find (b) the surface area, (c) the volume of the solid formed, by rotating the area about *YY*.

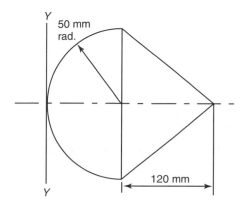

50 mm rad.

120 mm

Fig. 8.23

7. A dam wall of a reservoir has a cross-section shown in Fig. 8.24. The whole wall forms an arc of a circle, of radius 500 m, and subtends an angle of 120° at the centre. Find the volume of the dam wall.

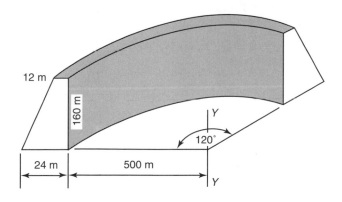

12 m

160 m

120°

24 m

500 m

Fig. 8.24

8. (a) In Fig. 8.25 find the centroid of the shaded area in relation to *XX*. Hence (b) find the volume generated when this area is rotated through 1 complete revolution about *XX*.

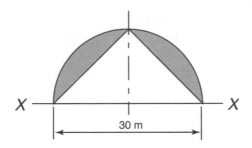

Fig. 8.25

8.9   Second moments of area and moments of inertia

(a) Second moment of area

Consider any lamina as shown in Fig. 8.26, divided into elementary areas δA_1, δA_2, ... at distances y_1, y_2, ... from an axis XX. The second moment of area of δA_1 about XX is $\delta A_1 y_1^2$. The second moment of area I_X of the whole lamina about XX is the sum of all the second moments of the elementary areas; that is,

$$I_X = \delta A_1 y_1^2 + \delta A_2 y_2^2 + \cdots$$

$$= \sum y^2 \delta A \tag{8.12a}$$

As $\delta A \to 0$, the summation becomes an integral.

$$\boxed{I_X = \int y^2 \mathrm{d}A} \tag{8.12b}$$

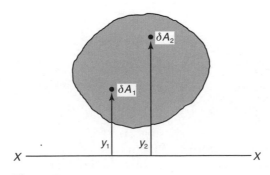

Fig. 8.26

From the second moment of area it is possible to obtain another quantity called the **radius of gyration**, K_X, which is defined by

$$K_X^2 = \frac{I_X}{A}$$

(8.13)

where A is the total area of the lamina.

(b) Moment of inertia

If any object is divided up into elementary masses m_1, m_2, m_3, ... at distances y_1, y_2, ... from an axis XX, then $m_1 y_1^2$, $m_2 y_2^2$, ... are the moments of inertia of m_1, m_2, ... about XX. The moment of inertia of the whole object is the sum of the moments of inertia of the elementary masses about XX.

$$I_X = \sum my^2$$

The radius of gyration is obtained in exactly the same way as in equation (8.13); that is,

$$K_X^2 = \frac{I_X}{M}$$

where M is the total mass of the object.

The significance of moment of inertia may be understood by reference to Fig. 8.27. Let the mass shown rotate through an angle θ about an axis XX. Any mass M_1 will sweep out an angle θ in the same time as any other mass M_2. However, M_1 is moving faster than M_2 since it has further to travel in

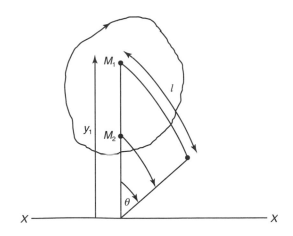

Fig. 8.27

the same time. The rate at which the angle is swept out, $\dfrac{\theta}{t}$, is the same for all masses, and is called the angular velocity ω.

Now in a time t, M_1 travels a distance l, where

$$l = y_1\theta$$

Dividing by t,

$$\frac{l}{t} = y_1\frac{\theta}{t}$$

But $\dfrac{l}{t}$ is the linear velocity of M_1, which we can call v_1.

Hence $v_1 = y_1\omega$ from equation (3.2)

The kinetic energy of all the masses is

$$E = \tfrac{1}{2}m_1v_1^2 + \tfrac{1}{2}m_2v_2^2 + \cdots$$
$$= \tfrac{1}{2}m_1\omega^2 y_1^2 + \tfrac{1}{2}m_2\omega^2 y_2^2 + \cdots$$
$$= \tfrac{1}{2}\omega^2 \sum my^2$$
$$= \tfrac{1}{2}\omega^2 I_X$$

Therefore it is seen that when an object rotates the kinetic energy involves the moment of inertia. Once the moment of inertia of a rotating object is known the kinetic energy can be determined in terms of the angular velocity.

(c) Second moment of area of a rectangle, about an axis along its base

Consider a small strip, width dy at a distance y from XX (dark strip in figure 8.28). Its area is

$$dA = B\,dy$$

Second moment of area about XX, using equation (8.12(b)), is

$$I_X = \int y^2 B\,dy$$

The positions of the strips vary from $y = 0$ to $y = D$, so that these are the limits of integration.

$$I_X = B\int_0^D y^2 dy$$

$$I_X = \tfrac{1}{3}BD^3 = \tfrac{1}{3}AD^2 \tag{8.14}$$

since area $A = BD$.

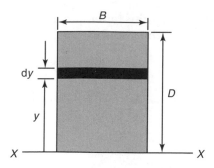

Fig. 8.28

The radius of gyration is given by

$$K_X^2 = \tfrac{1}{3}D^2$$

(d) Second moment of area of a rectangle, about an axis through the centre parallel to the base

The second moment of area in Fig 8.29 is obtained in exactly the same way as in section (c), except that the limits extend from $-\tfrac{1}{2}D$ to $+\tfrac{1}{2}D$.

$$I_X = B\int_{-\frac{1}{2}D}^{\frac{1}{2}D} y^2 \, dy$$

$$I_X = \frac{BD^3}{12} = \frac{AD^2}{12} \tag{8.15}$$

with $\quad K_Y^2 = \dfrac{D^2}{12}$

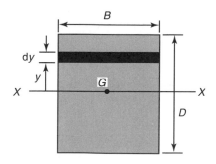

Fig. 8.29

(e) Second moment of area of a thin annulus, about an axis through the centre, perpendicular to the plane of the annulus

Consider a small element of the annulus d*l* (Fig 8.30). Its area is d*l* x d*t* and its second moment of area about ZZ is

$$\mathrm{d}l \, tR^2$$

For the whole annulus the second moment of area is

$$I_Z = tR^2 \int_0^L \mathrm{d}l$$

$$= tR^2 L$$

where L is the whole circumference $= 2\pi R$

$$I_Z = 2\pi tR^3 \tag{8.16a}$$

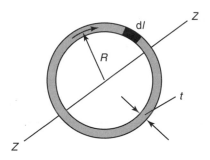

Fig. 8.30

(f) Second moment of area of a circle, about an axis through the centre, perpendicular to the plane of the circle

Consider an annular element, radius r, width dr (heavy shading in the figure 8.31). Using equation (8.16a), the second moment of area of this elementary annulus is

$$2\pi \, \mathrm{d}r \, r^3$$

The second moment of area of the whole circle is the sum of the moments of all these annular elements, extending from $r = 0$ to $r = R$.

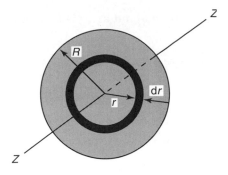

Fig. 8.31

$$I_Z = 2\pi \int_0^R r^3 \mathrm{d}r$$

$$= 2\pi \left[\tfrac{1}{4} r^4 \right]_0^R$$

$$I_Z = \tfrac{1}{2} \pi R^4 = \tfrac{1}{2} A R^2 \tag{8.16b}$$

since $A = \pi R^2$ for the circle.
The radius of gyration is given by

$$K_Z^2 = \tfrac{1}{2} R^2$$

(g) Second moment of area of a wide annulus,

The second moment of area (Fig. 8.32) is found as in section (f) above, except that now the limits of the integral are from $r = R_1$ to $r = R_2$. Therefore

$$I_Z = 2\pi \int_{R_1}^{R_2} r^3 \, \mathrm{d}r$$

$$= \tfrac{1}{2} \pi R_2^4 - \tfrac{1}{2} \pi R_1^4$$

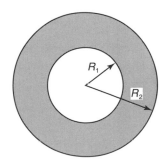

Fig. 8.32

(h) Second moment of area of a triangle about an axis through its apex parallel to the base

Using Fig. 8.11, second moment of area of the shaded strip about the axis XX is

$$\mathrm{d}A\, y^2$$

which is equal to $x\,\mathrm{d}y\, y^2$.
Therefore the second moment of area of the whole triangle about XX is

$$I_X = \int_0^h xy^2\, \mathrm{d}y$$

As in Section 8.4,

$$x = \frac{b}{h}y$$

Hence $$I_X = \int_0^h \frac{b}{h}y^3\, \mathrm{d}y = \frac{b}{h}\left[\tfrac{1}{4}y^4\right]_0^h$$

$$= \tfrac{1}{4}bh^3$$

8.10   Two theorems on second moments

(a) Perpendicular axes theorem

Theorem   If I_X, I_Y, I_Z are the second moments of area about the x, y and z axes, of any area A lying in the x–y plane, then

$$I_Z = I_X + I_Y \tag{8.17}$$

In Fig. 8.33 the area A is shown in the x–y plane. The three axes are mutually perpendicular and are called Cartesian axes. $\mathrm{d}A$ is an element of area such that its distance from the three axes is represented by lengths x, y and z as shown. Using equation (8.12), the second moment of area about the x-axis is

$$I_X = \int \mathrm{d}A\, y^2$$

Similarly

$$I_Y = \int \mathrm{d}A\, x^2$$

and

$$I_Z = \int \mathrm{d}A\, z^2$$

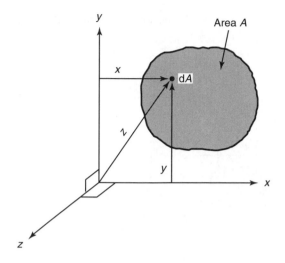

Fig. 8.33

Therefore

$$I_X + I_Y = \int dA(y^2 + x^2)$$

$$= \int dA\, z^2, \qquad \text{since } x^2 + y^2 = z^2$$
$$\text{(by Pythagoras' theorem)}$$

$$= I_Z$$

Note: This result is only valid for an area lying in the x–y plane.
   Iz is called the polar second moment of area.

(b) Parallel axes theorem

Theorem   If I_G is the second moment of area about an axis through the
centroid G of any area, and I_H the second moment of the area about any
parallel axis HH then

$$I_H = I_G + Ah^2 \tag{8.18}$$

where h is the distance between the two axes.

If we consider a small element dA, at a distance y from the axis through G
(Fig. 8.34), then its second moment of area is

$$dA\, y^2$$

Therefore, the total second moment of area of the area A about G is

$$I_G = \int y^2 \, \mathrm{d}A$$

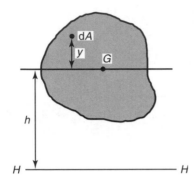

Fig. 8.34

The second moment of area of $\mathrm{d}A$ about HH is,

$$\mathrm{d}A(y+h)^2$$

so that the total moment of area of the area $\mathrm{d}A$ about HH is

$$I_H = \int (y+h)^2 \mathrm{d}A$$

$$= \int y^2 \, \mathrm{d}A + \int 2yh \, \mathrm{d}A + \int h^2 \, \mathrm{d}A$$

$$= I_G + 2h \int y \, \mathrm{d}A + h^2 \int \mathrm{d}A$$

From equation (8.4)

$$\int y \, \mathrm{d}A = \bar{y} \, \mathrm{d}A$$

where $\bar{y}$ is the distance of the centroid from the axis. In this case the axis passes through the centroid, so that $\bar{y} = 0$. Therefore

$$\int y \, \mathrm{d}A = 0$$

$$I_H = I_G + 0 + h^2 \int \mathrm{d}A$$

$$= I_G + Ah^2$$

Example 8.9   In Fig. 8.35, find the second moment of area about the axis *HH*. Find, also, the radius of gyration about *HH*.

From equation (8.15) for the second moment of area about an axis through the centroid,

$$I_G = \frac{BD^3}{12}$$

$$= \frac{5(12)^3}{12}$$

$$= 720 \text{ mm}^4$$

$$I_H = I_G + Ah^2$$

where from the diagram it is seen that $h = 8$ mm

$$I_H = 720 + 60(8)^4$$

$$= 4560 \text{ mm}^2$$

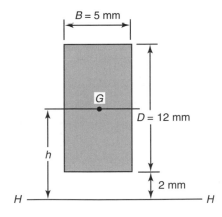

$B = 5$ mm

G

$D = 12$ mm

h

2 mm

H ——————————————— H

Fig. 8.35

Radius of gyration is given by

$$K_H^2 = \frac{I_H}{A}$$

$$= \frac{4560}{60}$$

$$= 76$$

$$K_H = \sqrt{76} = 8.72 \text{ mm}$$

Example 8.10   Find the second moment of area shown in Fig. 8.36 about the axis XX. The area is divided into three as shown.

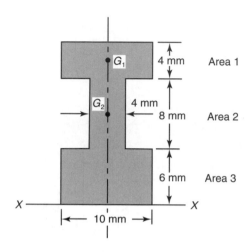

Fig. 8.36

Area 1: Let G_1 be its centroid. The second moment of area about an axis through G_1, parallel to XX, is

$$I_{G_1} = \frac{10(4)^3}{12} = 53.3 \text{ mm}^4$$

$$I_X = I_{G_1} + Ah^2$$

$$= 53.3 + 40(16)^2$$

$$= 10\,293 \text{ mm}^4$$

Area 2: Let G_2 be its centroid. The second moment of area about an axis through G_2 parallel to XX is

$$I_{G_2} = \frac{4(8)^3}{12} = 170.7 \text{ mm}^4$$

$$I_X = I_{G_2} + Ah^2$$

$$= 170.7 + 32(10)^2$$

$$= 3\,371 \text{ mm}^4$$

Area 3: Second moment of area about *XX* (which coincides with one of the sides) is

$$I_X \tfrac{1}{3} BD^3 = \tfrac{1}{3} 10(6)^3$$

$$= 720 \, \text{mm}^4$$

The second moment of area of the complete figure about *XX* is the sum of these results; that is,

$$I_X 10\,293 + 3\,371 + 720 = 14\,384 \, \text{mm}^4$$

8.11 Second moment of area of a circle

(a) About a diameter

In Fig. 8.37 the *x*- and *y*-axes lie along two perpendicular diameters, with the *z*-axis passing through the centre, perpendicular to the plane of the circle. The second moment of area of the circle, about the *z*-axis, is

$$I_Z = \tfrac{1}{2} \pi R^4 \qquad \text{(see equation 8.16b)}$$

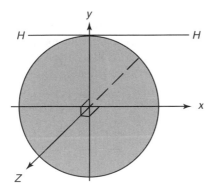

Fig. 8.37

By the perpendicular axes theorem

$$I_Z = I_X + I_Y$$

But because of symmetry

$$I_X = I_Y$$

Therefore

$$I_Z = 2I_X$$

or

$$I_X = \tfrac{1}{2}I_Z = \tfrac{1}{4}\pi R^4$$

(b) About a tangent

A tangent to the circle can always be chosen to be parallel to a diameter. Since the second moment of area has already been determined about the diameter which is the x-axis, we can choose the tangent HH parallel to the x-axis, which is at a distance R from the x-axis. Using the parallel axes theorem, since the x-axis passes through the centroid,

$$
\begin{aligned}
I_H &= I_X + Ah^2 \\
&= \tfrac{1}{4}\pi R^4 + \pi R^2 \times R^2 \\
&= \tfrac{5}{4}\pi R^4
\end{aligned}
$$

EXERCISE 8.5

1. In Question 1, Exercise 8.2, find the second moment of area of each figure about the axis YY. Hence find the second moment of area of each figure about an axis through the centroid, parallel to YY.
2. Find the second moment of area of a triangle about an axis along its base, if its height is h, and its base length b. Find also the second moment of area about an axis through the centroid, parallel to the base.
3. Determine the second moment of area of each of the areas in Fig. 8.38, (a) about the axis XX; (b) about an axis through the centroid parallel to XX.

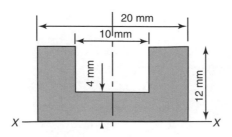

Fig. 8.38(a)

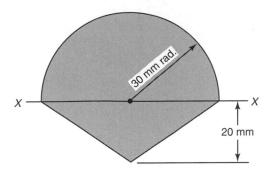

Fig. 8.38(b)

8.12   Second moment of area of an area beneath a curve between the limits $x = a$ and $x = b$

In Fig. 8.39, the second moment of the shaded area about the x-axis is required. The area is divided into a large number of elementary strips, of width dx. One such strip is shown heavily shaded. Since dx is infinitesimally small the strip approximates closely to a rectangle, thickness dx, length y. The second moment of area of the strip about OX using equation (8.14), is

$$\tfrac{1}{3}\,dx\,y^3$$

The second moment of area of the whole shaded area is the integral of this last expression between the limits $x = a$, $x = b$,

$$I_X = \int_a^b \tfrac{1}{3}y^3 \, dx$$

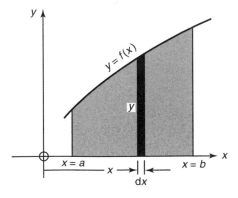

Fig. 8.39

The second moment of area of the strip about the y-axis is, from Section 8.9(a),

$$dA\, x^2 = y\, dx\, x^2$$

Since the strip approximates to a rectangle whose area is $dA = y\, dx$ the moment of the whole area about the y-axis is

$$I_Y = \int_a^b x^2 y\, dx$$

Example 8.11   Find the second moment of area under the curve $y = 3x^2$, between the limits $x = 1$ and $x = 2$ about the x- and y-axes.

From the results obtained above,

$$I_Y = \int_1^2 x^2 y\, dx$$

$$= 3 \int_1^2 x^4\, dx$$

$$= 3 \left[\frac{x^5}{5} \right]_1^2$$

$$= \tfrac{3}{5}(32 - 1)$$

$$= 18.6$$

$$I_X = \tfrac{1}{3} \int_1^2 y^3\, dx$$

$$= \tfrac{1}{3} \cdot 27 \int_1^2 x^6\, dx$$

$$= 9 \left[\frac{x^7}{7} \right]_1^2$$

$$= \tfrac{9}{7}(128 - 1)$$

$$= 163.3$$

EXERCISE 8.6

In Exercise 8.3, find the second moment of area about the x- and y-axes, in Questions 1–3, and 5.

8.13 Moment of inertia of a solid of revolution

The second moments treated in the previous paragraphs are only applicable to areas. In the case of a solid we have to consider, instead, the moment of inertia, which is defined by equation (8.12a) in which δA is replaced by the elementary mass m:

$$I_X = \sum my^2 \tag{i}$$

where y is the distance of a small mass m from the axis XX, as described in Section 8.9(a).

(a) Moment of inertia of a disc, radius R, thickness t, density ρ, about an axis through the centre perpendicular to the disc

The disc is shown in Fig. 8.40. Consider an elementary ring of width δr, at a radius r from the centre (shown as the shaded region in Fig. 8.40).

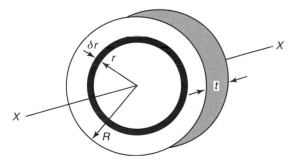

Fig. 8.40

Volume of ring = area of annulus × thickness t

$\qquad\qquad = 2\pi r\,\delta r \times t$

Mass of ring $\quad$ = volume × density

$\qquad\qquad = 2\pi r\,\delta r\,t \times \rho$

Moment of inertia of the ring = mass × r^2

$\qquad\qquad\qquad\qquad = 2\pi t\rho r^3\,\delta r$

Moment of inertia of the whole disc = sum of the moments of inertia

$\qquad\qquad\qquad\qquad\qquad$ of all the rings

$\qquad\qquad\qquad\qquad\qquad = \sum 2\pi t\rho r^3\,\delta r$

As $\delta r \to 0$, the summation can be replaced by integration.

Therefore

$$I_X = 2\pi t\rho \int_0^R r^3 dr$$

$$= 2\pi t\rho \left[\tfrac{1}{4}r^4\right]_0^R$$

$$I_X = \tfrac{1}{2}\pi t\rho R^4$$

But M the total mass of the disc = volume × density

$$= \pi R^2 t\rho \qquad\qquad\qquad\text{(ii)}$$

Therefore in equation (ii) $I_X = \tfrac{1}{2}MR^2$

(b) Moment of inertia of a solid of revolution

Consider any function $y = f(x)$, between the limits $x = a$ and $x = b$, as in Fig 8.41. If the curve is rotated it forms a surface of a solid of revolution. Let the solid so formed be subdivided into a large number of elementary discs of thickness dx, one of which is shown in the diagram. The moment of inertia of this particular disc about the x-axis is, using equation (ii),

$$\tfrac{1}{2}\pi \, dx \, \rho y^4$$

where ρ is the density.

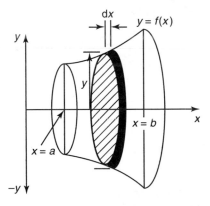

Fig. 8.41

The moment of inertia of the whole solid between $x = a$ and $x = b$ is

$$I_X = \tfrac{1}{2}\pi\rho \int_a^b y^4 dx \qquad\qquad\qquad\text{(iii)}$$

(i) The moment of inertia of a cone about its central axis

In Fig. 8.42a, if the line $y = (R/H)x$ is rotated about the x-axis, it produces the surface of a cone, with base radius R and height H. Therefore the moment of inertia about the x-axis is, using equation (iii),

$$I_X = \tfrac{1}{2}\pi\rho\left(\frac{R}{H}\right)^4 \int_0^H x^4 dx$$

$$= \tfrac{1}{2}\pi\rho\left(\frac{R}{H}\right)^4 \frac{H^5}{5}$$

$$= \tfrac{1}{10}\pi\rho R^4 H = (\tfrac{1}{3}\pi R^2 H\rho)\tfrac{3}{10}R^2$$

But the mass M of a cone is volume $\times$ density $= \tfrac{1}{3}\pi R^2 H\rho$. Therefore

$$I_X = \tfrac{3}{10}MR^2$$

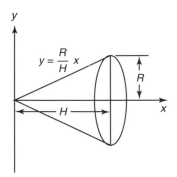

Fig. 8.42(a)

(ii) Moment of inertia of a sphere about a diameter

In Fig. 8.42b, a circle is shown, radius R, whose centre is at the origin. Its equation is

$$x^2 + y^2 = R^2$$

so that it cuts the x-axis at $-R$ and $+R$.

If this circle is rotated it produces a sphere. The moment of inertia about the x-axis is given by equation (iii) in which the limits are from $-R$ to $+R$; that is,

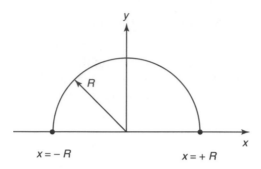

Fig. 8.42(b)

$$I_X = \tfrac{1}{2}\pi\rho \int_{-R}^{+R} (R^2 - x^2)^2 \, dx$$

$$= \tfrac{1}{2}\pi\rho \int_{-R}^{+R} (R^4 - 2R^2x^2 + x^4) dx$$

$$= \tfrac{1}{2}\pi\rho \left[R^4 x - 2R^2 \frac{x^3}{3} + \frac{x^5}{5} \right]_{-R}^{+R}$$

$$= \tfrac{1}{2}\pi\rho \tfrac{16}{15} R^5$$

$$= \tfrac{8}{15}\pi\rho R^5$$

$$= (\tfrac{4}{3}\pi R^3 \rho) \tfrac{2}{5} R^2$$

But the mass of a sphere $= \tfrac{4}{3}\pi R^3 \rho$ so that

$$I_X = \tfrac{2}{5} M R^2$$

MISCELLANEOUS EXERCISE 8

1. (a) Find, by integration, the second moment of area of a circle of radius r about an axis passing through its centre and perpendicular to its plane. Deduce the value of the second moment of area about a diameter of the circle.

   (b) The section shown in Fig. 8.43 consists of a rectangle $ABCD$ with two circular holes, centres E and F, cut from it. Calculate the second moment of area of the section about the axis AB.

2. Calculate the second moment of area of the section shown in Fig. 8.44 about
   (a) the edge AB,
   (b) the axis parallel to AB passing through the centroid of the section,
   (c) the neutral axis CD.

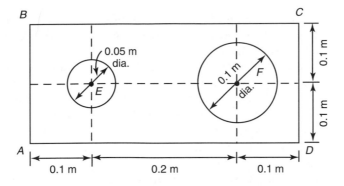

Fig. 8.43

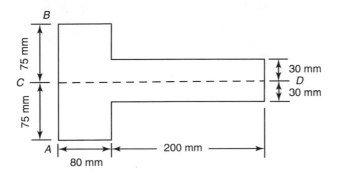

Fig. 8.44

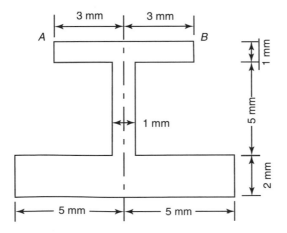

Fig. 8.45

3. Find the second moment of area of the section shown in Fig. 8.45 about an axis through the centroid of the section and parallel to AB.
4. A rectangular plate of the steel is in the shape of a rectangle of dimensions 250 mm by 150 mm, with a circular hole of radius 50 mm. The centre of the hole coincides with the centroid of the rectangle. Calculate the radius of gyration of the plate about an axis through the centroid perpendicular to the plate. Give your answer to the nearest millimeter.
5. Calculate the radius of gyration of a rectangular plate of length 1 m and width 0.6 m and mass 20 kg, about an axis through one vertex, perpendicular to the plane of the plate.
6. A regular hexagon has a side of length 2 units. Find the polar second moment of area through the centroid.
7. Determine, from first principles, the second moment of area of an equilateral triangle ABC of side 120 mm, about an axis along the side AB, and then use the theorem of parallel axes to find its second moment of area about an axis passing through the vertex C and parallel to the side AB.
8. A pulley wheel is produced by making a groove of triangular cross-section in a solid cylinder of diameter 180 mm and height 30 mm. The triangular cross-section is isosceles, of base 20 mm and height 18 mm. The bore of the pulley is of diameter 18 mm. Find the mass of the pulley if the density of the material is 7615 kg/m$^{3.}$
9. Find the centroid ordinate $\bar{y}$ for the area between the curve $y = x^2 + 4$, the x-axis and the ordinates $x = 0$ and $x = 4$. Determine the second moment of area about the x-axis.
10. Sketch, using the same axes, the straight line $y = x + 1$ and the curve $y = x^2 + 1$. Find the volume, expressed in terms of π, generated by rotating the area enclosed between the line and the curve about (a) the x-axis, (b) the y-axis.
11. Sketch the curve $y = 4 - x^2$ and calculate
    (a) the area in the first quadrant bounded by the curve and the coordinate axes;
    (b) the volume of the solid revolution formed by rotating the area about the x-axis;
    (c) the distance of the centroid of the area in (a) from the x-axis.
12. Sketch the curve $y = 3 \cos x$ between $x = -\frac{1}{2}\pi$ and $x = \frac{1}{2}\pi$. Determine.
    (a) the position of the centroid of the area between the curve and the x-axis;

    (b)   the root mean square value of the function over this interval.

13.   Find the distance from the y-axis of the centroid of the area enclosed between the curve $y = x^2 - 2x$ and the x-axis. Determine also the volume swept out when this area is rotated through one complete revolution about the x-axis.

14.   The velocity V of a body moving with simple harmonic motion at any time t is given by $V = a\omega \sin \omega t$, where a and ω are constants. Find the mean velocity between $t = 0$ and $t = \dfrac{\pi}{2\omega}$.

15.   Find the r.m.s. value of $1 - \sin 2x$ over the range $x = 0$ to $x = \pi/6$.

Chapter 9

Graphs

9.1 Representation of data

The relationship between two quantities in engineering or science can be expressed graphically, and usually by a formula or equation.

(a) Representation of data with a formula or equation

A formula or equation is a convenient way of expressing the relationship between two quantities. In equations such as $y = 3x^2 + 2x - 1$, x is called the **independent** variable. y is the **dependent** variable, so called because its value depends upon the x value. Such equations produce pairs of (x, y) values which can be used as Cartesian coordinates.

In equations such as $r = 2 \cos \theta$, θ is the independent variable and r the dependent variable, the pairs of (r, θ) values produced are called polar coordinates. Polar graphs are examined in Section 9.14.

(b) Graphical representation of two quantities related by an equation

For each value of x (or θ) a corresponding value of y (or r) is obtained from the equation. Each pair of values is used as the co-ordinates of a point on a plane. These points trace out a curve or straight line which is the graph representing the equation. The shape of the graph is a good indication of how one quantity depends on the other.

9.2 Cartesian and polar coordinates

To produce graphs and engineering drawings either by hand or by computer it is necessary to have a method of locating the positions of points on paper or on screen. At least two numbers, called **coordinates**, are required to locate a point in a plane. Two systems are used:

(a) Cartesian coordinates (*x*, *y*)

This is the most commonly used system. Two perpendicular datum lines are used, the horizontal line is called the **x-axis**, the vertical line is called the **y-axis**, as shown in Fig. 9.1. The point of intersection of the two axes is called the **origin** **O**. Any point P is located by its perpendicular distance from the two axes.

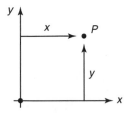

Fig. 9.1

(b) Polar coordinates (*r*, *θ*)

In this system a point *P* is located at a distance *r* along a line *OP* from a fixed point *O*, called the **pole**, as shown in Fig 9.2. *θ* is the angle that the line *OP* makes with the reference + *x*-axis. It is important to remember that *θ* is positive when *OP* rotates anticlockwise.

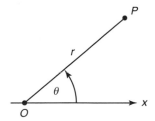

Fig. 9.2

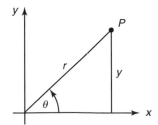

Fig. 9.3

Fig. 9.3 shows the relationship between both systems. Pythagoras's theorem and trigonometry can be used to change from one system to another.

(c) Conversion from Cartesian to polar coordinates

From Pythagoras $r = \sqrt{x^2 + y^2}$

and $\theta = \tan^{-1}\left(\dfrac{y}{x}\right)$

The smallest value of θ is usually quoted and can be positive or negative. The value of θ obtained must be checked so that it places P in the correct quadrant. This can be done by using a sketch to check the results, as shown in Example 9.1.

(d) Conversion from polar to Cartesian coordinates

From Fig. 9.3, using trigonometrical ratios in a right-angled triangle:

$x = r \cos \theta$
$y = r \sin \theta$

Example 9.1   The Cartesian coordinates of a point P are $(-4, -6)$. Convert these to polar coordinates.

The point P with these two coordinates is shown in Fig. 9.4.

$$r = \sqrt{(-4)^2 + (-6)^2} = \sqrt{16 + 36} = 7.2$$

$$\theta = \tan^{-1}\left(\frac{y}{x}\right) = \tan^{-1}\frac{-6}{-4}$$

$$= \tan^{-1} 1.5 = 56.3^\circ$$

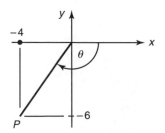

Fig. 9.4

The value of 56.3° is not in agreement with the position of P in the third quadrant. The smallest magnitude of θ is $180 - 56.3 = 123.7°$, and this is seen to be negative because it is in a clockwise direction.

The polar coordinates are $(7.2, -124°)$.

Example 9.2   Convert the polar coordinates $(18, 125°)$ to Cartesian coordinates.

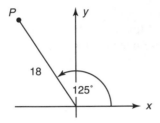

Fig. 9.5

From Fig. 9.5, $x = r \cos \theta = 18 \cos 125 = -10.3$

$y = r \sin \theta = 18 \sin 125 = 14.7$

The Cartesian coordinates are $(-10.3, 14.7)$.

9.3   The distance between two points

Fig. 9.6 shows two points $P(x_1, y_1)$ and $Q(x_2, y_2)$

Horizontal distance

$PR = x_2 - x_1$

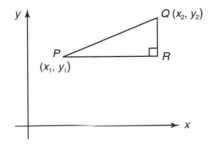

Fig. 9.6

Vertical distance

$$QR = y_2 - y_1$$

From Pythagoras,

$$PQ = \sqrt{PR^2 + QR^2}$$

$$= \sqrt{(x_2 - x_1)^2 + (y_2 - y_1)^2} \tag{9.1}$$

Example 9.3   The locations of two logic gates P and Q on a baseboard are given in polar coordinates as (80 mm, 60°) and (100 mm, 30°). Find the Cartesian coordinates and hence find the distance PQ.

At point P

$$x_1 = r \cos \theta = 80 \cos 60° = 40 \text{ mm}$$
$$y_1 = r \sin \theta = 80 \sin 60° = 69.3 \text{ mm}$$

At point Q

$$x_1 = r \cos \theta = 100 \cos 30° = 86.6 \text{ mm}$$
$$y_1 = r \sin \theta = 100 \sin 30° = 50 \text{ mm}$$

Distance PQ

$$= \sqrt{(x_2 - x_1)^2 + (y_2 - y_1)^2}$$

$$= \sqrt{(86.6 - 40)^2 + (50 - 69.3)^2}$$

$$= 50 \text{ mm}$$

EXERCISE 9.1

1.   Convert the following Cartesian coordinates into polar coordinates
     (a) (16, −20)   (b) (−2, 5)   (c) (−3, −6)
2.   Convert the following polar coordinates into Cartesian coordinates
     (a) (12, 30°)   (b) (20, −130°)   (c) (15, 160°)
3.   The location, in mm, of two components is shown in Fig. 9 7. Express the locations in polar coordinates. Find also the distance AB.
4.   Five holes are equally spaced on a circle of diameter 250 mm as shown in Fig. 9.8. Find the Cartesian coordinates for each of the holes, and hence calculate the distances AB and BC.

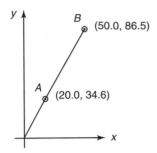

Fig. 9.7

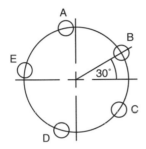

Fig. 9.8

5.   Three holes are to be drilled in a base-board as shown in Fig. 9.9.
Calculate the angle *BAC* and hence find the distances of the holes *B*
and *C* from the horizontal and vertical datum lines through *A*.

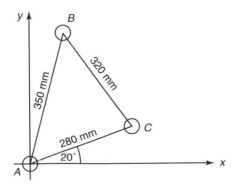

Fig. 9.9

9.4   Straight line graphs

(a) Gradient of a straight line

Consider the straight line passing through the two points A and B having coordinates (x_1, y_1) and (x_2, y_2) as shown in Fig. 9.10. Let the gradient of the line AB be m.

$$m = \frac{\text{change in } y}{\text{increase in } x} = \frac{DB}{AD}$$

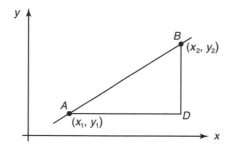

Fig. 9.10

But

$$DB = (y_2 - y_1) \qquad \text{and} \qquad AD = (x_2 - x_1)$$

Therefore

$$m = \frac{y_2 - y_1}{x_2 - x_1} \tag{9.2}$$

Example 9.4   A line passes through the points $(1, 4)$ and $(3, 1)$. Find the gradient of the line.

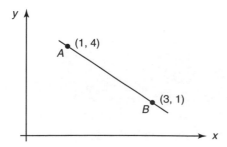

Fig. 9.11

In Fig. 9.11 we have, using equation (9.2)

$$m = \frac{1 - 4}{3 - 1} = -1.5$$

Gradient $m = -1.5$

Note: In this case since the value of y decreases as x increases the gradient is negative.

(b) Equation of a straight line

The equation representing a line is the equation relating the x and y coordinates of all points such as B (Fig. 9.12) that lie on it.

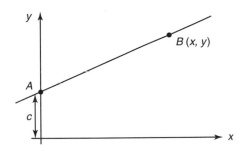

Fig. 9.12

Let the line cut the y-axis at a distance c from the origin at the point A. This distance is called the **intercept**. Let the coordinates of any point B be (x, y). The gradient m is given by equation (9.2),

$$m = \frac{y_2 - y_1}{x_2 - x_1} = \frac{y - c}{x - 0}$$

$$= \frac{y - c}{x}$$

The equation is

$$y = mx + c \tag{9.3}$$

Example 9.5 A line passes through two points A and B, whose coordinates are $(1, 5)$ and $(6, -7)$. Find the gradient and the equation of the line.

From equation (9.2) the gradient m is

$$m = \frac{y_2 - y_1}{x_2 - x_1} = \frac{-7 - 5}{6 - 1}$$

$$= -\frac{12}{5}$$

$$= -2.4$$

The equation of the line is $y = mx + c$. Since it passes through the point $(1, 5)$ we can substitute $x = 1$, $y = 5$, $m = -2.5$ into this equation to obtain the value of c.

$$5 = -2.4 \times 1 + c$$

$$c = 7.4$$

The equation is $y = -2.4x + 7.4$

Example 9.6   Find the equation of the straight line, gradient 2, which passes through the point $(3, 4)$.

Using equation (9.3):

$$y = mx + c$$

Substituting $x = 3$, $y = 4$, $m = 2$:

$$4 = 2 \times 3 + c$$

$$c = -2$$

The equation is $y = 2x - 2$

EXERCISE 9.2

Find the equation of the straight line in each of the following cases:

1.   Gradient is 5, passing through the point $(2, 6)$
2.   Gradient is -6, passing through the point $(-2, -5)$
3.   The tangent to a circle, gradient -2, passing through the point $(-4, 7)$
4.   The tangent at a point $(6, -2)$ on an ellipse if the gradient is -4
5.   The gradient is q, passing through the point $(q, 2q)$.
6.   Find the gradient and intercept of each of the following straight lines and hence sketch the graphs.
     (a) $y = 2x - 1$ (b) $y = -x + 4$ (c) $y = x + 1$
     (d) $y = 3x$ (e) $y = -x$ (f) $y = 2$

7. Results from a test on a machine are shown in Table 9.1

Table 9.1

Load lifted W (kN)	2.0	4.0	6.0	8.0	9.0
Applied force F (N)	0.81	1.22	1.58	2.04	2.17

Draw a graph of these results and hence find the law $F = aW + b$ of this machine.

8. A check carried out on a circuit showed when the voltage was 5 V the current was 4 A, and when the voltage was 11 V the current was 2 A. The relationship is known to be of the form $V = aI + b$. Find the values of a and b.

9. The resistance R (Ω) of a coil varies with temperature θ (°C) of the coil as shown in Table 9.2.

Table 9.2

Temperature θ (°C)	20	30	40	50	60	70
Resistance R (Ω)	60.4	65.4	69.2	75.6	79.8	84.8

Draw the graph and verify the law $R = a\theta + b$, and find the constants a and b.

(c) The coordinates of the mid point of a line AB in terms of the coordinates of A and B

In Fig 9.13 M is the mid-point of the line AB

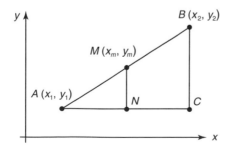

Fig. 9.13

Let the coordinates of A, B and M be, respectively (x_1, y_1), (x_2, y_2), (x_m, y_m). Then

$$AC = x_2 - x_1$$

N is the mid-point of AC because M is the mid-point of AB. Therefore

$$AN = \tfrac{1}{2} AC = \tfrac{1}{2}(x_2 - x_1)$$

so that

$$x_m = x_1 + \tfrac{1}{2}(x_2 - x_1)$$

$$= \tfrac{1}{2}(x_1 + x_2) \tag{9.4a}$$

Similarly

$$y_m = \tfrac{1}{2}(y_1 + y_2) \tag{9.4b}$$

Example 9.7   The coordinates of two points A and B are

(i)   (5, 7), (3, 4) respectively
(ii) (2, 3), (−6, −7) respectively

Find the coordinates of the mid-point of AB in each case.

Using the results obtained in equation (9.4) the coordinates of the mid-point of AB in each case are

(a) $\tfrac{1}{2}(5 + 3)$, $\tfrac{1}{2}(7 + 4)$ that is (4, 5.5)
(b) $\tfrac{1}{2}(2 - 6)$, $\tfrac{1}{2}(3 - 7)$ that is (−2, −2)

9.5   Perpendicular lines (normals)

Consider two lines AB and BD at right angles to one another (Fig. 9.14). AB is said to be **normal** to BD. Let θ be the inclination of BD to the horizontal. The inclination of AB to the horizontal is therefore $(90 - \theta)$. Let m_1 and m_2 be the gradients of BD and AB respectively. Therefore

$$m_1 = \frac{FD}{BF} = \tan \theta$$

The gradient of AB is negative, so that

$$m_2 = -\frac{AE}{EB}$$

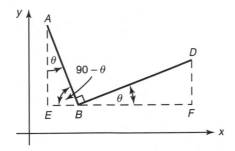

Fig. 9.14

But in triangle AEB,

$$\frac{AE}{EB} = \cot \theta = \frac{1}{\tan \theta}$$

Therefore

$$m_2 = -\frac{1}{\tan \theta}$$

Hence $$m_1 m_2 = \tan \theta \left(-\frac{1}{\tan \theta} \right)$$

that is $m_1 m_2 = -1$ (9.5)

Example 9.8 Find the equation of the line perpendicular to $y = 3x + 2$ which passes through the point $(3, 5)$.

Let the equation of the line by $y = m_2 x + c$. The gradient of $y = 3x + 2$ is 3; that is, $m_1 = 3$. Since the two lines are to be perpendicular

$m_1 m_2 = -1$

Therefore

$m_2 = -1/m_1 = -\frac{1}{3}$

The equation of the line becomes

$y = -\frac{1}{3} x + c$

Since the line passes through $(3, 5)$, $y = 5$ when $x = 3$
Therefore

$5 = -\frac{1}{3} \cdot 3 + c$

$c = 6$

The equation is

$$y = -\tfrac{1}{3}x + 6$$

EXERCISE 9.3

1. Determine the equation of the lines which are perpendicular to the lines represented by the following equations. The required line passes through the point shown in brackets in each case.
    (a) $y = 4x - 3$ (1, 2)
    (b) $y = -x + 7$ (2, −6)
    (c) $y = -5x - 1$ (−3, 4)
    (d) $y = -\tfrac{1}{2}x + 3$ (−2, −3)
    (e) $y = -x + 3$ (−4, 3)
2. The points A, B of the line AB have the following coordinates. Find the coordinates of the mid-point of AB in each case. Calculate also the gradient of the perpendicular to AB in each case:
    (a) (3, 4), (6, 7)   (b) (−6, 2), (7, −6)   (c) (7, −2), (−6, 5)
    (d) (8, 6), (14, 6)   (e) (−3, −4), (−6, −9)   (f) (2a, −4a), (−6a, 3a)
3. ABC is a triangle with coordinates $A(0, 0)$, $B(6, 0)$, $C(1, 8)$. Find the coordinates of the mid-point of each side.
    Find the equations of the perpendiculars to AB and AC which pass through these mid-points, respectively, and hence calculate the co-ordinates of the point of intersection of the two perpendiculars.
4. $ABCD$ is a parallelogram in which the coordinates of A, B and C respectively are (1, 2), (7, −1) and (1, −2).
    (a) Find the coordinates of the point D.
    (b) Find the equations of AD and CD.
5. The coordinates of A, B, C, D are respectively (−1, 3), (1, −2), (3, 1), (2, 4). Find the equations of AC and BD, and hence the coordinates of E their point of intersection.
6. Three points A, B, C have coordinates (−4, 8), (2, −1), (3, 4). Find
    (a) the equation of AB;
    (b) the equation of the line through C perpendicular to AB;
    (c) where the lines in (a) and (b) intersect.

9.6   The locus of a point

The term **locus** means the path over which a point will move when it has constraints imposed on it. When the path is formed between Cartesian axes it can be described by an equation relating the x- and y-coordinates. For example in Fig. 9.15, if a point P moves from the origin along a path with a

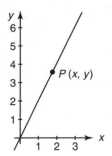

Fig. 9.15

constant gradient of 2 then the locus of the point is a straight line whose equation is $y = 2x$.

Example 9.9   A point P moves in such way that its distance from a point $A(0, 0)$ is equal to its distance from a point $B(4, -2)$. Find the locus of P, that is the equation which the point P traces out.

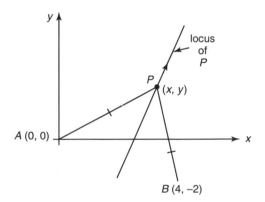

Fig. 9.16

Referring to Fig. (9.16) $AP = PB$.

   The lengths of AP and PB in terms of the coordinates are, using 9.1,

$$AP = [(x - 0)^2 + (y - 0)^2]^{\frac{1}{2}} = (x^2 + y^2)^{\frac{1}{2}}$$

$$PB = [(x - 4)^2 + (y - -2)^2]^{\frac{1}{2}}$$

Squaring both expressions we obtain

$$x^2 + y^2 = (x - 4)^2 + (y + 2)^2$$
$$0 = -8x + 16 + 4y + 4$$
$$y = 2x - 5$$

which is a straight line passing through the mid-point of AB and perpendicular to it.

9.7   The circle

A circle is the locus of a point which moves at a constant distance from a fixed point, as shown in Fig. 9.17. The fixed point is the centre of the circle and the constant distance is the radius of the circle.

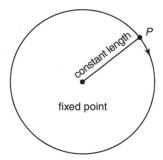

Fig. 9.17

Fig. 9.18 shows a circle whose centre is at the origin. P is any point (x, y) on the circle.

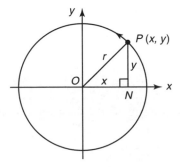

Fig. 9.18

Using the theorem of Pythagoras,

$$x^2 + y^2 = r^2$$

which is the equation of the circle.

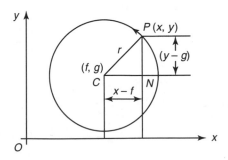

Fig. 9.19

When the centre of the circle is at a point (f, g) as in Fig. 9.19, then

$$CN = (x - f)$$
$$PN = (y - g)$$

From the theorem of Pythagoras

$$(x - f)^2 + (y - g)^2 = r^2$$

which expands to

$$x^2 + y^2 - 2fx - 2gy + f^2 + g^2 - r^2 = 0$$

The equation is usually written as

$$x^2 + y^2 + 2fx + 2gy + c = 0 \tag{9.6}$$

in which case the centre is at the point $(-f, -g)$ and $c = f^2 + g^2 - r^2$, from which

$$r^2 = f^2 + g^2 - c \tag{9.7}$$

For example, we can find the coordinates of the centre and the radius for the circle

$$x^2 + y^2 + 10x + 14y - 23 = 0$$

Comparing this with equation (9.6)

$$f = 5, \qquad g = 7, \qquad c = -23$$

The coordinates of the centre are $(-f, -g)$, that is $(-5, -7)$

Using the values of f and g in equation (9.7),

$$r^2 = f^2 + g^2 - c$$
$$= 5^2 + 7^2 - 23$$
$$= 51$$
$$r = 7.1$$

Example 9.10   Find the equation of a circle whose centre is at the point $(4, -3)$ and whose radius is 7.

From equation (9.6) the centre is at $(-f, -g)$, so that $-f = 4$, $-g = -3$

From equation (9.7),

$$r^2 = f^2 + g^2 - c$$
$$7^2 = (-4)^2 + 3^2 - c$$
$$c = 24$$

The equation is

$$x^2 + y^2 - 8x + 6y - 24 = 0$$

EXERCISE 9.4

Find the coordinates of the centre and the radius of each of the following circles.

1. $x^2 + y^2 - 6x - 8y - 56 = 0$ 2. $x^2 + y^2 - 4x + 12y + 24 = 0$
3. $x^2 + y^2 + 7x - 9y = 0$ 4. $x^2 + y^2 + 10x + 7 = 0$
5. $x^2 + y^2 - 8y - 20 = 0$

Find the equation of each of the following circles, having centre coordinates and radius as stated.

6. $(4, 6)$, 8   7. $(-3, -7)$, 6
8. $(-4, 5)$, 10   9. $(a, 2a)$, $5a$

9.8   The parabola, ellipse and hyperbola

In Fig. 9.20 a point P moves in such a way that its distance from a line AB, and its distance from a point S, are given by

$$SP = k\,PN$$

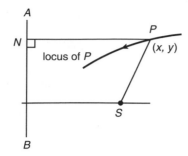

Fig. 9.20

When $k = 1$ the locus of P is a parabola.
When $k < 1$ the locus of P is an ellipse.
When $k > 1$ the locus of P is a hyperbola.

The fixed line AB is called the **directrix** and the point S is called the **focus**.

(a) The parabola

The locus, shown in Fig. 9.21, has an equation $y^2 = 4ax$ with respect to the Cartesian axes, which have their origin at the point where the curve cuts the x-axis. The distance from the origin to the focus is a.

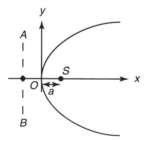

Fig. 9.21

In the equation $y^2 = 36x$, $a = 9$, so that the focus is at a distance 9 from the origin.

For each positive value of x, y has two values. For example, when $x = 1$,

$$y^2 = 36$$
$$y \pm 6$$

Therefore the curve is symmetrical about the x-axis.

When x is negative, say $x = -2$, then $y^2 = -72$

Since the square root of a negative quantity does not have a real value the curve cannot exist on the left of the origin.

If the parabola is rotated $90°$ anticlockwise about point O the formula will be $x^2 = 4ay$.

(b) The ellipse

The locus shown in Fig. 9.22 has the equation

$$\frac{x^2}{a^2} + \frac{y^2}{b^2} = 1 \tag{9.8}$$

where the centre of the ellipse is at the origin.

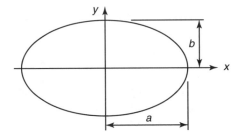

Fig. 9.22

From equation (9.8) when $y = 0$, $x = \pm a$ and when $x = 0$, $y = \pm b$

Therefore a and b are the lengths of the semi-axes. An ellipse is symmetrical about both axes. For an ellipse it can be proved by integration that

$$\text{area of the ellipse} = \pi ab$$
$$\text{perimeter of the ellipse} = \pi(a + b)$$

When $b = a$ the ellipse becomes a circle of radius a.

Example 9.11   Find the length of the semi-axes of the ellipse:

$3x^2 + 7y^2 = 42$

Divide throughout by 42 to obtain the ellipse in the form

$$\frac{x^2}{14} + \frac{y^2}{6} = 1$$

Therefore the semi-axes are $\sqrt{14}$ and $\sqrt{6}$.

(c) The hyperbola

The locus of the hyperbola shown in Fig. 9.23 is

$$\frac{x^2}{a^2} - \frac{y^2}{b^2} = 1 \tag{9.9}$$

In equation (9.9), when $y = 0$, $x = \pm a$, that is a is the distance from the origin to the points where the hyperbola cuts the x-axis. A hyberbola is symmetrical about both axes.

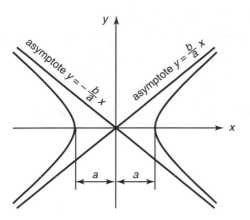

Fig. 9.23

Rewriting equation (9.9) gives

$$y^2 = b^2 \left(\frac{x^2}{a^2} - 1 \right)$$

It is seen that when x is very large x^2/a^2 is much greater than 1. Therefore

$$y^2 = \frac{b^2}{a^2} x^2$$

$$y = \pm \frac{b}{a} x$$

This means that as x gets very large the locus gets nearer and nearer to the two straight lines

$$y = \frac{b}{a} x \quad \text{and} \quad y = -\frac{b}{a} x$$

These two lines are called **asymptotes** as shown in Fig. 9.23.

If $b = a$ both asymptotes will be at $45°$ to the axes, that is, at $90°$ to each other. The equation of the hyperbola then becomes

$$x^2 - y^2 = a^2$$

and is called the **rectangular hyperbola**. If the asymptotes and the rectangular hyperbola are rotated about the origin through $45°$ in an anticlockwise direction the equation of the hyperbola becomes

$$xy = \text{constant}$$

A familiar application of this formula is Boyle's law, which is $pv = \text{constant}$, for the variation of pressure and volume of a fixed mass of gas at a constant temperature.

EXERCISE 9.5

1. Find the lengths of the semi-axes of the following ellipses:

   (a) $\dfrac{x^2}{25} + \dfrac{y^2}{9} = 1$

   (b) $\dfrac{x^2}{14} + \dfrac{y^2}{10} = 1$

   (c) $4x^2 + 7y^2 = 28$

   (d) $\dfrac{x^2}{4p^4} + \dfrac{y^2}{9m^2} = 1$

   (e) $25x^2 + 36y^2 = 1$.

2. A jet of water flows horizontally from an orifice in a tank and follows a parabolic path to hit the floor, which is 1.0 m below the tank, at a point 0.45 m horizontally from the point on the floor immediately below the orifice. Taking the origin of the Cartesian coordinates at the orifice find the equation for the parabola, and hence draw a diagram of the jet path to scale.

3. A receiving dish for transmission from outer space is in the form of a parabola. Given that the diameter of the dish is 400 cm and the depth of the dish at the centre is 20 cm, find the equation for the profile of the dish, assuming its axis of symmetry is horizontal.

4. An ellipse, whose semi-axes coincide with the Cartesian axes, passes through the points (1, 4) and (4, 2). Find the equation for the ellipse.

5. A round bar of diameter 100 mm is cut at an angle of $60°$ to the axis of the bar. Find the formula for the elliptical face so formed.

6. Draw the loci of the following hyperbolae and their asymptotes:

   (a) $\dfrac{x^2}{25} - \dfrac{y^2}{9} = 1$ (b) $x^2 - y^2 = 1$ (c) $xy = 20$.

Note: The curves dealt with in this section are referred to collectively as **conic sections**.

If a right circular cone is intersected by a flat plane, the intersection will produce a flat surface.
(a) If the plane is parallel to the base the intersection will be a circle.
(b) By varying the angle the plane makes with the base the intersection will be either a parabola, an ellipse, or a hyperbola.

9.9   Equation of the tangent and the normal at a point on a curve

The gradient of the tangent is equal to the gradient of the curve at the point of contact. The gradient of the curve can be found by differentiation.

The gradient of the normal can then be found using the formula $m_1 m_2 = -1$. In the case of the circle the normal passes through the centre of the circle. The procedure is shown in Example 9.12.

Example 9.12   Find the equation of the tangent and radius to the circle

$$x^2 + y^2 - 2x - 4y - 20 = 0$$

at the point in the first quadrant where $x = 4$.

To find the y coordinate substitute $x = 4$ into the equation of the circle.

$$4^2 + y^2 - 2(4) - 4y - 20 = 0$$
$$y^2 - 4y - 12 = 0$$
$$(y - 6)(y + 2) = 0$$
$$y = -2, \ 6$$

Therefore as seen in Fig. 9.24 there are two points P and Q on the circle with the same value of x.

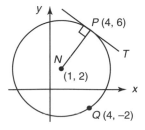

Fig. 9.24

To find the gradient at the point $P(4, 6)$, the equation of the circle is differentiated implicitly:

$$2x + 2y\frac{dy}{dx} - 2 - 4\frac{dy}{dx} = 0$$

$$\frac{dy}{dx} = \frac{2 - 2x}{2y - 4}$$

At the point $(4, 6)$ this gradient is $m = \dfrac{dy}{dx} = -\dfrac{3}{4}$

Equation of the tangent is

$$y = -\tfrac{3}{4}x + c$$

Since the tangent passes through the point $(4, 6)$

$$6 = -\tfrac{3}{4} \times 4 + c$$

$$c = 9$$

Equation of tangent PT is

$$y = -\tfrac{3}{4}x + 9$$

Gradient of normal (or radius)

$$m_2 = -\frac{1}{m_1} \qquad \text{where} \qquad m_1 = -\tfrac{3}{4}$$

$$m_2 = -\frac{1}{-\tfrac{3}{4}} = \tfrac{4}{3}$$

The equation of the radius is

$$y = \tfrac{4}{3}x + c$$

Since the radius passes through the point $(4, 6)$

$$6 = \tfrac{4}{3} \times 4 + c$$

$$c = \tfrac{2}{3}$$

The equation of the radius PN is $y = \tfrac{4}{3}x + \tfrac{2}{3}$

The tangent and radius are shown in Fig. 9.24. N is centre of circle.

A similar calculation may be made to find the equation of a tangent and normal to any curve.

EXERCISE 9.6

Each solution should be obtained by a calculation method and checked by drawing the graph.

1. Find the equation of the tangent and the normal at the stated point on each of the following circles:
    (a) $x^2 + y^2 = 25$ at $(3, 4)$
    (b) $x^2 + y^2 = 169$ at $(12, 5)$
    (c) $x^2 + y^2 = 85$ at $(7, 6)$
    (d) $x^2 + y^2 - 6x - 7y = 60$ at $(0, -5)$
    (e) $x^2 + y^2 - 4x + 12y = 274$ at $(-3, -23)$
2. Find the equation of the tangent and normal to the following curves at the stated point.
    (a) $3x^2 + 2y^2 = 30$ at the point $(2, 3)$
    (b) $2x^2 + 5y^2 = 47$ at the point $(1, 3)$
    (c) $7x^2 + 4y^2 = 128$ at the point $(2, 5)$
    (d) $4x^2 - 3y^2 = 24$ at the point $(3, 2)$
    (e) $8x^2 - 2y^2 = 10$ at the point $(1.5, 2)$
3. The normals to the ellipse $x^2 + 4y^2 = 100$ at the points $(6, 4)$ and $(8, 3)$ meet at a point A. Find the coordinates of A.
4. The normal to a parabola $y^2 = 16x$ at the point $(4, 8)$ crosses the curve again at a point P. What are the coordinates of P?

9.10   Quadratic graphs

Whereas in straight line graphs the two variables x and y have indices of 1, the quadratic graph is of the form

$$y = ax^2 + bx + c$$

where a, b, c are constants. For a particular graph b and/or c may be 0. The quadratic graph has a parabolic shape, and its width and position will depend on the values of a, b and c.

1. When a is large the graph is narrow, and when a is small the graph is wide. If a is positive the graph is $\cup$ shape, if a is negative the graph is $\cap$ shape.
2. A quadratic graph will have a maximum or minimum value through which passes the line of symmetry of the graph, as shown in Fig. 9.25.

    The position of this line is found by differentiating the equation

    $$y = ax^2 + bx + c$$

    $$\frac{dy}{dx} = 2ax + b$$

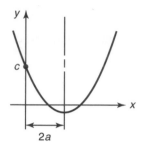

2a

Fig. 9.25

For a maximum or a minimum, $\dfrac{dy}{dx} = 0$ so that $x = -\dfrac{b}{2a}$

The line of symmetry is located at $x = -\dfrac{b}{2a}$

If $b = 0$, the line of symmetry is the y-axis.

3. c gives the value of the intercept.
4. A quadratic curve may cut the x-axis twice, just touch the x-axis, or not touch it at all. Therefore the equation $ax^2 + bx + c = 0$ may have two different real roots, two equal, real roots, or no real roots. (Note Section 1.1.)

Example 9.13   For the curve $y = -x^2 + 4x - 3$ find

(a)   the line of symmetry;
(b)   the maximum/minimum value of y;
(c)   the values of x which satisfy the equation $x^2 - 4x + 3 = 0$
      Sketch the graph using the above results.

(a)   The line of symmetry is at $x = -\dfrac{b}{2a} = -\dfrac{4}{2 \times -1} = 2$
(b)   Since a is negative the graph has a maximum value when
      $x = 2$, that is, $y - 2^2 + 4 \times 2 - 3 = 1$
      The intercept is $c = -3$.
(c)   When the graph cuts the x-axis, $y = 0$. Therefore

$$x^2 - 4x + 3 = 0$$
$$(x - 1)(x - 3) = 0$$
$$x = 1 \quad \text{and} \quad x = 3$$

With these results the curve can be sketched as shown in Fig. 9.26.

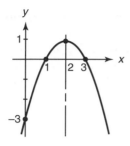

Fig. 9.26

A more accurate graph can be drawn using a table of values, but the example shows that a reasonably accurate sketch can be drawn using these four steps.

EXERCISE 9.7

1. Draw a graph of $y = 2x^2 + 3x - 15$, by first of all calculating a table of values of y for values of x between -4 and $+3$. Use the graph to find the solution of the equation $2x^2 + 3x - 15 = 0$. Calculate the position of the line of symmetry and read from the graph the minimum value of y.

2. The stress $S(MN/m^2)$ in a plate varies with the distance x (cm) from the centre and is given by $S = -x^2 + 9x - 14$. Draw the graph for the range $x = 1$ to $x = 8$ cm. Find (a) the greatest positive stress in the plate and the position at which it occurs, (b) the position(s) at which there is no stress in the plate.

9.11 Cubic graphs

Cubic graphs have equations which involve variables with the power of 3 such as

$$y = ax^3 + bx^2 + cx + d \qquad \text{where} \qquad a,\ b,\ c,\ d \text{ are constants.}$$

For a cubic equation b, c, or d may be equal to zero.

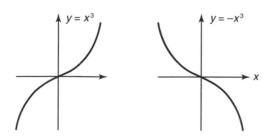

Fig. 9.27

The simplest cubic equations are $y = x^3$ and $y = -x^3$. The graphs of these functions are shown in Fig. 9.27.

Graphs of the more general equations are shown in Fig 9.28. These graphs have local maximum and minimum points or turning points. The value of the coefficient a determines the quadrant from which the graph commences on the left-hand side.

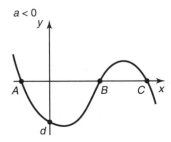

Fig. 9.28(a)

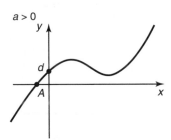

Fig. 9.28(b)

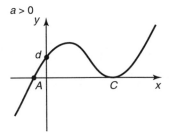

Fig. 9.28(c)

The cubic equation $ax^3 + bx^2 + cx + d = 0$ has

1. 3 real roots at A, B, C in Fig. 9.28(a), that is, where the graph cuts the x-axis, where $y = 0$;
2. 1 real root only at A in Fig. 9.28(b);
3. 1 real root at A and two equal real roots at B in Fig. 9.28(c).

9.12 Exponential graphs

An equation such as $y = 4^x$, where the variable x is an index, is called an exponential equation. 4^x is called a function of x; 4 is called the base number. In Chapter 1 it was stated that the base number widely used in engineering is the exponential number e, where e is the number 2.718..., being the value of the series

$$1 + \frac{1}{1} + \frac{1}{1.2} + \frac{1}{1.2.3} + \frac{1}{1.2.3.4} + \cdots$$

Exponential equations such as $y = Ae^{bx}$, $y = Ae^{-bx}$, $y = A(1 - e^{-bx})$, describe mathematically the behaviour of engineering situations, such as the cooling of a hot casting, the charging of a capacitor, etc.

The simplest of the exponential functions are $y = e^x$ and $y = e^{-x}$ and the graphs of the two functions are shown in Fig. 9.29.

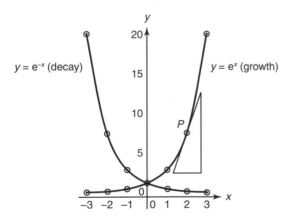

Fig. 9.29

The graphs show that for the function e^x, y increases exponentially with x, and for the function e^{-x}, y decreases exponentially with x. It is also seen that the two graphs are mirror images of one another.

The graphs have one property which is unique; the gradient of the curve at any point is always numerically equal to the height of the curve (the ordinate) at that point, that is, the rate at which the function is changing is always numerically equal to the value of the function at that point. This statement may be checked by finding the gradient at the point $P(x = 2)$, in Fig. 9.29. From the triangle, or by differentiation, the gradient at A is 7.39, which is the same as the value of y at P.

It is this property that makes the exponential function so important in technology. For example it is reasonable to expect that the rate at which the electric charge leaves a discharging capacitor will depend on the amount of charge left on the capacitor.

Example 9.14   The voltage V across the plates of a capacitor varies with the time $t(\mu s)$ according to the formula $V = 250e^{-0.4t}$. Draw a graph showing the voltage variation in the first $7\mu s$. From the graph find (a) the initial voltage when $t = 0$, (b) the rate at which the voltage is falling at $t = 3 \ \mu s$.

The values of V are shown in Table 9.3 and the graph is plotted in Fig. 9.30.

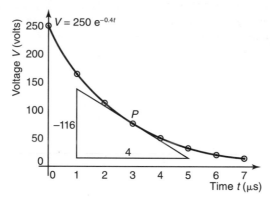

Fig. 9.30

Table 9.3

Time t (μs)	0	1	2	3	4	5	6	7
Voltage V (volts)	250	168	112	75.4	50.3	33.9	22.6	15.0

(a)   Initial voltage $= 250 \ V$
(b)   Rate of change of $V =$ gradient of tangent at P

$$= \frac{16 - 132}{5 - 1} = -29 \ V/\mu s$$

Note: The rate at which the voltage is changing at P is 29 V/μs, and this is not numerically equal to the voltage at P, which is 75 V. This will only be true for equations such as $y = e^x$ and $y = e^{-x}$. In an equation such as the one for voltage the rate of change of the voltage is **proportional** to the voltage, that is, if the voltage is halved the rate of change of voltage will be halved, etc.

Exponential curves have the following significant features:

1. Referring to the equations $y = Ae^{bx}$ and Ae^{-bx},
    A is the value of y where the curve crosses the y-axis.
    $+b$ gives a growth curve, $-b$ gives a decay curve.
    Increasing b increases the gradient of the curve at all points.
2. Referring to $y = A(1 - e^{-bx})$, the graph of which is shown in Fig 9.31, A is the maximum value attained by y.
    The larger the numerical value of b the quicker the curve will reach its maximum value.

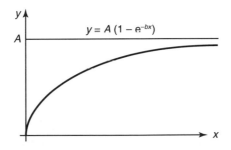

Fig. 9.31

EXERCISE 9.8

1. The voltage across the plates of a charging capacitor is given by
    $v = 30(1 - e^{-t/RC})$, where $R = 0.04\ M\Omega$, $C = 100\ \mu F$ and t is the time in microseconds. Plot a graph of v against t for the range $t = 0$ to $t = 15\ \mu s$. By drawing a tangent to the curve find the rate at which the voltage is increasing when $t = 5\ \mu s$. How does the rate of increase of voltage vary with time?
2. The formula for the force F in a drive belt is given by the formula
    $F = 800e^{-0.25\theta}$ where θ is the angle of contact in radians between the belt and the pulley. Draw the $F - \theta$ graph for the range $\theta = 0$ to $\theta = 3\pi/2$ rad, showing how the force in the belt depends upon the angle of contact with the pulley.

9.13   Logarithmic graphs

The graph of $y = \ln x$ is plotted in Fig. 9.32 for a range of values of x from 0.1 to 10. As discussed in Chapter 1 the base of the logarithm written in this form is e. For each of the values of x the value of y is obtained from a calculator by entering the value of x and pressing the $\boxed{\ln}$ key. These values are shown in Table 9.4.

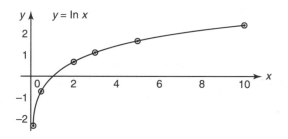

Fig. 9.32

Table 9.4

x	0.1	0.5	1	2	3	5	10
$y = \ln x$	−2.30	−0.69	0	0.69	1.10	1.61	2.30

In Chapter 1 it was seen that $x = \ln y$ is the logarithmic form of $y = e^x$. For this reason a logarithmic graph is a mirror image of the exponential graph. Similarly with an exponential graph representing a practical problem, such as a hot casting which is cooling according to Newton's law of cooling, $\theta = \theta_0 e^{-kt}$, where k is a constant and θ is the temperature of the casting above the surroundings after a time t and θ_0 is the initial temperature above the surroundings. The graph of θ (vertically) against t (horizontally) will give an exponential decay graph. The graph of t (vertically) against θ (horizontally) will give a logarithmic graph.

9.14   Polar graphs

In many engineering situations, such as drilling a set of holes in a circle, or measuring the intensity of light around a lamp, it is more convenient to use polar coordinates in place of the Cartesian coordinates. Again, shapes such as cams have complicated Cartesian formulae but much simpler polar formulae. The application of polar graphs is shown in the following two examples.

Example 9.15 The shape of the ground area illuminated by a motorway lamp is given by $r = 20(1 + \cos \theta)$, where r is the distance along the ground from the base of the lamp standard, and θ is the angle made by r from a fixed reference line. Plot the shape of the illuminated area.

The values of r in Table 9.5 are calculated from the equation.

Table 9.5

$\theta°$	0	30	60	90	120	150	180	210	240	270	300	330	360
r (m)	40	37.3	30	20	10	2.68	0	2.68	10	20	30	37.3	40

The illuminated area is shown in Fig. 9.33

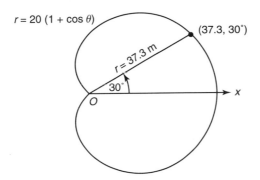

Fig. 9.33

Example 9.16 The output channel from a water turbine has a profile shape given by $r = 200\theta$ mm. Draw the profile for values of θ from $0°$ to $270°$

θ must be expressed in radians which have no units. If θ were expressed in degrees in the equation then the distance r would be expressed in degrees which is impossible. The (mm) unit is contained within the coefficient 200.

Table 9.6

$\theta°$	0	10	20	30	90	120	180	240	270
θ rad	0	0.175	0.350	0.524	1.571	2.094	3.142	4.189	4.712
r (mm)	0	35.0	70.0	104.8	314.2	418.8	628.4	837.8	942.4

Thus r is calculated using θ in radians, but for convenience the graph is plotted with θ in degrees. Table 9.6 is obtained from the formula and the profile plotted in Fig. 9.34.

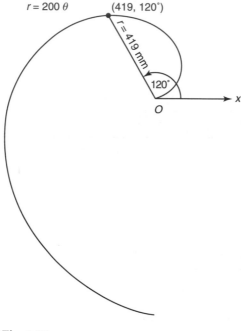

Fig. 9.34

EXERCISE 9.9

1. Draw the graphs of (a) $r = 2 \cos \theta$ (b) $r = 2 \sin \theta$ for values of θ from $0°$ to $360°$.
2. A steel plate is to be marked out with the curve $r = 10 + 50\theta$ mm. Plot this curve for θ from 0 to 3π radians.
3. The equation for the profile of a cam is $r = 50 + 40 \cos \theta$ (mm). Draw this profile for the range of θ from $0°$ to $360°$.
4. A display panel is to have the shape $r = 1 + \cos \theta$ (mm), for θ from $0°$ to $360°$. Draw the panel to scale.

9.15   Laws of experimental data

In experimental work it is often necessary to determine whether a set of values of x and y obeys a particular equation or formula. If the y values are plotted against x the points may, or may not, lie on, or close to, a straight line.

(a) Linear law

If the points so plotted lie on a straight line the values obey a linear law $y = ax + b$. The values of the gradient a and the intercept b can then be obtained from the graph, so that the particular law for the data can be found.

(b) Non-linear laws

If the values so plotted do not lie on a straight line it is possible that the values obey non-linear laws, such as,

$$y = ax^2 + b$$

$$y = a\sqrt{x} + b$$

$$y = \frac{a}{x} + b$$

$$y = ax^2 + bx$$

$$y = ax^b$$

$$y = ab^x$$

It is not generally possible to decide from a curved graph to which law the data will apply. The technique is to modify the values plotted on each axis to produce a straight line. From the straight line so plotted it is possible to calculate the values of a and b. Each of the above equations can be reduced to give a linear graph $Y = mX + c$ as follows:

$$y = ax^2 + b$$

Comparing with $Y = mX + c$ we see that plotting Y against X, that is, y against x^2, will produce a straight line, if the values obey this law. From the graph the gradient is a and the intercept is b.

$$y = a\sqrt{x} + b$$

Comparing with $Y = mX + c$ we see that plotting Y against X, that is, y against $x^{1/2}$ will produce a straight line, if the values obey this law. From the graph the gradient is a and the intercept is b.

$$y = \frac{a}{x} + b$$

Comparing with $Y = mX + c$ we see that plotting Y against X, that is, y

against $1/x$ will produce a straight line, if the values obey this law. From the graph the gradient is a and the intercept is b.

$$y = ax^2 + bx$$

Divide throughout by x to give

$$\frac{y}{x} = ax + b$$

Comparing with $Y = mX + c$, we see that plotting Y against X, that is, y/x against x, will produce a straight line, if the values obey this law. From the graph the gradient is a and the intercept is b.

$$y = ax^b$$

Take logarithms to base 10 of both sides of the equation and apply the laws of logarithms

$$\log y = \log(ax^b)$$
$$= \log a + \log(x^b)$$
$$= b \log x + \log a$$

Comparing with $Y = mX + c$ we see that plotting Y against X, that is, $\log y$ against $\log x$ will produce a straight line, if the values obey this law. From the graph the gradient is b and the intercept is $\log a$, from which a is found from the inverse logarithm.

$$y = ab^x$$

Take logarithms to base 10 on both sides of the equation and apply the laws of logarithms:

$$\log y = \log(ab^x)$$
$$= \log a + \log(b^x)$$
$$= x \log b + \log a$$

Comparing with $Y = mX + c$ we see that plotting Y against X, that is, $\log y$ against x will produce a straight line, if the values obey this law. From the graph the gradient is $\log b$ and the intercept is $\log a$, from which a and b are found.

This law is an exponential and will generally be found in the form $y = ae^{bx}$.

It should be noted that because experimental or actual data are subject to errors the various points are unlikely to lie exactly on a straight line. It is important, therefore, to choose the best fit straight line. The procedure is to calculate the gradient of the line from the graph to obtain the first constant.

The intercept will not give the second constant if the X-axis scale does not start at 0. In such a case the second constant is obtained by calculation from the original formula.

Example 9.17   The set of results shown in Table 9.7 was obtained in a practical test to find how the stress y (MN/m²) in a casting varied with the flange thickness x (cm). It is thought that the results obey approximately the law $y = ax^2 + b$. Verify that this is so by plotting a suitable graph and hence find the values of a and b. Calculate the maximum percentage error between any of the experimental results and the law obtained.

Table 9.7

x (cm)	1.0	2.0	3.0	4.0	5.0
y (MN/m²)	7.0	15	32	50	80

As described above if y plotted against x^2 produces a linear graph then the law is verified. Therefore we produce a table of results of y against x^2, and plot these as in Fig. 9.35.

Table 9.8

x (cm)	1.0	2.0	3.0	4.0	5.0
x^2	1.0	4.0	9.0	16.0	25.0
y (MN/m²)	7.0	15	32	50	80

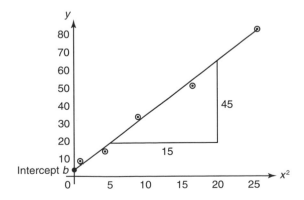

Fig. 9.35

From Fig. 9.35 it is seen that the readings approximate closely to a straight line. Therefore the results obey the law. From the graph

gradient $a = 45/15 = 3.0$
intercept $b = 4.0$

so that the law is

$$y = 3x^2 + 4$$

The largest deviation between a result and the straight line occurs when $x^2 = 16$. The true value from the graph is 52, the actual value is 50. Therefore

$$\text{maximum percentage error} = \frac{2}{50} \times 100 = 4\%$$

Example 9.18   In an electrical circuit the results in Table 9.9 were obtained.

Table 9.9

Current I (A)	3.0	2.0	1.0	0.71	0.55
Resistance R (Ω)	2.0	4.0	10	15	20

Verify that these results obey the law $R = a/I + b$, and if so, find the constants a and b.

As shown above this law reduces to a straight line when R is plotted against $1/I$. Therefore the table is recalculated to obtain values for $1/I$ (Table 9.10).

Table 9.10

Resistance R (Ω)	2.0	4.0	10	15	20
$1/I$	0.33	0.5	1.0	1.41	1.82

The graph is shown in Fig. 9.35. It is a straight line which verifies that the results obey the law. From the graph, gradient $a = 12$ and intercept $b = -2$ so that the law is

$$R = \frac{12}{I} - 2$$

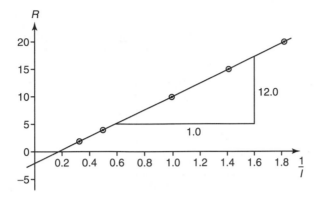

Fig. 9.36

Example 9.19 Table 9.11 shows the deflection x of a beam under given loads N.

Table 9.11

x (mm)	6.0	10.0	12.0	15.0	17.8	24.0
L (kN)	4.0	11.2	16.0	25.0	42.2	64.0

Verify that the results obey the law $L = ax^b$ and hence find a and b.

The linear form of the law is $\log L = b \log x + \log a$.

This equation should produce a straight line graph when $\log L$ is plotted against $\log x$. The table is recalculated (Table 9.12) and the graph plotted as in Fig. 9.37.

Table 9.12

log x	0.78	1.00	1.08	1.18	1.25	1.38
log L	0.60	1.05	1.20	1.40	1.63	1.81

With the exception of one point, which is probably caused by experimental error, the results obey the law closely. From the graph

Gradient $b = 0.80/0.4 = 2$

The intercept cannot be obtained from the graph because the scale on the horizontal axis does not start at the origin. We choose a

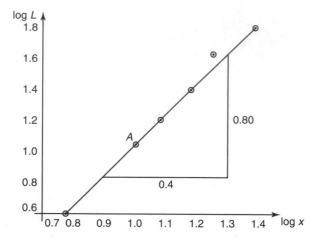

Fig. 9.37

known point, such as A, where $L = 11.2$ and $x = 10$ and substitute these readings, as well as $b = 2$, into the equation $L = ax^b$. Therefore

$$11.2 = a\,10^2 = 100a$$
$$a = 0.112$$

The law is

$$L = 0.11t^2$$

Example 9.20   Verify that the law $L = ab^x$ holds for the maximum load L that can be carried by an alloy-steel bracket containing varying percentages x of a certain chemical (Table 9.13). Hence calculate the values of a and b.

Table 9.13

x (%)	1	2	3	4	5
L (kN)	0.95	1.42	2.13	3.20	4.80

The linear form of this law is $\log L = x \log b + \log a$.

To verify this law $\log L$ is plotted against x. Recalculating the table gives Table 9.14.

Table 9.14

x (%)	1	2	3	4	5
log L	−0.02	0.15	0.33	0.51	0.68

The graph of log L against x is shown in Fig. 9.38 where the points lie on a straight line, thus verifying the law. From the graph

Gradient $= \log b = 0.53/3 = 0.177$

$$b = \text{inv log } 0.177 = 1.50$$

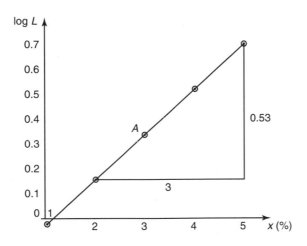

Fig. 9.38

The intercept cannot be obtained from the graph because the horizontal scale does not start at the origin. Using the values at the point A on the graph, where $L = 2.13$ and $x = 3$, we obtain

$$L = ab^x$$

$$2.13 = a \times 1.50^3$$

$$a = 0.63.$$

The law is

$$L = 0.63 \times 1.50^x$$

EXERCISE 9.10

1. The stress σ in an alloy steel plate was recorded at different distances x and the results given in Table 9.15 were obtained.

   Table 9.15

Distance x (cm)	4.0	7.0	8.0	10.0	11.0	12.0
Stress σ (MN/m^2)	20.8	102	141	238	296	360

   Plot a suitable graph to show that the results follow the law $\sigma = ax^2 + bx$, and hence find the value of the constants a and b.

2. The readings of an ammeter were taken in an electrical circuit as the resistance changed, and Table 9.16 was generated.

   Table 9.16

R (Ω)	51.4	53.8	56.0	58.0	61.0	64.0
I (A)	7.00	4.99	3.96	3.31	2.70	2.26

   It is thought that R and I are related approximately by the law $R = a/I + b$. Plot a straight line to verify that this is so and find the constants a and b.

3. The results in Table 9.17 were obtained experimentally and are thought to be related by the law $T = aL^2 + b$.

   Table 9.17

Load L (kN)	1.0	1.5	2.0	2.3	2.5	2.7
Torque T (kNm)	3.1	4.2	6.0	7.1	8.1	9.0

   Plot a suitable graph to check that this is so and find the constants a and b. Use the equation to estimate the torque with the load of 3.0 kN.

4. Show graphically that the values of the area of cross-section A of a beam, and the radius r of holes drilled in the beam, shown in Table 9.18, obey the law $A = ar^b$, where a and b are constants.

   Find the values of a and b and use the formula to find the area when the radius of the hole required is 1.2 cm.

Table 9.18

Radius r (cm)	1.6	1.8	2.0	2.2	2.4	2.6
Area A (cm^2)	2.48	2.06	1.78	1.55	1.36	1.20

5. The temperature of a metal plate cooling in a wind tunnel varies with time according to Table 9.19, where θ is the temperature above its surroundings and t is the time in minutes.

Table 9.19

Time t (min)	0	1.0	2.0	3.0	4.0	5.0
Temperature θ (°C)	500	260	120	65	30	17

Show that these results approximately obey the law $\theta = ab^t$ and find the constants a and b.

6. Table 9.20 shows the current flowing in a circuit at different times.

Table 9.20

Time t (μs)	1.0	2.0	3.0	4.0	5.0	6.0	7.0
Current i (A)	2.02	1.36	0.904	0.607	0.405	0.271	0.181

Show that these values fit the formula $i = ae^{bt}$ and find the constants a and b.

9.16 Logarithmic graph paper

In Examples 9.19 and 9.20 graphs were plotted of the logarithms of experimental readings. To avoid calculating logarithms the results may be plotted directly on logarithmic paper. With normal graph paper the linear scaled axes have distances marked out directly proportional to the size of the numbers. For example, the length of axis between 1 and 2 is the same as the length of axis between 2 and 3. On a logarithm scale the length of axis between two numbers is proportional to the logarithm of those numbers, that is, logarithm scales are not linear. For example the distances between 1 and 2 and between 2 and 3 are not the same. Furthermore, logarithm scales repeat themselves in cycles, since $\log 0.1 = -1$, $\log 1 = 0$, $\log 10 = 1, \log 100 = 2$, etc.

Two types of logarithmic paper are available, and both are available with 1, 2, or 3 cycles:

1. log/log paper, in which both axes have logarithmic scales, which would be used to verify the law $y = ax^b$.
2. log/linear paper, in which one axis only has a logarithmic scale, the other scale being linear, which would be used to verify the law $y = ab^x$.

Examples 9.19 and 9.20 are now repeated using log/log and log/linear paper respectively.

1. In Example 9.19 the law is $L = ax^b$, the linear form being $\log L = b \log x + \log a$. In this case both axes need the logarithmic scale so that log/log paper is used. Also since both variables are within the $1-10$ and $10-100$ bands, two-cycle paper will be required. The graph is shown in Fig. 9.39.

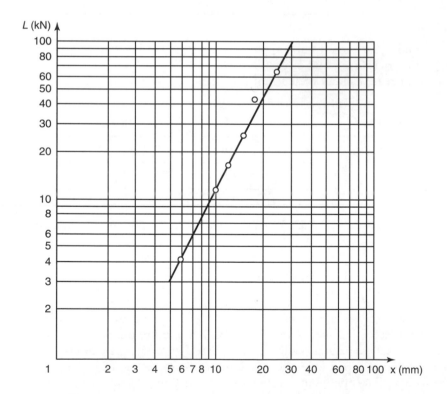

Fig. 9.39

It is seen from the graph that the points lie on a straight line, thus verifying the law. The constants a and b are now determined by substituting values of x and L of two points *on the line*, into $\log L = b \log x + \log a$, and solving two simultaneous equations.

2.   In Example 9.20 the law is $L = ab^x$, the linear form being $\log L = x \log b + \log a$. In this case $\log L$ is plotted against x so that log/linear paper is required. Since L is within the $0.1 - 1$ and $1 - 10$ bands a 2-cycle paper is required as shown in Fig. 9 40.

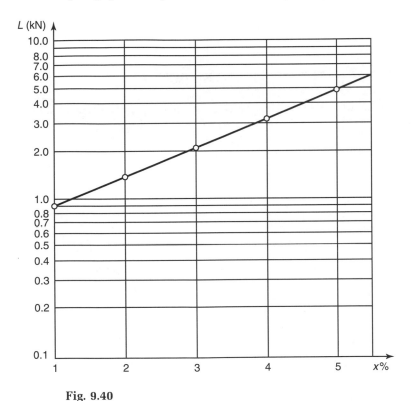

Fig. 9.40

The graph is a straight line and thus the law is verified. Again to find the constants a and b two points are considered *on the line* and values of L and x used in a calculation with two simultaneous equations.

9.17   The graphical solution of equations

Consider the graph of $y = x^2 - 3x + 2$. The coordinates are given in Table 9.21(a).

The graph is shown in Fig. 9.41. It can be seen from the graph that the curve cuts the x-axis at A and B, where $x = 1$ and $x = 2$ respectively. At A

Table 9.21

x	(a) $y = x^2 - 3x + 2$	(b) $y = x - 1$
-2	12	-3
-1	6	
0	2	-1
1	0	
2	0	
3	2	
4	6	
5	12	4

and *B* the value of *y* is zero. Therefore $x = 1$ and $x = 2$ are the solutions of the equation

$$x^2 - 3x + 2 = 0$$

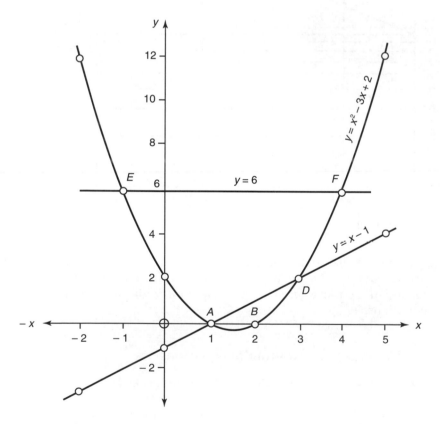

Fig. 9.41

Fig. 9.41 also shows the intersection of the curve $y = x^2 - 3x + 2$ with the line $y = 6$, at the points E and F, where $x = -1$ and $x = 4$. These are the values of x at which the function equals 6: that is, $x = -1$ and $x = 4$ are the solutions of the equation

$$x^2 - 3x + 2 = 6$$

that is of

$$x^2 - 3x - 4 = 0$$

Again, Fig. 9.41 shows the intersection of the line $y = x - 1$ with the curve $y = x^2 - 3x + 2$. The graph of $y = x - 1$ is plotted using the values determined in Table 9.21(b). They intersect at A and D, where $x = 1$ and $x = 3$. At these two values of x it is obvious that the two functions have the same values of y. Hence, $x = 1$ and $x = 3$ are the solutions of

$$x^2 - 3x + 2 = x - 1$$

which reduces to

$$x^2 - 4x + 3 = 0.$$

This last discussion shows that an equation can be solved graphically by plotting, on the same axes, graphs of the functions that appear on either side of the equation, and determining their points of intersection.

This method is extremely useful for solving more complicated equations, as shown in Example 9.21.

Example 9.21   Find graphically the solution of

$$4 \sin x = x^2$$

In this equation there are two distinct parts. The graphs of $y = 4 \sin x$ and $y = x^2$ are plotted on the same axes, as shown in Fig. 9.42. In the function $y = 4 \sin x$ the values of x must be expressed in radians because x appears on the right-hand side. In obtaining values of sin x it must be remembered that when

$$\pi > x > \frac{\pi}{2} \text{ (see Chapter 3) sin } x = \sin(\pi - x). \text{ The values to be}$$

plotted are given in Table 9.22.

Table 9.22

x	0	0.2	0.4	0.6	0.8	1.0	1.2	1.4	1.6	1.8	2.0
$y = 4 \sin x$	0	0.80	1.56	2.26	2.87	3.37	3.73	3.94	4.00	3.89	3.64
$y = x^2$	0	0.04	0.16	0.36	0.64	1.00	1.44	1.96	2.56	3.24	4.00

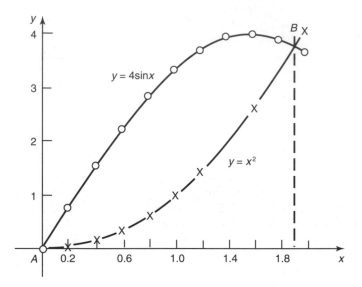

Fig. 9.42

The points of intersection are at A and B, where

$$x = 0 \quad \text{and} \quad x = 1.93$$

which are the solutions of the equation.

EXERCISE 9.11

1. Solve graphically the equation $x^3 - 9x^2 = 0$.
2. By plotting the graphs of $y = x^2$ and $y = x + 6$ on the same axes, in the range $x = -3$ to $x = 4$, solve graphically the equation

   $$x^2 - x - 6 = 0$$

3. Plot the graph of $y = x^2 + 2x - 3$. Use the graph to solve the equations

   (a) $x^2 + 2x - 5 = 0$
   (b) $x^2 + 3x - 5 = 0$

4. Plot the graph of $y = x^3 + x^2 - 8x - 5$ for values of x from -4 to $+3$, and hence solve the equation

   $$x^3 + x^2 - 8x - 5 = 0$$

   By drawing a portion of the graph to a larger scale find the positive root of this equation correct to two decimal places.

5. By plotting suitable graphs on the same axes, for values of x from 1 to 7, solve the equation

    $\ln x^2 = (x - 4)^2$

6. Plot the graphs of $y = x^3$, $y = 15x - 2$ on the same axes, taking values of x from -4 to $+4$, and using the two graphs solve the equation

    $x^3 - 15x + 2 = 0$

7. Solve the following equation graphically, for values of x between $-\frac{1}{2}\pi$ and $+\frac{1}{2}\pi$,

    $e^x = 2\cos x$

8. Solve, graphically, the equation,

    $\sin x = \frac{1}{3}x$

    for values of x between $-\frac{1}{2}\pi$ and $+\frac{1}{2}\pi$.

9. Plot the graphs of $x^2 + y^2 = 16$ and $y = -3x + 3$. Hence find the coordinates of the points of intersection to solve these simultaneous equations.

10. Plot the graphs of $(x^2/25) + (y^2/9) = 1$ and $xy = 2$. Hence find the coordinates M and N in the first quadrant where the rectangular hyperbola cuts the ellipse. Find the equation of the straight line MN.

Note: If the graphical solution is required more accurately the portion of the graph about the point of intersection is plotted on a larger scale, as in Question 4.

9.18 Curve sketching

It is often important to find the general form of a graph without going to the extent of plotting the curve exactly. In such cases a sketch of the curve showing its significant features will be sufficient. We have come across this idea already with turning points and in dealing with the main features of different graphs such as quadratic, cubic, logarithmic, exponential, trigonometrical graphs etc. A systematic approach to curve sketching is described below.

(a) Find the value of y when x = 0, or as x → 0

In an equation such as

$$y = \frac{x^2 + 3x + 3}{x - 1}$$

the value of y can readily be obtained when $x = 0$. It is $\dfrac{3}{-1} = -3$.

This gives one point on the graph immediately.

In other equations, such as, for example,

$$y = 4 + \frac{1}{x} \tag{i}$$

when $x \to 0$ the value of $y \to$ a large value that cannot be placed on the graph. Therefore, x is allowed to tend to zero, so that y tends to a large value, which can be shown on the graph.

As $x \to 0$ from positive value $y \to +\infty$
As $x \to 0$ from negative value $y \to -\infty$

The sketch of this equation is shown in Fig. 9.43.

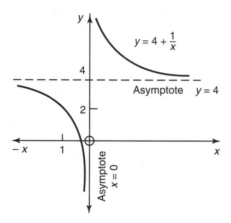

Fig. 9.43

From the figure it is seen that the curve gets closer and closer to the y axis (that is $x = 0$) as the value of y gets larger. The line forming the y-axis is called an **asymptote** in this case. An asymptote is defined as the straight line to which the curve gets closer and closer, as it stretches to infinity.

(b) Find the value of.y as $x \to \pm\infty$

Referring again to equation (i), as $x \to \infty$, $\frac{1}{x} \to 0$, so that for large values of x the curve approximates to

$$y = 4$$

The line $y = 4$ is therefore an asymptote to the curve.
As $x \to +\infty$, $y \to 4^+$, that is, y tends to 4 from values greater than 4.
As $x \to -\infty$, $y \to 4^-$, that is, y tends to 4 from values less than 4. The

sketch of the function for large values of x is shown in Fig. 9.19. The points discussed in sections (a) and (b) are used to sketch the curve in Example 9.22.

(c) Find and distinguish between maximum and minimum points

This method has already been described in Section 5.4.

Example 9.22   Sketch the curve $y = x + \dfrac{1}{x}$

(i)   As $x \to 0$ from positive values, $y \to +\infty$.
As $x \to 0$ from negative values, $y \to -\infty$.
Therefore $x = 0$ is an asymptote to the curve, as shown in Fig 9.44

(ii)   As $x \to \infty$, $\dfrac{1}{x} \to 0$, so that for large values of x the function

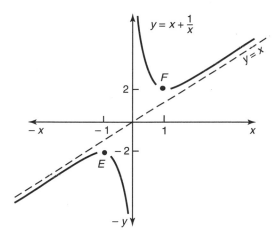

Fig. 9.44

approximates to $y = x$

The line $y = x$ by the above definition, is an asymptote.

(iii)   Differentiating: $\dfrac{dy}{dx} = 1 - \dfrac{1}{x^2}$

At the turning points, $\dfrac{dy}{dx} = 0$.

$$\frac{dy}{dx} = 0 \qquad \text{when} \qquad 1 - \frac{1}{x^2} = 0$$

$$x^2 = 1$$

$$x = \pm 1$$

Therefore $x = \pm 1$ denote the positions of the maximum or minimum points.

When $x = 1$, $y = 2$ and $x = -1$, $y = -2$.

Because of the information already gained about the curve in (i) and (ii) point $E(-1, -2)$ must be a maximum point and point $F(1, 2)$ must be a minimum point. This avoids the necessity of using d^2y/dx^2 to distinguish between them.

(d) Asymptotes as a result of the denominator tending to zero

Consider the function

$$y = \frac{1}{x - 1}$$

When $x = 1$, the denominator is zero, so that y becomes infinite.
As $x \to 1$ from values greater than 1, $y \to +\infty$.
As $x \to 1$ from values less than 1, $y \to -\infty$.
Therefore $x = 1$ is an asymptote, as shown in Fig. 9.45.
Also as $x \to \pm\infty$, $y \to 1/\pm\infty \to 0$ so the x axis is also an asymptote.
When $x = 0$, $y = -1$.

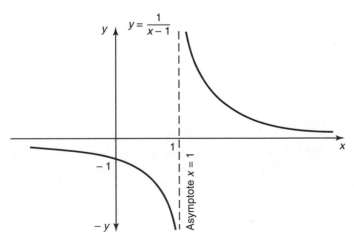

Fig. 9.45

If the denominator of a function becomes zero at a certain value of x, the function has an asymptote there provided the numerator does not become zero at the same value of x.

Example 9.23   Sketch the curve

$$y = \frac{1}{(x-2)(x-6)}$$

The denominator becomes zero at $x = 2$ and $x = 6$. Therefore, these two lines are asymptotes to the curve, as shown in Fig. 9.46.

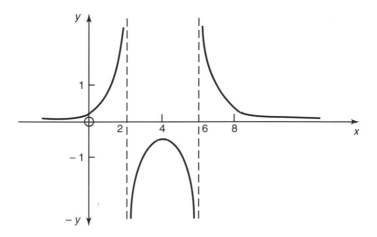

Fig. 9.46

As $x \to 2$ from values less than 2, $y \to +\infty$, and as $x \to 2$ from values greater than 2, $y \to -\infty$.
As $x \to 6$ from values less than 6, $y \to -\infty$, and as $x \to 6$ from values greater than 6, $y \to +\infty$.
If $2 < x < 6$ then y is always negative; therefore any maximum point will be below the x-axis. By differentiating it can be shown that the maximum point is at $(4, -\frac{1}{4})$.
Also as $x \to \pm\infty$, $y \to 1/+\infty \to 0$ so the x-axis is an asymptote.
When $x = 0$, $y = \frac{1}{12}$.

$$y = \frac{1}{(x-2)(x-6}$$

(e) Symmetry

There is one other property which can be useful in curve sketching and that is the property of symmetry.

If x can be replaced by $-x$, without changing the equation, then the curve is symmetrical about the y-axis. $y = x^2$ is such a function, since $(-x)^2 = x^2$. The sketch of this function is shown in Fig. 9.47(a).

If y can be replaced by $-y$ without changing the equation then the curve is symmetrical about the x-axis. $y^2 = x$ is such a function. A sketch of this function is shown in Fig. 9.47(b).

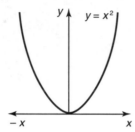

Fig. 9.47(a)

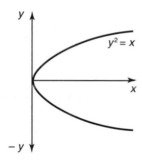

Fig. 9.47(b)

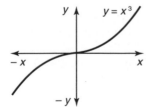

Fig. 9.47(c)

If y can be replaced by $-y$ and x replaced by $-x$, without changing the equation then the function is **skew symmetrical**, and the curve is symmetrical about the origin. $y = x^3$ is such a function. A sketch is shown in Fig. 9.47(c).

Example 9.24 Sketch the curve $y^2 = 4 + 2x$.

(i) When $x = 0$, $y^2 = 4$ so that $y = +2$ or -2. This determines two points on the curve immediately.
(ii) As $x \to \infty$, $y \to \pm\infty$.
(iii) Differentiating as an implicit function:

$$2y \frac{dy}{dx} = 2$$

Thus dy/dx cannot be zero for any finite value of x so that there are no turning points.
(iv) There are no asymptotes.
(v) y can be replaced by $-y$ without changing the equation, so that the curve is symmetrical about the x-axis.

Finally, because of y^2 the curve cannot exist if the right-hand side is negative, because the square root of a negative quantity does not exist. Therefore the curve will not exist in the region where $x < -2$. The curve is sketched in Fig. 9.48

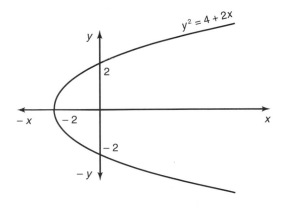

Fig. 9.48

Example 9.25 Sketch the curve $y = \dfrac{2x}{x - 4}$.

(i) When $x = 0$, $y = 0$.

(ii)   When $x \to \pm\infty$, $x - 4$ is almost equal to x so that the curve approximates to

$$y = \frac{2x}{x} = 2$$

Therefore, $y = 2$ is an asymptote.

(iii) Differentiating:

$$\frac{dy}{dx} = \frac{(x-4).2 - 2x}{(x-4)^2}$$

$$= \frac{-8}{(x-4)^2}$$

The differential coefficient cannot be zero for any finite value of x. Therefore, the graph will not have any turning points.

(iv)   The denominator becomes zero when $x = 4$, which is therefore an asymptote.

(v)   There is no symmetry about either axis.

The sketch is shown in Fig. 9.49.

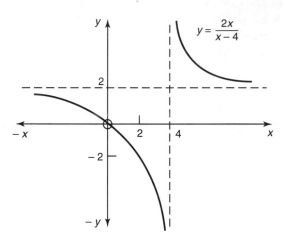

Fig. 9.49

EXERCISE 9.12

Sketch the following curves:

1. $y^2 = 6x - 16$ 2. $y = x^2 - 9$

3. $y = \dfrac{1}{2x - 5}$ 4. $y = \dfrac{1}{(x+1)(x-1)}$

5. $\quad y = \dfrac{1}{x} + 2x$ **6.** $\quad y = 4x + \dfrac{1}{x-1}$

7. $\quad y = \dfrac{1}{x^2}$ **8.** $\quad y = \dfrac{1}{x^2} + 8x$

9. $\quad y = \dfrac{x}{(x+1)(x-1)}$ **10.** $\quad y = \dfrac{1}{x^3}$

11. $\quad y = -x^3 + 8x^2 - 19x + 12$ **12.** $\quad xy = 10$

MISCELLANEOUS EXERCISES 9

1. Plot the curve $y = x + 4e^{-x}$ for values of x between 1 and 2 and hence give the value of x for which y is a minimum.

2. Plot on the same axes, the two functions

(a) $\quad y = \dfrac{1}{x}$

(b) $\quad y = 1 + \dfrac{1}{2-x}$

for values of x, at intervals of 0.2, between 0 and 4. Hence, give two solutions of the equation

$$1 + \frac{1}{2-x} = \frac{1}{x}$$

3. Sketch the curves, showing salient values:
(a) $\quad y = (x-1)^2$
(b) $\quad y^2 = 1 - x$
(c) $\quad y^2 = 4 - x^2$

4. Express in polar coordinates the points represented by (4, 3) and (−3, 4). Given the lines joining each of these points to the origin represent forces acting outwards from the origin, express in polar coordinates the direction and magnitude of the equilibrant.

5. Find the equation of the line joining the points $A(-2, 6)$ and $B(3, -4)$. Show that the curve $y = x^2 - 8x + 11$ passes through B and that the line AB is a tangent to the curve at this point.

6. (a) Mark on a sketch the points $A(4, 3)$ and $B(-3, -1)$. Find the equation of the line AB. Find the polar coordinates of A.
(b) Plot the graph of $y = x^2 + 4x$ and use it to solve the equation $x^2 + 4x + 3 = 0$.

7. Find the equation of the chord joining the points on the curve $y = 10x - x^2$, whose x-coordinates are 2 and 4 respectively.

8. Plot the graphs of $y = x^3$ and $y = 5x - 4$ for values of x from -3 to $+3$. Hence solve the equation

$$x^3 - 5x + 4 = 0.$$

9. Plot the graph of $y = x^3 + 2x^2 - x - 2$ for values of x between -3 and $+2$ in $\frac{1}{2}$ unit increments.
    Use the graph to solve the following equations:
    (a) $x^3 + 2x^2 - x - 2 = 0$
    (b) $x^3 + 2x^2 - x = 0$
    (c) $x^3 + 2x^2 - x - 2 = x - 1$

10. In a test on a rectifier the results shown in Table 9.23 were obtained when measuring the variation of current I (amperes) with the applied voltage V (volts).

Table 9.23

I	0.039	0.048	0.061	0.075	0.090
V	6.5	7.6	9.1	10.6	12

By drawing a suitable graph show that I and V are related by a law of the form

$$I = mV + cV^2$$

and find the appropriate values of the constants m and c.

11. The variables s and t are related by the law $s = at^2 - 5t + b$. Verify graphically that the experimental figures shown in Table 9.24 substantiate this statement and obtain values for the unknown constants a and b.

Table 9.24

t	1	2	3	4	5	6
s	3.9	1.5	0.8	1.9	4.8	9.3

Hence, or otherwise, estimate the value of t at which s has its minimum value.

12. The mass m mg of a radioactive substance remaining at times t min from the first observation are given in Table 9.25.

Table 9.25

Time (t min)	10	20	30	40	50	60	70
Mass remaining (m mg)	181	163	148	134	121	110	99

By plotting ln m against t verify that the radioactive decay satisfies the relationship $m = Me^{-kt}$, where M and k are constants, and from your graph determine their approximate values. What meaning can be given to the value of M?

Assuming that the law holds for all values of t and m calculate the time taken, to the nearest minute, for the initial mass to decrease to 10 mg.

13. The mass M which can safely be distributed uniformly on a girder of span L varies as in Table 9.26.

Table 9.26

M (kg)	184	140	103	68	40
L (metres)	1.6	2.0	2.5	3.5	4.0

Plot a graph of ML against L^2 and show that this is a straight line. Explain why this verifies the relationship

$$M = aL + \frac{b}{L}$$

and find the constants a and b.

14. To drill a hole of diameter d mm with a given torque, the maximum permitted feed F mm/rev of a machine is given by Table 9.27.

Table 9.27

d	8	12	16	24	28
F	1.472	0.576	0.294 4	0.115 2	0.08

Show by drawing a suitable graph that F and d are connected by a law of the form $Fd^n = c$, where n and c are constants. Find the best values of n and c. Find the maximum feed for a hole of diameter 20 mm.

15. Table 9.28 shows the deflection d (mm) of a beam under given loads L (N).

Table 9.28

d	7.58	10.8	14.5	18.6	23.0	27.7
L	20	25	30	35	40	45

The equation connecting L and d is of the form $L = Kd^n$, where K and n are constants. Verify this by drawing the appropriate straight line graph and determine the values of K and n. Use the graph to find the deflection produced by a load of 38 N.

16. The points $P(0, 2)$ and $Q(3, 2)$ lie on the curve $y = x^2 - 3x + 2$. Find the equation of the normal to the curve at each of the points P and Q.

Find the coordinates of R, the point where the normals meet.

The curve cuts the x-axis at the points A and B, A being the point nearer to the origin. Obtain the coordinates of A and B, and sketch of the curve.

17. The tangents to the parabola $y^2 = x$ at the variable points $P(p^2, p)$ and $Q(q^2, q)$ intersect at N. Given that M is the mid-point of PQ, prove that

$$MN = \tfrac{1}{2}(p - q)^2.$$

18. $ABCD$ is a rectangle in which the coordinates of A and C are $(0, 4)$ and $(11, 1)$ respectively and the gradient of the side AB is -5.
    (a) Find the equations of the sides AB and BC.
    (b) Show that the coordinates of B are $(1, -1)$.
    (c) Calculate the area of the rectangle.

19. A point P moves in such a way that its distance from the point $(1, 2)$ is equal to its distance from the line $2y = 3$. Find
    (a) the equation of the locus of P,
    (b) the equation of the tangent to the locus at the point where $x = 3$.
    Draw a graph of the locus.

20. Prove that the equation of the tangent at the point where $x = h$ on the curve $y = 2x^2 + 18$ is $y = 4hx + 18 - 2h^2$. Hence, or otherwise, find the equations of the two tangents to the curve which pass through the origin, and the coordinates at which the tangents touch the curve.

21. The closed curve $4x^2 + y^2 = 100$ is symmetrical about the x-axis and about the y-axis. It passes through the points whose coordinates are $(a, 0)$, $(0, b)$, $(c, 6)$, $(d, 8)$, where a, b, c, d are positive integers. Find the values of a, b, c, d and sketch the curve.

22. A point P moves in such a way that

$$AP^2 - BP^2 = 5$$

where A and B are the points $(1, 2)$ and $(4, 3)$ respectively. Show that the locus of P is a straight line.

23. A point $P(x, y)$ moves in such a way that

$$\frac{AP}{BP} = \frac{3}{2}$$

where A and B have coordinates $(-4, -2)$ and $(1, 3)$ respectively.
   (a) Find the equation of the locus of P and show that it is a circle.
   (b) Find the coordinates of its centre and show that its radius is $6\sqrt{2}$.

Chapter 10

Series

10.1 Sequences and series

A sequence of numbers is a set of numbers which follow a pattern or rule. Examples of sequences are:

(a) 1, 3, 5, 7, ..., (b) 1, 4, 9, 16, ..., (c) 16, 8, 4, 2, ...

Any number in a sequence can be predicted from the previous one once the rule has been determined.

Each number in a sequence is called a **term**. When the terms of a sequence are linked together algebraically it is called a **series**. Examples of series are:

(a) $1 + 4 + 7 + 10 + \cdots$.
(b) $16 - 8 + 4 - 2 \cdots$.

A series can be **finite** or **infinite** depending upon whether it terminates after a definite number of terms or not.

10.2 Arithmetic progression (AP)

A series of numbers is said to be in arithmetic progression if the difference between consecutive terms is constant. This difference is called the **common difference**. Examples of APs are

(i) $3 + 7 + 11 + 15 + \cdots$ common difference $= 7 - 3 = 15 - 11 = 4$
(ii) $27 + 25 + 23 + 21 + \cdots$ common difference $= 25 - 27 = 23 - 25 = -2$

In general terms

$a =$ first term of an AP
$d =$ common difference
$n =$ number of terms
$S_n =$ sum of n terms

Therefore the sum of n terms using these definitions is

$$S_n = a + [a + d] + [a + 2d] + [a + 3d] + \cdots + \underset{n\text{th term}}{[a + (n - 1)d]}$$

Writing this sum in reverse

$$S_n = [a + (n-1)d] + [a + (n-2)d] + \cdots + a$$

Adding both expressions together

$$2S_n = [2a + (n-1)d] + [2a + (n-1)d] + \cdots + [2a + (n-1)d]$$

The right-hand side is the sum of n equal terms

$$2S_n = n \times [2a + (n-1)d]$$
$$S_n = \frac{n}{2}[2a + (n-1)d]$$

This may be written as

$$S_n = \frac{n}{2}[a + a + (n-1)d]$$
$$= \frac{n}{2}[\text{first term} + \text{last term}]$$

Example 10.1   Find the tenth term and the sum of the series $1 + 4 + 7 + 10 + 13 + \cdots$ to 10 terms.

For this series $a = 1$, $d = 3$, $n = 10$

tenth term $= a + (10 - 1) \times 3 = 28$

$$S_{10} = \frac{10}{2}[2 \times 1 + 9 \times 3] = 145$$

Example 10.2   The fourth term of an AP is 21 and the fourteenth term is -9. Find the first term and the common difference.

The nth term $= a + (n-1)d$

Therefore

fourth term $= a + 3d = 21$ (i)

fourteenth term $= a + 13d = -9$ (ii)

Subtracting (i) from (ii):

$10d = -30$

$d = -3$

Substituting into equation (i):

$a = 30$

EXERCISE 10.1

1. The third term of an AP is 28 and the eighth term is 53. Find the first term and the common difference.
2. An AP has a first term of 16 and a common difference of −2. Find the value of the twentieth term and the sum of the first 20 terms.
3. How many terms must be in the series $10 + 14 + 18 + 22 + \cdots$ to have a sum greater than 2170?
4. A machine is to operate at six possible speeds, ranging from 600 rev/min to 1800 rev/min. If the speeds form an AP calculate the other speeds.

10.3   Geometric progression (GP)

A series of numbers is said to be in geometrical progression if the ratio of each term of the series to its preceding term is a constant value. This ratio is known as the **common ratio**. Examples of GP are

(a) $9 + 18 + 36 + 72 + \cdots$ Common ratio $= \dfrac{18}{9} = \dfrac{36}{18} = 2$

(b) $30 - 15 + 7.5 - 3.75 + \cdots$ Common ratio $= \dfrac{-15}{30} = \dfrac{7.5}{-15} = -\frac{1}{2}$

In general terms

$$a = \text{first term of a GP}$$
$$r = \text{common ratio}$$
$$n = \text{number of terms}$$
$$S_n = \text{sum of } n \text{ terms}$$

then

$$S_n = a + ar + ar^2 + \cdots + ar^{n-1} \qquad \text{(i)}$$
$$\text{(nth term)}$$

Multiply (i) by r

$$rS_n = ar + ar^2 + \cdots + ar^{n-1} + ar^n \qquad \text{(ii)}$$

Subtract (ii) from (1)

$$S_n - rS_n = a - ar^n$$

Therefore

$$S_n = a\frac{(1 - r^n)}{1 - r}$$

Example 10.3 Find the sum of eight terms and the eighth term the series $2 - 4 + 8 - 16 + \cdots$.

From the series it is seen that $a = 2$, $n = 8$, $r = \dfrac{-4}{2} = -2$.

Sum of eight terms $\quad S_8 = a\dfrac{(1 - r^8)}{1 - r}$

$$= 2\dfrac{(1 - [-2]^8)}{1 - [-2]}$$

$$= \tfrac{2}{3}(1 - 256)$$

$$= -170$$

Eighth term $\quad = ar^7$

$$= 2(-2)^7$$

$$= -256$$

Example 10.4 Given that the sixth term of a GP is 160 and the third term is 20, find the tenth term.

Sixth term:	$ar^5 = 160$	
Third term:	$ar^2 = 20$	
By division	$r^3 = 8$	$r = 2$
Now	$ar^2 = 20$	
Therefore	$a2^2 = 20$	$a = 5$
Tenth term:	$ar^9 = 5 \times 2^9$	
	$\quad = 2560$	

Example 10.5 The value of a machine depreciates each year by 12% of its value at the commencement of that year. If its original value was £5000, determine

(a) its value after five years;
(b) how long it takes before the value falls below £2000.

At the beginning of the first year, machine value = £5000.
At the beginning of the second year, machine value = 5000 × 0.88.
At the beginning of the third year machine value = 5000 × 0.88².
Therefore the values form a GP with $a = 5000$, $r = 0.88$.

(a) Value after five years is the 6th term of the GP.

$$= ar^5$$

$$= 5000 \times 0.88^5$$

$$= £2639$$

(b) Let the machine take n years to depreciate to £2000. This is the $(n+1)$th term of the GP.

Therefore

$$2000 = ar^n = 5000 \times 0.88^n$$

$$0.88^n = 0.4$$

Take logarithms of both sides:

$$n \log 0.88 = \log 0.4$$

$$n = \frac{\log 0.4}{\log 0.88}$$

$$= \frac{-0.397\,9}{-0.055\,5}$$

$$= 7.2 \text{ years, i.e. 8 years.}$$

EXERCISE 10.2

1. In each of the following progressions determine the term and sum of terms requested.
    (a) $1 - 2 + 4 - 8 + \cdots$ 8th term, sum of 10 terms.
    (b) $1 + 3 + 9 \cdots$ 9th term, sum of 7 terms.
    (c) $2 + 1\frac{1}{3} + \frac{8}{9} + \frac{16}{27} \cdots$ 10th term, sum of 5 terms.
    (d) $x + x^2 + x^3 + \cdots$ 11th term, sum of 18 terms.
2. The first term of a GP is 3 and the sixth term is 96. Find the common ratio, the tenth term and the sum of the first 10 terms.
3. The mass of a radioactive pellet decreases each year by 8% of its mass at the start of that year. Find the time taken, in complete years, before the pellet loses one half of its mass.
4. If a machine is to have 8 speeds ranging from 100 to 800 rev/min in geometrical progression, determine these speeds to the nearest revolution.
5. A spring vibrates 30 mm in the first oscillation and subsequently 90% of its previous value in each succeeding oscillation. How many oscillations will it take before the vibration falls below 10 mm and how far will it have vibrated in this time?
6. The third and seventh term of a GP are 486 and 6 respectively. What are the first six terms of this series?

7. Given that £100 is invested at compound interest of 5% per annum determine
    (a)   the capital value after 17 years,
    (b)   the time it takes to reach £200.

10.4   Infinite series

Consider the sum of the terms of the following infinite series:

(a) $1 + 2 + 3 + 4 + 5 + \cdots$;

(b) $1 + \frac{1}{2} - \frac{1}{2} + \frac{1}{2} - \frac{1}{2} + \cdots$;

(c) $1 - \frac{1}{4} + \frac{1}{16} - \frac{1}{64} + \cdots$;

(d) $1 + \frac{1}{2} + \frac{1}{3} + \frac{1}{4} + \cdots$.

For series (a), the sum of an infinite number of terms is obviously infinitely large, that is $S_\infty - \infty$. For series (b), the sum will always fluctuate between 1 and $1\frac{1}{2}$. For series (c), the sum of the first n terms, S_n, will increase as the number of terms increases, but it will be found that this sum will approach a limiting value of $\frac{4}{5}$; that is,

$$S = \lim_{n \to \infty} S_n = \frac{4}{5}$$

For series (d), it appears that S_n approaches a limiting value, but in fact $S_\infty = \infty$. If the sum of a series approaches a limiting value as $n \to \infty$, the series is said be **convergent**. If the sum does not approach a limiting value the series is said to be **divergent**. In the above series only (c) is convergent.

The terms of a convergent series must get progressively smaller but this is no proof of convergency, as was seen in series (d). There are various tests for proving convergency including the following ratio test. Consider the series $u_1 + u_2 + u_3 + \cdots + u_n + \cdots$. If

$$\lim_{n \to \infty} \left| \frac{u_{n+1}}{u_n} \right| < 1 \quad \text{series is convergent}$$

$$> 1 \quad \text{series is divergent}$$

$$= 1 \quad \text{further tests required}$$

The lines $|\ |$ signify that the signs of the terms are ignored. For example $|x| < 1$ means $-1 < x < 1$ and $|x|$ is referred to as the **modulus** of x.

Example 10.6   Check the following series for convergency:

(a) $1 - \dfrac{1}{2!} + \dfrac{1}{3!} - \dfrac{1}{4!} + \cdots$ (where $n! = 1 \times 2 \times \cdots \times n$)

(b) $1 + 2x + 3x^2 + 4x^3 + \cdots$

(a)   using the nth term $\qquad |u_n| \quad = \left| \dfrac{1}{n!} \right|$

using the $(n+1)$th term $|u_{n+1}| = \left| \dfrac{1}{(n+1)!} \right|$

Therefore $\qquad\qquad \left| \dfrac{u_{n+1}}{u_n} \right| = \left| \dfrac{1}{(n+1)!} \middle/ \dfrac{1}{n!} \right| = \left| \dfrac{1}{n+1} \right|$

Therefore $\displaystyle\lim_{n \to \infty} \left| \dfrac{u_{n+1}}{u_n} \right| < 1$

Therefore the series is convergent.

(b) $|u_{n+1}| = |(n+1)x^n|$

$\quad |u_n| = |nx^{n-1}|$

Therefore $\qquad\qquad \left| \dfrac{u_{n+1}}{u_n} \right| = \left| \dfrac{(n+1)x^n}{nx^{n-1}} \right| = \left| \dfrac{(n+1)}{n}x \right|$

Therefore $\displaystyle\lim_{n \to \infty} \left| \dfrac{u_{n+1}}{u_n} \right| = \lim_{n \to \infty} \left| \left(1 + \dfrac{1}{n}\right)x \right|$

$$= |x| \quad \text{since} \quad \dfrac{1}{\infty} = 0$$

Therefore the series is convergent if $|x| < 1$

and divergent if $|x| > 1$

In addition if $x = 1$ the series is obviously divergent. It is recommended that the student checks the convergency of series (a) to (d) above and the other series in this chapter.

(a) Sum of an infinite GP

An infinite GP will be convergent if $|r| < 1$. This can be proved using the ratio test or from the formula

$$S_n = a\frac{(1 - r^n)}{1 - r}$$

If $|r| < 1$ then as $n \to \infty$ $r^n \to 0$. For example: $(\tfrac{1}{3})^2 = \tfrac{1}{9}$, $(\tfrac{1}{3})^6 = \tfrac{1}{729}$, etc.

Therefore $\qquad S_\infty = \displaystyle\lim_{n \to \infty} S^n$

$$= \frac{a}{1 - r}$$

Example 10.7 Find the sum of an infinite number of terms of the series $16 + 8 + 4 + 2 + 1 + 0.5 \ldots$

The series is a GP with $r = \frac{8}{16} = 0.5$.

Sum to infinity $S_\infty = \dfrac{a}{1 - r} = \dfrac{16}{1 - 0.5} = 32.$

EXERCISE 10.3

1. Determine the sum to infinity of
   (a) $1 + \frac{1}{4} + \frac{1}{16} + \cdots$ (b) $0.2 + 0.02 + 0.002 + \cdots$
   (c) $1 - \frac{1}{5} + \frac{1}{25} \cdots$ (d) $9 - 3 + 1 \cdots$
   (e) $1 - \frac{1}{2} + \frac{1}{4} - \frac{1}{8} + \cdots$

2. The sum of an infinite GP is 15. Given that the common ratio is $\frac{1}{3}$, find the first term.

3. A contract is signed giving a person £1000 profit in the first year. The profit falls by 8% of the previous year's profit for each subsequent year. What is the total possible profit?

4. The spring in Question 5, Exercise 10.2 is allowed to vibrate until it eventually stops. Find the total distance moved by the spring before stopping.

5. Express the recurring decimals (a) $0.333\ldots$ (b) $0.353\,5\ldots$ as geometrical progressions and hence evaluate them as vulgar fractions.

10.5 Power series

A series containing terms in ascending powers of x is called a power series. An example of such a series is $16 + 7x + 8x^2 + 9x^3 + \cdots$.

(a) Maclaurin's theorem

This states that a function $f(x)$ can be represented in the form of a power series provided that, at $x = 0$,

1. the function is finite;
2. the derivatives are finite;
3. the series obtained is convergent.

If $f(0)$ is the value of the function $f(x)$ at $x = 0$ and $f'(0), f''(0)$, etc., are the values of the first, second, etc., derivatives at $x = 0$, then, by Maclaurin's theorem,

$$f(x) = f(0) + xf'(0) + \frac{x^2}{2!}f''(0) + \frac{x^3}{3!}f'''(0) + \cdots .$$

(Note that $2! = 1 \times 2$, $3! = 1 \times 2 \times 3$, $4! = 1 \times 2 \times 3 \times 4$, etc.) Maclaurin's expansion will now be used to obtain series for various functions.

If functions can be expressed as convergent series then these series can be programmed into calculators or computers to provide a means by which a machine can calculate values such as sin 0.2, $e^{1.3}$, ln 6.2 etc. Evaluating functions manually using the power series is tedious but the examples are included to show how the series are used.

(i) Sine series

$$f(x) = \sin x \qquad\qquad \text{therefore} \quad f(0) = \sin 0 = 0$$
$$f'(x) = \cos x \qquad\qquad\qquad\quad f'(0) = \cos 0 = 1$$
$$f''(x) = -\sin x \qquad\qquad\qquad\quad f''(0) \qquad\quad = 0$$
$$f'''(x) = -\cos x, \text{ etc.} \qquad\qquad f'''(0) \qquad\quad = -1$$

Therefore

$$\sin x = 0 + 1.x + \frac{0.x^2}{2!} - \frac{1.x^3}{3!} + \cdots$$

$$= x - \frac{x^3}{3!} + \frac{x^5}{5!} - \frac{x^7}{7!} + \cdots \qquad\qquad (10.1)$$

This series is convergent for all values of x.
Note: The angle x must be measured in radians.

Example 10.8 Use the sine series to find the value of sin 1 rad correct to four decimal places.

$$\sin 1 = 1 - \frac{1^3}{3!} + \frac{1^5}{5!} - \frac{1^7}{7!} + \cdots$$

$$= 1 - 0.166\,667 + 0.008\,333 - 0.000\,198 + \cdots$$

$$= 0.841\,468 \ldots$$

$$= 0.841\,5 \text{ correct to four decimal places.}$$

(ii) Cosine series

This can be obtained either by using Maclaurin's theorem or by differentiation of the sine series.

$$\frac{d}{dx}(\sin x) = 1 - \frac{3x^2}{3!} + \frac{5x^4}{5!} - \frac{7x^6}{7!}$$

Therefore

$$\cos x = 1 - \frac{x^2}{2!} + \frac{x^4}{4!} - \frac{x^6}{6!} + \cdots \qquad (10.2)$$

This series is convergent for all values of x, x being measured in radians.

Example 10.9   Evaluate cos 0.6 rad, correct to four significant figures, by means of the cosine series.

$$\cos 0.6 = 1 - \frac{0.6^2}{2!} + \frac{0.6^4}{4!} - \frac{0.6^6}{6!} + \cdots$$

$$= 1 - 0.180\,000 + 0.005\,400 - 0.000\,065 + \cdots$$

$$= 0.825\,335...$$

$$= 0.825\,3 \text{ correct to four significant figures.}$$

EXERCISE 10.4

1.   Use the appropriate series to evaluate sin 0.5 rad and cos 0.2 rad, correct to four decimal places.
2.   Express $\cos 2x$, $\sin \frac{x}{2}$ and $\tan ax$ as power series in ascending powers of x as far as the 4th term.
3.   Expand
     (a) $(1 - 2x^2) \sin 3x$ as a power series to x^5 term
     (b) $(1 + x) \sin x^2$ as a power series to x^{10} term.

(iii) Logarithmic series

$$f(x) = \ln x \qquad \text{therefore} \qquad f(0) = \ln 0 = -\infty$$

$$f'(x) = \frac{1}{x} \qquad\qquad\qquad f'(0) = \frac{1}{0} = \infty$$

$$f''(x) = -\frac{1}{x^2} \qquad\qquad\qquad f''(0) = -\frac{1}{0} = -\infty$$

The function and the derivatives at $x = 0$ are not finite, therefore $\ln x$ cannot be expressed as a power series. However, consider

$$f(x) = \ln(1 + x) \qquad \text{therefore} \qquad f(0) = \ln 1 = 0$$

$$f'(x) = \frac{1}{1 + x} \qquad\qquad\qquad f'(0) = \frac{1}{1} = 1$$

$$f''(x) = -\frac{1}{(1 + x)^2} \qquad\qquad\qquad f''(0) = \frac{-1}{1} = -1$$

Therefore

$$\ln(1 + x) = x - \frac{x^2}{2!} + \frac{2x^3}{3!} - \frac{6x^4}{4!} + \cdots$$

$$= x - \frac{x^2}{2} + \frac{x^3}{3} - \frac{x^4}{4} + \cdots$$

A series can be obtained for $\ln(1 - x)$ by substituting $-x$ for x in the above series

$$\ln(1 - x) = -x - \frac{x^2}{2} - \frac{x^3}{3} - \frac{x^4}{4} + \cdots$$

Both these series are convergent if x is in the range $-1 < x < 1$ or $|x| < 1$.

In their present form these series are of limited use because they can only be used to evaluate logarithms of numbers less than 2.

In addition, unless x is very small, these series will converge extremely slowly and many terms of the series must be evaluated even for relatively few figures accuracy.

These series will now be combined to produce other series without these two limitations.

$$\ln\left(\frac{1 + x}{1 - x}\right) = \ln(1 + x) - \ln(1 - x)$$

$$= \left(x - \frac{x^2}{2} + \frac{x^3}{3} - \frac{x^4}{4} + \cdots\right) - \left(-x - \frac{x^2}{2} - \frac{x^3}{3} - \frac{x^4}{4} \cdots\right)$$

$$\ln\left(\frac{1 + x}{1 - x}\right) = 2\left(x + \frac{x^3}{3} + \frac{x^5}{5} + \cdots\right)$$

This series is convergent if $|x| < 1$. Let

$$\frac{1 + x}{1 - x} = \frac{y + 1}{y}$$

therefore by rearrangement

$$x = \frac{1}{2y + 1}$$

Substitution in the above series gives

$$\ln\left(\frac{y + 1}{y}\right) = 2\left[\frac{1}{(2y + 1)} + \frac{1}{3(2y + 1)^3} + \frac{1}{5(2y + 1)^5} + \cdots\right]$$

This series is convergent if $y > 0$.

The following example illustrates how these series can be used.

Example 10.10   Calculate (a) ln 2, (b) ln 3, correct to three decimal places.

(a)   ln 2, using the first series let $\dfrac{1+x}{1-x} = 2$. Therefore $x = \tfrac{1}{3}$.

$$\ln\left(\frac{1+x}{1-x}\right) = 2\left(x + \frac{x^3}{3} + \frac{x^5}{5} + \cdots\right)$$

$$\ln 2 = 2[\tfrac{1}{3} + \tfrac{1}{3}(\tfrac{1}{3})^3 + \tfrac{1}{5}(\tfrac{1}{3})^5 + \cdots]$$

$$= 2[0.333\,33 + 0.012\,34 + 0.000\,82 + 0.000\,07 + \cdots]$$

$$\simeq 2(0.346\,56)$$

$$= 0.693$$

(b)   ln 3, using the second series let $y = 2$.

$$\ln(y+1) = \ln y + 2\left[\frac{1}{(2y+1)} + \frac{1}{3(2y+1)^3} + \frac{1}{5(2y+1)^5} + \cdots\right]$$

if $y = 2$,

$$\ln 3 = \ln 2 + 2\left[\frac{1}{5} + \frac{1}{3 \times 125} + \frac{1}{5 \times 3\,125} + \cdots\right]$$

$$= 0.693\,12 + 2(0.2 + 0.002\,67 + 0.000\,06 + \cdots)$$

$$\simeq 1.098\,58$$

$$= 1.099$$

Example 10.11   Express $\sin x \ln(1 + 2x)$ as a power series in ascending powers of x as far as the term containing x^4. For what values of x is this series valid (i.e. convergent)?

Comparison of $\ln(1 + x)$ with $\ln(1 + 2x)$ shows that $2x$ must be substituted for x in the logarithmic series; that is, in

$$\ln(1 + x) = x - \frac{x^2}{2} + \frac{x^3}{3} - \frac{x^4}{4}$$

$$\ln(1 + 2x) = 2x - \frac{(2x)^2}{2} + \frac{(2x)^3}{3} - \frac{(2x)^4}{4}$$

$$= 2x - 2x^2 + \frac{8x^3}{3} - 4x^4$$

Therefore

$$\sin x \ln(1 + 2x) = \left(x - \frac{x^3}{6}\right)\left(2x - 2x^2 + \frac{8x^3}{3} - 4x^4\right)$$

$$= 2x^2 - 2x^3 - \frac{x^4}{3} + \frac{8x^4}{3} \cdots$$

$$= 2x^2 - 2x^3 + \frac{7x^4}{3} \cdots$$

The series for $\ln(1 + x)$ is valid for $|x| < 1$. Therefore the series for $\ln(1 + 2x)$ is valid for $|2x| < 1$ or $|x| < \frac{1}{2}$. The sine series is valid for all values of x. The combined series for $\sin x \ln(1 + 2x)$ is valid for the smallest validity range, namely $|x| < \frac{1}{2}$.

EXERCISE 10.5

1. Calculate ln 1.1 and ln 0.9 using the series for $\ln(1 + x)$ and $\ln(1 - x)$, giving the values correct to four decimal places.
2. Show, by using the series for

    $$\ln(1 + x) \quad \text{and} \quad \ln\left(\frac{1 + x}{1 - x}\right)$$

    to evaluate ln 1.5 correct to five decimal places, that the latter series is a better one to use.
3. Given that $\ln 5 = 1.609\,437$ use a suitable logarithmic series to evaluate ln 6 and ln 7 correct to four decimal places.
4. Expand the following expressions as power series as far as the fourth term in ascending powers of x. State the range of validity values for each series:
    (a) $\ln(1 - 3x)$
    (b) $\ln 3(1 - x)$
    (c) $\ln(1 - x)^2$
    (d) $\ln\left(\frac{1 - 3x}{1 + x}\right)$
    (e) $\cos x \ln(1 + x)$
    (f) $\ln(x^2 + 5x + 4)$
    (g) $\ln \sqrt{(1 + 5x)}$

10.6   Binomial series

Binomial expressions (that is, those containing two separate terms) of the type $(a + x)^2$, $(a + x)^4$ etc. can be expanded by multiplication:

$$(a + x) \qquad\qquad\qquad = \qquad\qquad a + x$$
$$(a + x)^2 = (a + x)(a + x) \qquad = \qquad a^2 + 2ax + x^2$$
$$(a + x)^3 = (a + x)(a + x)(a + x) \qquad = \qquad a^3 + 3a^2x + 3ax^2 + x^3$$
$$(a + x)^4 = (a + x)(a + x)(a + x)(a + x) = a^4 + 4a^3x + 6a^2x^2 + 4ax^3 + x^4$$

If the coefficients are examined it will be seen that they form a predictable pattern (Fig 10.1). This pattern is referred to as **Pascal's triangle**.

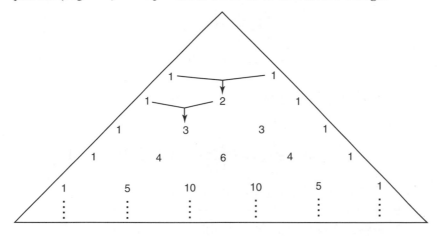

Fig. 10.1

Apart from the unity coefficients, any coefficient can be obtained by adding the two adjacent coefficients immediately above in the previous row (see Fig. 10.1).

This triangle can be used to expand any binomial expression of the type $(a + x)^n$ where n is a positive integer. This result is known as the **binomial theorem** and it states that

$$(a + x)^n = a^n + na^{n-1}x + \frac{n(n-1)}{2!}a^{n-2}x^2$$
$$+ \frac{n(n-1)(n-2)}{3!}a^{n-3}x^3 + \cdots + x^n$$

When n is a positive integer this series is finite and terminates after $(n + 1)$ terms.

Example 10.12 Expand completely (a) $(1 + 2x)^4$, (b) $(1 - x)^5$.

Each expression must be compared with the binomial theorem. In both cases $a = 1$ and the theorem is now read as

$$(1 + x)^n = 1 + nx + \frac{n(n-1)}{2!}x^2 + \frac{n(n-1)(n-2)}{3!}x^3 + \cdots + x^n$$

(a) $(1 + [2x])^4 = 1 + 4[2x] + \frac{4 \cdot 3}{1 \cdot 2}[2x]^2 + \frac{4 \cdot 3 \cdot 2}{1 \cdot 2 \cdot 3}[2x]^3 + [2x]^4$

$\qquad = 1 + 8x + 6[4x^2] + 4[8x^3] + [16x^4]$

$\qquad = 1 + 8x + 24x^2 + 32x^3 + 16x^4$

(b) $(1 + [-x])^5 = 1 + 5[-x] + \dfrac{5.4}{1.2}[-x]^2 + \dfrac{5.4.3}{1.2.3}[-x]^3$

$\qquad\qquad + \dfrac{5.4.3.2}{1.2.3.4}[-x]^4 + [-x]^5$

$\qquad = 1 - 5x + 10x^2 - 10x^3 + 5x^4 - x^5$

Example 10.13   Expand $(2x + y)^4$ using the binomial theorem.

$(2x + y)^4 = ([2x] + [y])^4$

$\qquad = [2x]^4 + 4[2x]^3 y + \dfrac{4.3}{1.2}[2x]^2 y^2 + \dfrac{4.3.2}{1.2.3}[2x]y^3 + y^4$

$\qquad = 16x^4 + 32x^3 y + 24x^2 y^2 + 8xy^3 + y^4$

Example 10.14   What is the tenth term in the expansion of $(1 - \frac{1}{2}x)^{15}$?

$(1 - \frac{1}{2}x)^{15} = (1 + [-\frac{1}{2}x])^{15}$

Examination of the terms of the binomial series shows that the tenth term of the expansion is

$$\dfrac{n(n-1)(n-2)\ldots(n-8)}{1.2.3\ldots9}x^9$$

Therefore the tenth term of the expansion

$$\dfrac{15.14.13.12.11.10.9.8.7}{1.2.3.4.5.6.7.8.9}[-\tfrac{1}{2}x]^9$$

$$= \dfrac{-5005}{412}x^9$$

(a) Binomial series for any index

By considering $f(x) = (a + x)^n$, Maclaurin's theorem can be used to prove the binomial theorem in the form

$$(a + x)^n = a^n + na^{n-1}x + \dfrac{n(n-1)}{1.2}a^{n-2}x^2 + \dfrac{n(n-1)(n-2)}{1.2.3}a^{n-3}x^3 + \cdots$$

It is therefore true for any value of n.

It was seen that when n was a positive integer the series terminated after $(n + 1)$ terms with the x^n term. However, if n is fractional or negative the series will be infinite. The expansion will be valid (convergent) if $|x| < a$.

Example 10.15 Use the binomial theorem to express $\dfrac{1}{(1-x^3)^2}$ as a power series in ascending powers of x as far as the term containing x^9. For what values of x is this series valid?

The expression must first be rearranged:

$$\frac{1}{(1-x^3)^2} = (1-x^3)^{-2} = (1+[-x^3])^{-2}$$

Using the binomial theorem when $a = 1$,

$$(1+x)^n = 1 + nx + \frac{n(n-1)}{1.2}x^2 + \frac{n(n-1)(n-2)}{1.2.3}x^3 + \cdots$$

$$(1+[-x^3])^{-2} = 1 + (-2)[-x^3] + \frac{(-2)(-3)}{1.2}[-x^3]^2$$

$$+ \frac{(-2)(-3)(-4)}{1.2.3}[-x^3]^3 + \cdots$$

$$= 1 + 2x^3 + 3x^6 + 4x^9 + \cdots$$

This expansion is valid, that is, convergent, provided

$$|x^3| < 1 \quad \text{i.e.} \quad |x| < 1$$

Example 10.16 If x^3 and higher powers of x can be neglected, express

$$\frac{\sqrt[3]{(1+\tfrac{1}{2}x)}}{\sqrt{(1-2x)}}$$

as a power series. Give the range of values of x for which this series is convergent.

$$\frac{\sqrt[3]{(1+\tfrac{1}{2}x)}}{\sqrt{(1-2x)}} = \frac{(1+\tfrac{1}{2}x)^{\frac{1}{3}}}{(1-2x)^{\frac{1}{2}}} = (1+\tfrac{1}{2}x)^{\frac{1}{3}}(1-2x)^{-\frac{1}{2}}$$

$$= (1+[\tfrac{1}{2}x])^{\frac{1}{3}}(1+[-2x])^{-\frac{1}{2}}$$

Each binomial expression is now expanded separately:

$$(1+[\tfrac{1}{2}x])^{\frac{1}{3}} = 1 + \tfrac{1}{3}[\tfrac{1}{2}x] + \frac{\tfrac{1}{3}(\tfrac{1}{3}-1)}{1.2}[\tfrac{1}{2}x]^2 + \cdots$$

$$= 1 + \tfrac{1}{6}x + \tfrac{1}{3}\left(\frac{-2}{3}\right) \cdot \tfrac{1}{2} \cdot \tfrac{1}{4}x^2 + \cdots$$

$$= 1 + \tfrac{1}{6}x - \tfrac{1}{36}x^2 + \cdots$$

This is valid for $|\frac{1}{2}x| < 1$, i.e. $|x| < 2$

$$(1 + [-2x])^{-\frac{1}{2}} = 1 + (-\frac{1}{2})[-2x] + \frac{(-\frac{1}{2})(-\frac{1}{2} - 1)}{1.2}[-2x]^2 + \cdots$$

$$= 1 + x + (-\frac{1}{2})\left(\frac{-3}{2}\right)\frac{1}{2}.4x^2 + \cdots$$

$$= 1 + x + \frac{3}{2}x^2 + \cdots$$

This is valid for $|2x| < 1$, i.e. $|x| < \frac{1}{2}$

$$\frac{\sqrt[3]{(1 + \frac{1}{2}x)}}{\sqrt{(1 - 2x)}} = (1 + \frac{1}{6}x - \frac{1}{36}x^2 + \cdots)(1 + x + \frac{3}{2}x^2 + \cdots)$$

$$= 1 + \frac{1}{6}x + x + \frac{1}{6}x^2 - \frac{1}{36}x^2 + \frac{3}{2}x^2 + \cdots)$$

$$= 1 + \frac{7}{6}x + \frac{59}{36}x^2$$

This series is convergent if $-\frac{1}{2} < x < \frac{1}{2}$.

EXERCISE 10.6

1. Expand completely using the binomial theorem:
   (a) $(1 + x)^7$ (b) $(1 - x)^4$ (c) $(1 - x^2)^6$ (d) $(1 - \frac{1}{2}x)^4$.

2. Express as power series, in ascending powers of x, as far as the fourth term:

   (a) $(1 + x)^{-3}$ (b) $(1 - 2x)^{-5}$ (c) $\dfrac{1}{1 - 3x}$ (d) $(1 + x)^{\frac{2}{3}}$

   (e) $\dfrac{1}{(1 + 3x)^2}$ (f) $\sqrt[3]{(1 + x^2)}$ (g) $\dfrac{1}{\sqrt{(1 + 2x)}}$ (h) $\left(1 - \dfrac{1}{x}\right)^{-4}$.

   State the validity values of x for each series.

3. Expand completely   (a) $(2 - 3x)^5$,   (b) $(3a + 2b)^4$.

4. Use the binomial theorem to expand the following to the fourth term in ascending powers of x:

   (a) $\sqrt{(4 - x)}$ (b) $\dfrac{1}{\sqrt{(4 + x)}}$ (c) $\dfrac{1}{2 + 5x^2}$ (d) $\left(x + \dfrac{1}{x}\right)^{\frac{1}{3}}$.

   For what values of x are these series convergent?

5. (a) Find the sixth term in the expansion of $(2a + 1/a^2)^{12}$
   (b) Find the middle term in the expansion of $(3x^2 - y^2/3)^{16}$

6. Expand $\sqrt{(1 + 2x + x^2)}$ using the binomial theorem, as far as the x^3 term.

7. Expand by the binomial theorem, in ascending powers of x,
   (a) $(1 + 2x)^{\frac{1}{2}}$, (b) $(1 - x)^{-\frac{1}{2}}$ as far as the term in x^2. State, in each case the range of values of x for which the expansion is valid. Given that x

is so small that terms containing x^3 and higher powers may be neglected show that

$$\sqrt{\left(\frac{1+2x}{1-x}\right)} = 1 + \frac{3x}{2} + \frac{3x^2}{8}$$

8.   Express the following as power series in ascending powers or x as far as the x^2 term. Give the validity values for each series:

(a) $\dfrac{1+x}{(1-3x)}$ (b) $\dfrac{(1+2x)^2}{(1-3x)^2}$ (c) $\sqrt{\left(\dfrac{1+x}{1-x}\right)}$

(d) $\dfrac{(1+x)(1-3x)^{\frac{2}{3}}}{(1+x^2)^2}$

10.7   Applications of the binomial theorem

Binomial expansions are useful for numerical approximations, for calculations with small variations and errors and also in probability theory (see Chapter 13).

(a) Approximations

Powers and roots of numbers can be obtained to any desired degree of accuracy by arranging the calculation in the form $(1+x)^n$ where $|x| < 1$.

The accuracy will depend upon the number of terms of the binomial series that are evaluated. The smaller the x term the quicker the series will converge, that is less terms will be required for a given accuracy.

Example 10.17   Evaluate $(1.002)^4$, correct to five decimal places.

$(1.002)^4 = (1 + [0.002])^4$ i.e. $x = 0.002$

$$= 1 + 4[0.002] + \frac{4.3}{1.2}[0.002]^2 + \cdots$$

$$= 1 + 0.008 + 0.000\,024 + \cdots$$

$$= 1.008\,02$$

Example 10.18   Find the approximate value of $1/0.98$.

$$\frac{1}{0.98} = 0.98^{-1} = (1 + [-0.02])^{-1}$$

$$= 1 + (-1)[-0.02] + \frac{(-1)(-2)}{1.2}[-0.02]^2$$

Since x is very small the third term need not be used to find an approximate value.

$$= 1 + 0.02$$

$$= 1.02$$

(b) Small variations and errors

The next two examples show how easy it is to use the binomal theorem to assess the approximate effect of small numerical changes on a calculation.

Example 10.19   The time t_p of one complete oscillation of a simple pendulum of length L is given by $t_p = 2\pi\sqrt{\left(\dfrac{L}{g}\right)}$. If the length is decreased by 2% what is the percentage change in t_p?

$$\text{New length} = L - \frac{2}{100}L = L(1 - 0.02)$$

$$\text{Therefore new time} = 2\pi\sqrt{\left[\frac{L(1 - 0.02)}{g}\right]}$$

$$= 2\pi\sqrt{\left(\frac{L}{g}\right)}\sqrt{(1 - 0.02)}$$

$$= t_p(1 - 0.02)^{\frac{1}{2}}$$

$$= t_p(1 - \tfrac{1}{2} \times 0.02\ldots)$$

$$= t_p(1 - 0.01)$$

that is, the time of one oscillation will decrease by 1% approximately.

Example 10.20   What is the error involved in calculating the volume of a cone if the radius of the base is measured 1% too large and the height is measured 1.5% too small?

If correct radius $= r$ then measured radius $= r(1 + 0.01)$.
If correct height $= h$ then measured height $= h(1 - 0.015)$.

$$\text{Correct volume} = V = \tfrac{1}{3}\pi r^2 h$$

$$\text{Therefore calculated volume} = V_1 = \tfrac{1}{3}\pi r^2(1 + 0.01)^2 h(1 - 0.015)$$

$$\approx \tfrac{1}{3}\pi r^2 h(1 + 2 \times 0.01)(1 - 0.015)$$

$$= V(1 + 0.02)(1 - 0.015)$$

$$= V(1 + 0.02 - 0.015)$$

$$= V(1 + 0.005)$$

$$\text{Therefore error in volume} \qquad = V_1 - V$$

$$= 0.005V$$

that is, the calculated volume is 0.5% too large.

The above approach has also been used in Chapter 2 to deal with the accuracy of numerical calculations.

EXERCISE 10.7

1. Use the binomial theorem to find the approximate values of

$$1.02^4, \quad 0.996^8, \quad 1.005^{\frac{1}{2}}, \quad \frac{1}{1.003}, \quad \frac{1}{0.998}, \quad \frac{1}{0.995^3}$$

2. Find the value of the following:

    (a) $\dfrac{1}{\sqrt{1.009}}$ (five decimal places);

    (b) $\dfrac{\sqrt{0.993}}{\sqrt[3]{1.004}}$ (four decimal places).

3. Given that an error of 1% is made when measuring the diameter of a sphere, determine the approximate errors in the calculated surface area and volume of the sphere.

4. Use the binomial theorem to evaluate

    (a) $\sqrt{35}$, correct to five decimal places,

    (b) $\dfrac{1}{\sqrt{50}}$, correct to four significant figures,

    (c) $\sqrt[4]{82}$, correct to four decimal places.

5. The natural frequency f of a vibrating shaft is given by

    $$f = \frac{1}{2\pi} \sqrt{\left(\frac{\text{stiffness}}{\text{inertia}}\right)}$$

    Calculate the percentage error in determining f if the stiffness is measured 2% too small and the inertia is measured 1% too large.

6. The maximum bending stress σ in a cantilever is given by $6WL/BD^2$. Find the percentage increase in σ if W is increased 0.5%, B is reduced 1% and D is reduced 1.5%.

7. Evaluate $(3.02)^{10}$ to the nearest integer using the binomial theorem.

8. The period of oscillation T of a bar magnet in the Earth's magnetic field is given by

    $$T = 2\pi \sqrt{\left(\frac{I}{MH}\right)}$$

    Find the approximate error in T if I is recorded 1% high, H 4% high and M 2% low.

10.8 Exponential series

Consider the function $f(x)$ such that

$$f(x) = 1 + x + \frac{x^2}{2!} + \frac{x^3}{3!} + \cdots \tag{i}$$

This infinite series is covergent for all values of x. Substituting z for x gives

$$f(z) = 1 + z + \frac{z^2}{2!} + \frac{z^3}{3!} + \cdots$$

Therefore

$$f(x)f(z) = \left(1 + x + \frac{x^2}{2!} + \cdots\right)\left(1 + z + \frac{z^2}{2!} + \cdots\right)$$

$$= 1 + (x + z) + \frac{(x + z)^2}{2!} + \cdots$$

If $z = -x$, then $f(x)f(-x) = 1$.

This can only be true if $f(x) = a^x$, since $a^x a^{-x} = 1$ where the base a is not yet determined.

Equation (i) now reads

$$a^x = 1 + x + \frac{x^2}{2!} + \frac{x^3}{3!} \cdots$$

If $x = 1$,

$$a = 1 + 1 + \frac{1}{2!} + \frac{1}{3!} + \cdots$$

This series is the one that defines the exponential constant e; that is, $a = $ e.

Therefore $f(x) = a^x$

$$= e^x = 1 + x + \frac{x^2}{2!} + \frac{x^3}{3!} + \cdots$$

Example 10.21   Use the exponential series to evaluate $e^{0.1}$ and $e^{-0.1}$ correct to five significant figures.

$$e^{0.1} = 1 + 0.1 + \frac{0.1^2}{2} + \frac{0.1^3}{6} + \frac{0.1^4}{24} + \cdots$$

$$= 1.1 + 0.005 + 0.000\,167 + 0.000\,004$$

$$= 1.105\,171$$

$$= 1.105\,2$$

$$e^{-0.1} = 1 - 0.1 + 0.005 - 0.000\,167 + 0.000\,004$$

$$= 0.904\,837$$

$$= 0.904\,84$$

Example 10.22 Express $e^x(1+2x)^2$ as a power series in ascending powers of x as far as the x^3 term.

$$e^x(1+2x)^2 = \left(1 + x + \frac{x^2}{2} + \frac{x^3}{6} + \cdots\right)(1 + 4x + 4x^2)$$

$$= 1 + (1+4)x + (\tfrac{1}{2}+4+4)x^2 + (\tfrac{1}{6}+2+4)x^3 + \cdots$$

$$= 1 + 5x + \tfrac{17}{2}x^2 + \tfrac{37}{6}x^3 + \cdots$$

EXERCISE 10.8

1. Use the series for e^x to evaluate $e^{0.09}$ and $e^{-0.1}$ correct to five significant figures.
2. Use the exponential series to express (a) $(1 + 3x + x^2)e^x$ and (b) $(1 - 2x)e^{-x}$ as a power series in ascending powers of x with four terms.
3. What is the expansion of $e^x \sin x$ in ascending powers of x as far as the x^4 term?
4. Express the following as power series:
   (a) $\tfrac{1}{2}(e^x - e^{-x})$ (b) $\tfrac{1}{2}(e^x + e^{-x})$ (c) e^{-x^2}
   (d) $e^{2x}\ln(1-x)$ (c) $e^{-x}\ln(1+x)$

MISCELLANEOUS EXERCISE 10

1. The values of five resistors are in geometric progression. Given that the least is 100 ohm and the greatest is 220 ohm, find the values of the three intermediate resistors, each correct to two significant figures.
2. The initial cost of a machine was £8000. It is expected that this cost will increase yearly by 15% of the previous year's value. Calculate the expected cost of the machine after six years, correct to the nearest £100.
3. Due to improving conditions, it is expected that the annual cost of combating damage by atmospheric pollution will form the geometric series £600 + £540 + £486 + · · ·. Find, to the nearest £ the expected expenditure over the first seven years. What would be the total expenditure if these conditions were to continue indefinitely?
4. (a) The first term of a geometric progression is 2 and the fifth term is 162. Find the common ratio, the seventh term and the sum of the first six terms.
   (b) Using the binomial theorem, find the value of $\frac{1}{\sqrt{9.0300}}$ correct to four decimal places.

5. Show that if x^3 and higher powers of x can be neglected

$$\frac{\sqrt{(1+x)}}{(1-x)^2} = 1 + \tfrac{5}{2}x + \tfrac{31}{8}x^2$$

Use this expression to evaluate $\sqrt{1.02}/(0.98)^2$ correct to four decimal places.

6. (a) Using the binomial theorem, expand $\dfrac{1}{(3+2x)^2}$ as a series in ascending powers of x as far as the term in x^3. State for what values of x the expansion is valid.

   (b) Show that

$$\sin\left(\frac{\pi}{6}+x\right) = \tfrac{1}{2}\cos x + \frac{\sqrt{3}}{2}\sin x$$

   Using the series for $\sin x$ and $\cos x$, show that

$$\sin\left(\frac{\pi}{6}+x\right) = \tfrac{1}{2} + \frac{\sqrt{3}}{2}x - \tfrac{1}{4}x^2 - \frac{\sqrt{3}}{12}x^3 + \cdots$$

7. Expand $(1-x^2)^{-\frac{1}{2}}$ as far as the term in x^4, stating the range of values of x for which the expansion is valid. Deduce the expansion of $e^x/\sqrt{(1-x^2)}$ as far as the term in x^4.

8. Give, with each term in its simplest form, the first six terms of the binomial expansion $(2+\tfrac{1}{2}x)^8$. Hence obtain the value of $(2.05)^8$, correct to three decimal places.

9. (a) Write down the first four terms of the expansion of $\sqrt{(1+x)}$ in ascending powers of x. Hence calculate $\sqrt{(4.064)}$ to five places of decimals.

   (b) The equation $y = WL^3/D^4$ refers to the deflection y of a metal rod. Using the binomial theorem, show that the deflection y increases by approximately 18.45% when L increases by 3% and D decreases by 2%. (*Note*: In this part the binomial expansions should include three terms.)

10. (a) Expand $(x+x^2)^5$, using the binomial theorem.

    (b) Determine, using the binomial theorem, the approximate percentage change in the volume of a sphere, radius R, if the radius is increased by 2%.

    (c) If x is so small that x^3 and higher powers can be neglected, show that

$$\frac{1}{\sqrt[3]{(1-x)}} - \sqrt[5]{(1+x)} = \frac{x^2}{5}$$

    (d) Use a binomial approximation to evaluate $\sqrt[3]{(1.06)}$ correct to two decimal places.

11. Two bank accounts are opened and a principal sum of £100 deposited in each. One account, A, pays 7% simple interest and the other account, B, pays $5\frac{1}{2}$% compound interest. Show that the total sum of principal plus interest after 10 years is greater for account B than for account A.

12. The periodic time T of a simple pendulum of length L is given by

$$T = 2\pi\sqrt{\left(\frac{L}{g}\right)}$$

If the percentage errors in L and g are 1% high and 0.5% low respectively, determine the percentage error in T correct to two decimal places.

13. The formula for the specific heat of a gas is

$$\left\{1 + \frac{c_1}{p}\right\}^{\gamma-1} = \left\{1 + \frac{c_2}{p}\right\}^{\gamma}$$

If $p \gg c_1$ or c_2, so that the third and all subsequent terms in the expansion of both sides may be neglected, derive an equation relating $\dfrac{c_1}{c_2}$ to γ. ($\gg$ means 'very much greater than').

Chapter 11

Numerical methods

11.1 Introduction

Many numerical methods are available for the solution of mathematical problems that arise in technology. These methods, which are basically very simple but repetitive, can involve long and tedious numerical work.

Electronic calculators and computers can process large quantities of numerical data accurately and rapidly. These machines are well suited for numerical analysis techniques.

Numerical methods may also be the only methods of solution if an analytical approach is impossible, as for many complicated integrals and equations.

11.2 Solution of equations with one variable

It was seen in Chapter 9 that equations of the type $f(x) = 0$ can be solved graphically by plotting the curve $y = f(x)$. Solutions will be located at the point or points at which the curve cuts the x-axis.

The accuracy of these solutions is limited by the scale of the graph. A greater accuracy can be attained by plotting the curves on a larger scale in the vicinity of the roots.

A method which enables the roots to be calculated to any desired accuracy is **Newton's method**.

(a) Newton's method

This states that if x_1 is an approximate value for a real root of the equation $f(x) = 0$, then a closer approximation x_2 may be obtained for this root using the formula

$$x_2 = x_1 - \frac{f(x_1)}{f'(x_1)}$$

where $f'(x_1)$ is the gradient of the curve at the point where $x = x_1$.

The formula can be verified in the following way. Figure 11.1 shows the portion of the graph $y = f(x)$ in the region of a real root $x = a$ of the equation $f(x) = 0$.

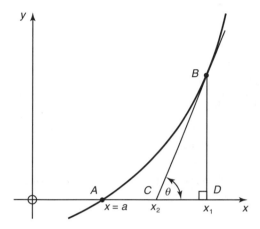

Fig. 11.1

Gradient of the tangent $BC = \tan \theta = f'(x_1)$. In triangle BCD,

$$CD = \frac{BD}{\tan \theta}$$

$$= \frac{f(x_1)}{f'(x_1)}$$

where $f(x_1)$ is the y ordinate at $x = x_1$. But

$$OC = OD - CD$$

that is

$$x_2 = x_1 - \frac{f(x_1)}{f'(x_1)}$$

This formula can be used repeatedly to find successively closer approximations, x_3, x_4, etc., to the root $x = a$. This is represented graphically in Fig. 11.2.

Newton's formula can generally be used for any polynomial or transcendental (i.e. non-polynomial) function. First approximations can be obtained by drawing the graph or examining the function. The first approximation x_1 should be chosen so that the tangent to the curve at this point cuts the x-axis at a point nearer to the real root, thus giving a more accurate second approximation x_2. In general, the closer the first

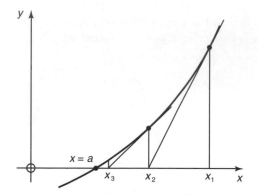

Fig. 11.2

approximation is to the real root the sooner the series formed from the terms $x_1, x_2, x_3 \ldots$ will converge towards the root.

Example 11.1   The equation $x^3 - 3x - 3 = 0$ has one real root. Use Newton's method to determine its value correct to four significant figures.

Let

$$f(x) = x^3 - 3x - 3$$

At $x = 2$,

$$f(2) = 2^3 - 3(2) - 3 = -1$$

At $x = 3$,

$$f(3) = 3^3 - 3(3) - 3 = 15$$

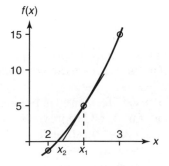

Fig. 11.3

Since $f(2)$ and $f(3)$ are of opposite sign the root must be between $x = 2$ and $x = 3$ (see Fig. 11.3).

At $x = 2.5$, $f(2.5) = 5.125$

From Fig. 11.3 $x_1 = 2.5$ is a suitable first approximation since the tangent at this point gives a better second approximation x_2. Since

$$f(x) = x^3 - 3x - 3$$

then

$$f'(x) = 3x^2 - 3$$

Substituting these in Newton's formula gives

$$x_2 = x_1 - \frac{x_1^3 - 3x_1 - 3}{3x_1^2 - 3}$$

$$= \frac{x_1(3x_1^2 - 3) - (x_1^3 - 3x_1 - 3)}{3x_1^2 - 3}$$

$$= \frac{2x_1^3 + 3}{3x_1^2 - 3}$$

Take $x_1 = 2.5$ as a first approximation.

Second approximation $= \dfrac{2(2.5)^3 + 3}{3(2.5)^2 - 3} = \dfrac{34.25}{15.75} = 2.1746$

Third approximation $= \dfrac{2(2.1746)^3 + 3}{3(2.1746)^2 - 3} = \dfrac{23.567}{11.187} = 2.1068$

Fourth approximation $= \dfrac{2(2.1067)^3 + 3}{3(2.1067)^2 - 3} = \dfrac{21.703}{10.316} = 2.1038$

Fifth approximation $= \dfrac{2(2.1038)^3 + 3}{3(2.1038)^2 - 3} = \dfrac{21.623}{10.278} = 2.1038$

The root is 2.104 (correct to four significant figures). It will be noticed that the successive approximations after the first are calculated to an increased number of significant figures until the final desired accuracy is attained.

(b) Iteration

A process such as Newton's method, which repeatedly uses the same formula to obtain successively closer approximations to a solution, is known as an **iterative** process. Such a process can never produce an exact solution but the process can be continued to obtain an approximate answer to any desired degree of accuracy.

Iterative methods are well suited to machine calculations. Once the iterative formula has been established the machine can perform the repetitive calculations very quickly. In addition a computer can be programmed (i.e. instructed) to terminate the program automatically when the desired accuracy has been attained.

Figure 11.4 shows the algorithm as a flow chart to carry out the necessary calculations for determining the root in Example 11.1. The iterative formula for this example, in general terms, is

$$x_{r+1} = \frac{2x_r^3 + 3}{3x_r^2 - 3}$$

where $r = 1, 2, 3 \ldots$

x_r and x_{r+1} are the rth and $(r+1)$th approximations respectively.

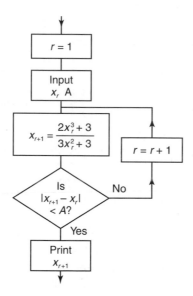

Fig. 11.4

Note: A is an accuracy value. For example, if $A = 0.001$ the answer will be accurate to 3 decimal places or, in this case, to 4 significant figures.

11.3   Evaluation of N^k

Numbers such as $2.4^{1.3}$, 0.82^{-5}, $7.9^{0.5}$, $1/2.3$ may be evaluated by machine using an iterative process such Newton's method, described earlier in the chapter. This can be done by arranging the number in the form $f(x) = 0$, as follows.

Let

$$x = N^k$$

Therefore

$$x^{1/k} = N \qquad \text{(that is, taking the kth root on either side)}$$

$$x^{1/k} - N = 0$$

Let

$$f(x) = x^{1/k} - N$$

Differentiating $f'(x) = \dfrac{1}{k} x^{(1/k) - 1}$

Therefore using Newton's formula

$$x_2 = x_1 - \frac{x_1^{1/k} - N}{\dfrac{1}{k} x_1^{(1/k) - 1}}$$

A suitable flow chart is shown in Fig. 11.5.

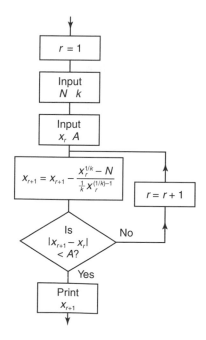

Fig. 11.5

Example 11.2 Evaluate

(a) $\sqrt[3]{19}$ correct to 5 decimal places

(b) $\dfrac{1}{2.8}$ correct to 6 decimal places

(a) $\sqrt[3]{19} = 19^{1/3}$; therefore $N = 19$, $k = \frac{1}{3}$ that is, $1/k = 3$.
Let the first approximation be $x_1 = 2.5$.

$$x_2 = 2.5 - \frac{2.5^3 - 19}{3 \times 2.5^{3-1}} \qquad\qquad = 2.68$$

$$x_3 = 2.68 - \frac{2.68^3 - 19}{3 \times 2.68^2} \qquad = 2.6684518$$

$$x_4 = 2.6684518 - \frac{2.6684518^3 - 19}{3 \times 2.6684518^2} = 2.6684016$$

$$x_5 = 2.6684016 - \frac{2.6684016^3 - 19}{3 \times 2.6684016^2} = 2.6684016$$

$\sqrt[3]{19} = 2.66840$ correct to 5 decimal places

(b) $\dfrac{1}{2.8} = 2.8^{-1}$ therefore $N = 2.8$, $k = -1$

Let the first approximation be $x_1 = 0.3$. Then

$$x_2 = 0.3 - \frac{0.3^{(1/-1)} - 2.8}{\left(\dfrac{1}{-1}\right).0.3^{(1/-1)-1}} = 0.3 - \frac{0.3^{-1} - 2.8}{-0.3^{-2}} = 0.348$$

$$x_3 = 0.348 - \frac{0.348^{-1} - 2.8}{-0.348^{-2}} \qquad = 0.356\,908\,8$$

Continuing

$x_4 = 0.357\,142\,7$

$x_5 = 0.357\,142\,9$

$x_6 = 0.357\,142\,9$

$\dfrac{1}{2.8} = 0.357\,143$ correct to 6 decimal places.

Note: The iterative process requires a first approximation to start it going.

EXERCISE 11.1

Flow charts should be drawn when answering these questions to gain further practice in their usage.

1. Use Newton's method to solve the equation $x^3 - 2x - 3 = 0$ correct to four decimal places.
2. Use Newton's method to determine the root of the equation $e^x + x = 0$ which lies in the region -0.5 to -0.7. Give your answer correct to three decimal places.
3. The solution of the equation $3 \ln x = x$ lies in the range 4 to 5. Find this solution correct to three decimal places.
4. Use Newton's method to solve the following equations to the accuracy given:
    (a) $\cos x = 1.5 \tan x$ (four decimal places, in range $0°$ to $90°$)
    (b) $x^4 + x = 16$ (four significant figures)
    (c) $2 \sin x + 2x = 1$ (three decimal places in range 0 to $\pi/2$).
5. Use an iterative process to evaluate the following roots to the accuracy indicated:
    (a) $\sqrt{1462}$ (4 sig. figs)   (b) $\sqrt{2.33}$ (5 sig. figs)
    (c) $\sqrt[3]{0.3247}$ (5 sig. figs)   (d) $\sqrt[4]{846}$ (4 dec. places)
    (e) $\sqrt[5]{9794.8}$ (5 dec. places).
6. By iteration determine the reciprocals of
    (a) 3.49   correct to three decimal places;
    (b) 139   correct to six significant figures;
    (c) 0.002 134   correct to four significant figures.

11.4   Linear simultaneous equations

The usual method of solving these equations by successive elimination of the unknowns was discussed in Chapter 1.

In Chapter 16 a matrix method is utilised which is suitable for machine solution.

Two further methods, also suitable for machine solution, are discussed below.

These methods can be used for any set of n linear equations containing n unknowns. For simplicity, however, the examples have been chosen with three unknowns.

(a) Pivotal elimination method

This is an elimination method modified for machine computation. Errors due to rounding off are minimised if the term with the largest coefficient is always chosen for elimination. Such a coefficient is termed a **pivot**.

The pivotal equation (i.e. that containing the pivot) is *divided* throughout by the pivot. This reduces the tendency to build up large coefficients when solving a large set of equations. The method is illustrated in Example 11.3 below. In the three equations the largest coefficient (i.e. the pivot) is 7.6 Therefore x is the first unknown to be eliminated. Equations (6) and (9) in Table 11.1 are then obtained containing two unknowns. The largest coefficient now is $-6.143\,8$. Therefore z is the next unknown to be eliminated.

Example 11.3 Solve the simultaneous equations

$7.6x + 1.2y - 1.4z = 10$

$1.9x + 1.0y - 3.1z = 8.2$

$0.82x - 2.4y + 3.7z = 12$

Give your answer correct to two decimal places.

The procedure is shown in table 11.1.
From equ (11) in the table

$0.307\,5y = -3.752\,2$

$y = -12.202$

Substitute in equ (10):

$-0.254\,5y + z = -2.072\,7$

$$z = -2.0727 + 0.254\,5 \times -12.202$$

$$= -5.178$$

Substitute in equ (1):

$7.6x + 1.2y - 1.4z = 10$

$$x = \frac{1}{7.6}(10 + 1.4z - 1.2y)$$

$$= \frac{1}{7.6}(10 + 1.4x - 5.178 - 1.2x - 12.202)$$

$$= 2.288\,5$$

Therefore, the solutions correct to two decimal places are

$x = 2.29, \qquad y = -12.20, \qquad z = -5.18$

(b) Iterative method

This method is not suitable for all sets of linear simultaneous equations because successive approximations do not always converge to the required

Table 11.1

Operation	Coefficient of			Constant	Current summation	Equation
	x	*y*	*z*			
In (1), (2), (3) **pivot = 7.6**	7.6	1.2	−1.4	10	17.4	(1)
	1.9	1.0	−3.1	8.2	8.0	(2)
	0.82	−2.4	3.7	12	14.12	(3)
(1) ÷ 7.6	1	0.157 9	−0.184 2	1.315 8	2.289 5	(4)
(2) ÷ 1.9	1	0.526 3	−1.631 6	4.315 8	4.210 5	(5)
Eliminate *x* **in (1) and (2)** (4) − (5)		−0.368 4	1.447 4	−3.000 0	−1.921 0	(6)
(2) ÷ 1.9	1	0.526 3	−1.631 6	4.315 8	4.2105	(5)
(3) ÷ 0.82	1	−2.926 8	4.512 2	14.634 1	17.219 5	(7)
Eliminate *x* **in (2) and (3)** (5) − (7)		3.453 1	−6.143 8	−10.318 3	−13.009 0	(8)
In (6) and (8) **pivot = 6.143 8** (8) ÷ 6.143 8		0.562 0	−1	−1.679 5	−2.117 4	(9)
(6) ÷ 1.447 4		−0.254 5	1	−2.072 7	−1.327 2	(10)
Eliminate *z* **in (6) and (8)** (9) + (10)		0.307 5		−3.752 2	−3.444 6	(11)

solution. An iterative method will always produce convergence when the set of equations obey the following conditions:

1. the largest coefficients are on the diagonal drawn through the equations (as in example 11.4);
2. in each equation the largest coefficient exceeds the sum of the other coefficients.

The iteration relationships are so arranged that each subsequent new value of the variable calculated makes use of the best available approximation for the other two.

Example 11.4   Find the simultaneous solutions (correct to two decimal places) for the following equations:

$$5x + y + z = 8$$
$$x + 10y + z = 10$$
$$x + y + 15z = 40$$

Note that conditions (1) and (2) do apply.
The equations are rearranged, using the largest coefficient in each equation, to obtain the iterative relationships:

$$z_r = \tfrac{1}{15}(40 - x_r - y_r)$$
$$x_{r+1} = \tfrac{1}{5}(8 - y_r - z_r)$$
$$y_{r+1} = \tfrac{1}{10}(10 - x_{r+1} - z_r)$$

For three unknowns x, y and z, the first approximations can be $x_1 = 0$, $y_1 = 0$, with z then determined using the first equation.

First approximations

$$\text{if} \quad x_1 = 0$$
$$\text{and} \quad y_1 = 0$$
$$\text{then} \quad z_1 = \tfrac{1}{15}(40 - 0 - 0) \qquad = 2.67$$

Second approximations

$$x_2 = \tfrac{1}{5}(8 - 0 - 2.67) \qquad = 1.066$$
$$y_2 = \tfrac{1}{10}(10 - 1.066 - 2.67) \quad = 0.626$$
$$z_2 = \tfrac{1}{15}(40 - 1.066 - 0.626) = 2.554$$

The table shows the successive approximations.

x	0	1.066	0.964	0.958 6	0.958 5
y	0	0.626	0.648	0.648 2	0.648 2
z	2.67	2.554	2.559 2	2.559 5	2.559 5

The required solutions are $x = 0.96$, $y = 0.65$, $z = 2.60$. The flow chart for this example is shown in Fig. 11.6.

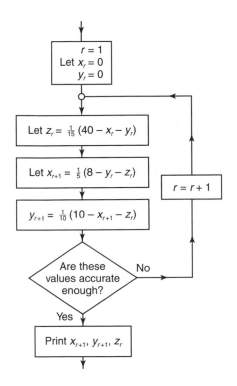

Fig. 11.6

EXERCISE 11.2

Solve the following simultaneous equations by the method indicated and to the accuracy specified:

1. $6x + 2y - z = -5$ elimination
   $2x - 3y + 4z = -4$ two significant figures
   $3x + 2y + 2z = 4$

2. $4.63x + 3.82y - 3.41z = 3.87$ elimination
   $3.01x - 2.16y - 3.16z = 0.642$ three significant figures
   $2.96x - 3.93y + 3.27z = 1.28$

3. $9x + y + z = 18$ iteration
    $2x + 18y + z = 50$ two decimal places
    $x + y + 25z = 82$
4. $2x + y - 2z = 2.17$ elimination
    $4x - y + 3z = 7.15$ three decimal places
    $x + y - 6z = -3.82$
5. $x - y - 4z = -8$ iteration
    $x - 8y + 2z = -5$ four decimal places.
    $22x - 2y + 3z = -20$

11.5   Difference tables

It is very often necessary to compile a set of values of a function at regular intervals of the variable. A function that is extensively used in numerical analysis is the polynomial function. The reasons for this are (1) it is relatively easy to find a polynomial to correlate closely with most sets of numerical values and (2) it is an easy function to work with. The evaluation of a known polynomial is more easily done by first rearranging the polynomial. Consider the evaluation of a polynomial

$$f(x) = 2x^3 + 6x^2 - 4x + 5$$

in intervals of 1 unit from $x = 0$ to $x = 10$.

Since repeated values of the polynomial are required $f(x)$ can be grouped or 'nested' as shown below to make the numerical work easier:

$$f(x) = x[2x^2 + 6x - 4] + 5$$
$$= x[2x(x + 3) - 4] + 5$$

The value of the polynomial for various values of x are calculated as follows: For $x = 2$

$$f(x) = 2[2 \times 2(2 + 3) - 4] + 5$$
$$= 2[2 \times 2 \times 5 - 4] + 5$$
$$= 2[16] + 5$$
$$= 37$$

Nesting provides a smoother calculation process.

The difference table shown as Table 11.2 lists values of $f(x)$ calculated in this manner, and the following differences:

(a)   1st differences: differences between successive values of $f(x)$.
(b)   2nd differences: differences between successive 1st differences.
(c)   3rd differences: differences between successive 2nd differences.

Table 11.2

x	f(x)	First differences	Second differences	Third differences
0	5			
1	9	4		
2	37	28	24	
3	101	64	36	12
4	213	112	48	12
5	385	172	60	12
6	629	244	72	12
7	957	328	84	12
8	1 381	424	96	12
9	1 913	532	108	12
10	2 565	652	120	12

It is seen that with a polynomial of degree 3 the third differences are constant. For a polynomial of degree n the nth differences are constant.

If values of an unknown function $f(x)$ are known at regular intervals of x, the difference tables can be used to:

1. find the best polynomial expression for $f(x)$. With experimental values the final differences are unlikely to be constant but become nearly constant;
2. **extrapolate** further values of $f(x)$, i.e. to find values of $f(x)$ outside the given range for x;
3. **interpolate** further values of $f(x)$, i.e. to find values of $f(x)$ at values of x in between the given regular intervals.

Note: With experimental results extrapolated and interpolated values must be treated with caution. There is no guarantee that the form of the function is always the same. This is especially true with extrapolation.

Example 11.5 Stress concentration factors $f(x)$ measured at distances x metres from one edge of an aeroplane wing are given in Table 11.3

Table 11.3

x	0.5	0.6	0.7	0.8	0.9	1.0
f(x)	3.75	3.88	4.07	4.32	4.63	5.0

Determine

(a) the form of the function $f(x)$;

(b)   the likely stress factor $f(1.2)$;
(c)   the likely stress factor at $x = 0.65$ m.

From the results the difference table can be completed down to the broken line (Table 11.4).

Table 11.4

x	$f(x)$	First difference	Second difference
0.5	3.75		
0.6	3.88	0.13	
0.7	4.07	0.19	0.06
0.8	4.32	0.25	0.06
0.9	4.63	0.31	0.06
1.0	5.00	0.37	0.06
1.1	5.43	0.43	0.06
1.2	5.92	0.49	0.06

(a)   Since the second differences are constant then $f(x)$ is of degree 2, i.e. $f(x) = ax^2 + bx + c$ where a, b and c are constants. From the difference table,

when $x = 0.5$ $f(x) = 3.75$

therefore $f(x) = a(0.5)^2 + b(0.5) + c = 3.75$

$$0.25a \ + 0.5b \ + c = 3.75 \qquad (1)$$

when $x = 0.8$ $0.64a \ + 0.8b \ + c = 4.32 \qquad (2)$

when $x = 1$ $a \ + \ b \ + c = 5.00 \qquad (3)$

Solving these three simultaneous equations gives $a = 3$, $b = -2$, $c = 4$.

Therefore $f(x) = 3x^2 - 2x + 4$

(b) $f(1.2)$ may be extrapolated either by completing the difference table below the broken line or by substituting $x = 1.2$ in $3x^2 - 2x + 4$. The difference table is completed by continuing the second difference column down and then working back through the columns to the left.

$$f(1.2) = 5.92$$

(c) $f(0.65) = 3(0.65)^2 - 2(0.65) + 4$
$$= 3.97$$

EXERCISE 11.3

1. Evaluate the following polynomials using the nesting technique:
   (a) $6x^2 + 2x - 10$: $(x = 2.9)$ (b) $3x^3 - x^2 + 5x - 16$: $(x = 1.5)$.
2. Construct difference tables for the following:
   (a) $2x^3 - 3x^2 + 4x + 6$ for $x = 0$ to $x = 6$ in unit intervals.
   (b) $x^4 - 2x^3 - x + 9$ for $x = 0$ to $x = 5$ in unit intervals.
   In both cases use your difference table to evaluate $f(7)$ and check your answers by using the polynomials.
3. In each case find the polynomial to fit the results:

(a)

x	0	1	2	3	4	5	6
$f(x)$	-7	-2	7	20	37	58	83

(b)

t	-2	-1	0	1	2	3
$f(t)$	-21	-7	-3	-3	-1	9

(c)

θ	0	0.1	0.2	0.3	0.4	0.5
$f(\theta)$	0	0.099	0.192	0.273	0.336	0.375

Use your tables to extrapolate $f(8)$ in (a), $f(-4)$ in (b), $f(-0.1)$ in (c). Use the polynomials to interpolate $f(1.5)$, in (a), $f(-0.6)$ in (b), $f(0.48)$ in (c).

11.6   Numerical integration

It is not possible to integrate all functions analytically. However, it is known that the definite integral gives the area under a curve. Evaluation of an area by Simpson's rule can therefore provide a numerical method of determining a definite integral. To prove Simpson's formula the boundary curve is assumed to be a quadratic function. Therefore, Simpson's rule provides an exact solution in this case only. However, it is possible to obtain answers with an accuracy within $\pm 1\%$ for non-quadratic functions.

Simpson's rule

The area under the curve between the two points A and B is divided into an *even* number of strips of equal width h (Fig. 11.7). Normally eight to fourteen strips are chosen. In this case eight strips have been chosen and nine ordinates y_1 to y_9 drawn.

The curve at the top of each pair of strips is considered to be of the form $y = ax^2 + bx + c$.

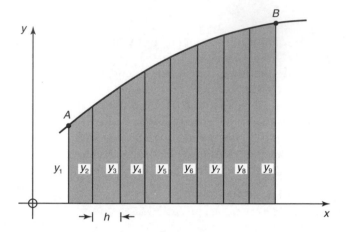

Fig. 11.7

Consider one pair of strips and let them be displaced horizontally to the y-axis as shown in Fig. 11.8. This does not affect the area but simplifies the proof.

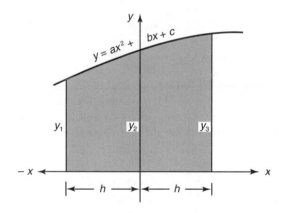

Fig. 11.8

$$y = ax^2 + bx + c$$

$$\text{at} \quad x = -h, \quad y_1 = ah^2 - bh + c \tag{1}$$

$$x = 0, \quad y_2 = \qquad c \tag{2}$$

$$x = h, \quad y_3 = ah^2 + bh + c \tag{3}$$

Now area of first pair of strips $= \int_{-h}^{h} (ax^2 + bx + c) \, dx$

$$= \left[\frac{ax^3}{3} + \frac{bx^2}{2} + cx \right]_{-h}^{h}$$

$$= \frac{h}{3} [2ah^2 + 6c]$$

But from equations (1), (2) and (3)

$$y_1 + 4y_2 + y_3 = 2ah^2 + 6c$$

Therefore area of 1st pair of strips $= \dfrac{h}{3} [y_1 + 4y_2 + y_3]$

Similarly, area of 2nd pair $\qquad = \dfrac{h}{3} [y_3 + 4y_4 + y_5]$

area of 3rd pair $\qquad = \dfrac{h}{3} [y_5 + 4y_6 + y_7]$

area of 4th pair $\qquad = \dfrac{h}{3} [y_7 + 4y_8 + y_9]$

By addition,

$$\text{total area} = \frac{h}{3} [(y_1 + y_9) + 4(y_2 + y_4 + y_6 + y_8) + 2(y_3 + y_5 + y_7)]$$

For an even number of strips where

$\qquad A = $ sum of first and last ordinates;

$\qquad B = $ sum of even ordinates;

$\qquad C = $ sum of odd ordinates excluding the first and last,

then

$$\textbf{Area under curve} = \frac{h}{3} [A + 4B + 2C]$$

Example 11.6   Use Simpson's rule to evaluate $\displaystyle\int_{0}^{0.8} \sqrt{(1 - x^2)} \, dx$
correct to five significant figures.
Consider eight strips, i.e. $h = 0.1$. Table 11.5 shows the values
required for the calculation.

Table 11.5

x	x^2	$1-x^2$	$\sqrt{(1-x^2)} = y$		$\times$ Factor	
0	0	1	1	y_1	$\times 1 =$	1
0.1	0.01	0.99	0.994 99	y_2	$4 =$	3.979 96
0.2	0.04	0.96	0.979 80	y_3	$2 =$	1.959 60
0.3	0.09	0.91	0.953 94	y_4	$4 =$	3.815 76
0.4	0.16	0.84	0.916 52	y_5	$2 =$	1.833 03
0.5	0.25	0.75	0.866 03	y_6	$4 =$	3.464 10
0.6	0.36	0.64	0.800 00	y_7	$2 =$	1.600 00
0.7	0.49	0.51	0.714 14	y_8	$4 =$	2.856 57
0.8	0.64	0.36	0.600 00	y_9	$1 =$	0.600 00
					Total $=$	21.109 0

$$\text{Area} = \frac{h}{3}\left[A + 4B + 2C\right]$$

$$= \frac{0.1}{3} \times 21.109\ 0 = 0.703\ 63 \text{ (correct to 5 significant figures).}$$

That is

$$\int_0^{0.8} \sqrt{(1-x^2)}\,dx = 0.70363$$

This particular integral could have been solved analytically. The result obtained by the above method is then found to be only 0.002% in error, showing that Simpson's rule can provide quite accurate results with very few strips.

EXERCISE 11.4

1. Evaluate

    (a) $\displaystyle\int_1^2 e^{-\frac{1}{2}x^2}\,dx$, correct to four decimal places

    (b) $\displaystyle\int_0^{0.5} \frac{\sin x}{x}\,dx$, correct to five decimal places.

    Check your answer by expanding $(\sin x)/x$ as a power series and then integrating analytically. (*Hint*, See Section 4.13 for $(\sin 0)/0^0$)

    (c) $\displaystyle\int_{0.1}^{0.8} \frac{\sqrt{(1-x^2)}}{x}\,dx$, correct to four significant figures

    (d) $\displaystyle\int_0^1 \frac{1}{\sqrt{(1+x^2)}}\,dx$; correct to three decimal places

    (e) $\displaystyle\int_1^6 \ln x\,dx$, correct to three significant figures.

2.   Find the approximate value of

(a) $\displaystyle\int_0^{0.12} \sqrt{(1 - \sin x)}\, dx$ (b) $\displaystyle\int_0^{\pi/2} \sqrt{(1 + 2\cos\theta)}\, d\theta$

3.   Evaluate $\displaystyle\int_0^6 \sqrt{(2x + 1)}\, dx$ using Simpson's rule with twelve strips.

Also find the value of this integral by an analytical method to find the percentage error involved using the approximation method. In both cases work to eight decimal places.

4.   The gas in an engine expands from a volume of 1 litre to 9 litres according to the law $pv^{1.4} = $ constant. The pressure of the gas is $90\,kN/m^2$ when the volume is 1 litre. Given that the work done by the gas during the expansion is $\displaystyle\int_{v_1}^{v_2} p\, dv$, use Simpson's rule with eight strips to find the work done. Check your answer by an analytical method.

11.7   Differential equations (Euler's method)

In Chapter 7 differential equations were solved analytically to find a general solution in the form $y = f(x)$ from a differential equation $dy/dx = f(x)$. From this analytical solution, provided boundary conditions are known particular solutions can be determined.

With a method of numerical analysis, such as Euler's method, the only solution that can be found is an approximation to the particular value of y at a given value of x.

Consider a differential equation $dy/dx = f(x, y)$, for which the boundary conditions (x_0, y_0) are known. The task is to find the value of y_p at the point $x = x_p$. Assume that the solution to this differential equation is $y = F(x)$ (although, of course, this solution is unknown). Let the curve of this solution be represented by AP in Fig. 11.9.

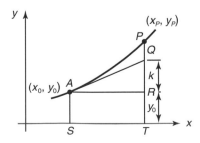

Fig. 11.9

Since the actual value y_p is unknown an approximate value is given by TQ, that is

$$y_p \fallingdotseq y_0 + k$$

Let AQ be the tangent to the curve at the point A. Therefore from triangle AQR

$$k = AR \tan RAQ$$
$$= (x_p - x_0)(\mathrm{d}y/\mathrm{d}x_A) = (x_p - x_0)f(x_0, y_0)$$

where $(\mathrm{d}y/\mathrm{d}x)_A$ the gradient of the curve at $A = f(x_0, y_0)$
   Therefore the estimated value of y_p is

$$y_0 + (x_p - x_0)f(x_0, y_0) \tag{i}$$

Unless the curve is nearly a straight line this is a poor estimate of y_p, since it is in error by an amount PQ. An improved estimate for y_p can be obtained by dividing ST into a number of small intervals as shown in Fig. 11.10.

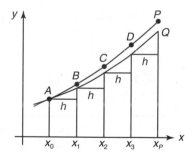

Fig. 11.10

The calculations are then carried out for each interval, using equation (i), with $(x_p - x_0)$ replaced by h.
   Therefore the estimated value of

$$y_B = y_A + hf(x_0, y_0)$$
$$y_C = y_B + hf(x_1, y_1)$$
$$y_D = y_C + hf(x_2, y_2)$$
$$y_p = y_D + hf(x_3, y_3)$$

It is seen in Fig. 11.10 that these estimated values are getting further from the true values. The reason is that the gradient of the curve at the beginning of the interval is being used to estimate the y value at the end of the interval. A further improvement in the accuracy can be gained by using an average value of the gradient of the curve across the interval.

Fig. 11.11 refers to the first interval in Fig. 11.10.

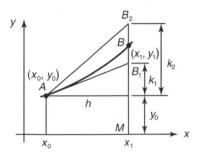

Fig. 11.11

The y coordinate at B_1, $MB_1 = y_0 + k_1 = y_0 + hf(x_0, y_0)$.
The coordinates at $B_1(x_0 + h, y_0 + k_1)$, which are known, can be used to obtain an estimate of the gradient of the curve at B_1, that is, $f(x_0 + h, y_0 + k_1)$. AB_2 is drawn with a gradient of this value.
The y coordinate of B_2, $MB_2 = y_0 + k_2 = y_0 + hf(x_0 + h, y_0 + k_1)$.
The estimated y coordinate of B, MB, is the average of MB_1 and MB_2 that is,

$$y_B = y_0 + \tfrac{1}{2}(k_1 + k_2)$$

This process is repeated for each interval in Fig. 11.10. In summary

$$\boxed{\begin{aligned} k_1 &= hf(x_0, y_0) \\ k_2 &= hf(x_0 + h, y_0 + k_1) \\ y_1 &= y_0 + \tfrac{1}{2}(k_1 + k_2) \end{aligned}}$$ (ii)

This method is now illustrated in Example 11.7.

Example 11.7 Use a numerical method to solve the differential equation $dy/dx = 3xy$, to find the value of y when $x = 1$, given that $y = 2$ when $x = 0$.

If the range of x (0 to 1) is divided into 10 intervals then $h = 0.1$. For the first interval $x_0 = 0$ to $x_1 = 0.1$, using (ii)

$$k_1 = h \times 3x_0 y_0 \qquad\qquad = 0.1 \times 3 \times 0 \times 2 \qquad = 0$$

$$k_2 = h \times 3(x_0 + h)(y_0 + k_1) = 0.1 \times 3(0 + 0.1)(2 + 0) = 0.06$$

$$y_1 = y_0 + \tfrac{1}{2}(k_1 + k_2) \qquad = 2 + 0.5(0 + 0.06) \qquad = 2.03$$

For the second interval, using (ii) again,

$$k_1 = h \times 3x_1 y_1 \qquad\qquad = 0.1 \times 3 \times 0.1 \times 2.03$$

$$= 0.0609$$

$$k_2 = h \times 3(x_1 + h)(y_1 + k_1) = 0.1 \times 3 \times (0.1 + 0.1)(2.03 + 0.0609)$$

$$= 0.125454$$

$$y_2 = y_1 + \tfrac{1}{2}(k_1 + k_2) \qquad = 2.03 + 0.5(0.0609 + 0.125454)$$

$$= 2.123177$$

This process can be continued for the 10 increments giving $y_{10} = 8.9235772$ which is an estimate of y when $x = 1$.

This differential equation is of the variables separable type and can be integrated analytically. The solution is

$$y = e^{(1.5x^2 + 0.69315)}$$

When $x = 1$, $y = 8.96338$. It is seen that the numerical solution is close to the analytical solution, with an error of -0.44%

EXERCISE 11.5

1. Solve each of the following differential equations at the respective value of x using Euler's method and check your solutions, if possible, with the analytical method.
    (a) $dy/dx = 2 + x$ at $x = 1$; the boundary condition is $y = 0$ when $x = 0$
    (b) $dy/dx = -2y$ at $x = 1$; the curve passes through the point $(0, 2.7183)$
    (c) $dy/dx = \cos 2x$ at $x = 2$; given that $y = 2$ when $x = 1$
2. An object is moving according to the equation

    $$dx/dt = 1 + t \sin x$$

    Given that its position x is 0.5 m when the time is zero, find its position when $t = 0.5$ s.

MISCELLANEOUS EXERCISE 11

1. Show that the equation $e^{-2x} - x^2 = 0$ has a root between 0.5 and 0.6. Use Newton's method to obtain the root correct to 2 places of decimals.

2. Use an iterative formula to find $\sqrt{N}$, correct to three decimal places, when $N = 20$. Take as the first approximation $x_1 = 4.4$.

3. The equation $x^3 + x - 1 = 0$ has a root in the vicinity of $x = 0.7$. By using a simple iterative process, find the value of this root correct to three decimal places.

4. Solve the following equations by successive elimination, working to three decimal places, setting out the working clearly in tabular form and including current sum checks:

    $45.3p + 15.7q = 62.6$

    $11.2p + 44.9q = 15.8$

    Check the solutions and round them off to two decimal places.

5. Use an iterative formula to find $1/N$, correct to five decimal places, when $N = 5.6$. Take as a first approximation $x_1 = 0.2$.

6. Solve the following simultaneous equations by successive elimination, setting out the working clearly in tabular form and including current sum checks:

    $3x + 2y - z = \quad 5.9$

    $4x + 3y + 4z = -17.2$

    $2x + 2y - 5z = \quad 27.3$

    Check the solutions.

7. Solve the following equations by successive elimination, setting out the working clearly in tabular form and including current sum checks:

    $a + \quad b + 2c = -10$

    $3a - \quad b - 2c = \quad 13$

    $7a + 5b - \quad c = \quad 1$

    Check the solutions.

8. Plot the curve $y = \frac{1}{2}\sqrt{(4 - x)}$ for values of x between 0 and 4 at intervals of 0.5. The second moment of area about the

y-axis of the region bounded by the curve and the

coordinate axes is $\int_0^4 x^2 y \, dx$. Evaluate this integral using

Simpson's rule.

9. Given that $f(x)$ is a smooth function use a difference method to complete the Table 11.6.

Table 11.6

x	0	1	2	3	4	5	6	7	8	9
$f(x)$	2	4	8	14	22	32	44			

What is the polynomial for $f(x)$? Hence obtain $f(3.4)$. Determine the error in determining $f(3.4)$ by linear interpolation.

10. Using Simpson's rule with eight strips, evaluate $\int_0^4 y \, dx$, if $y = \sqrt{(16 - x^2)}$.
    Calculate the mean value of y over this range.

11. Use Simpson's rule to evaluate $\int_1^5 \ln x \, dx$.

12. Use Simpson's rule, with four strips, to obtain the approximate value for the integral $\int_0^\pi x \sin x \, dx$.

13. Use Simpson's rule with six strips to evaluate $\int_{\pi/4}^{\pi/3} \tan x \, dx$.

14. Use Euler's method to solve the differential equations below, and check the results using the analytical method.
    (a) Given that $dy/dx = y \sin x$, find the value of y when $x = \pi$ given that $y = 2$ when $x = \pi/2$
    (b) Given that $dy/dx = x e^{2y}$ and $y = 0$ when $x = 1$, find y when $x = 0.2$

Chapter 12

Statistics

12.1 Introduction

The branch of mathematics that deals with the collection and analysis of data is called **Statistics**. A typical statistical survey consists of the following main steps:

1. collection of the data – this involves sampling techniques;
2. data presentation;
3. data analysis;
4. conclusions and decisions based on the analysis.

The data collected is usually in the form of sets of values of a particular variable; for example, the lengths of steel bars. The variable may be of two types:

(a) discrete – a variable consisting of separate values; for example, the number of bolts in a packet. There may be, say, 8 or 9 bolts but there cannot be any number in between 8 and 9.
(b) continuous – a variable which may have any value. For example, the diameters of steel bars produced by a machine. Any diameter is possible within the allowable tolerance to which the machine is set.

12.2 Sampling

When a statistical survey is carried out, it is not usual to examine every case. This could make the survey very lengthy and costly and, in many cases, completely impractical. For instance, it would be impossible for a factory to check the life of all the light bulbs it produces, otherwise it would never sell any.

For these reasons a sample must be chosen. This sample should be representative of the complete set of values from which it has been chosen. The complete set is called the **population**.

Although it can never be guaranteed that any sample is completely representative it is usually best to try to choose an **unbiased** sample. To be unbiased every possible sample must have an equal chance of being chosen.

This condition is likely to be satisfied if the sample is chosen at **random**; that is, if there is no order in the way the sample is chosen.

Such a sample is called a **random sample** and the larger this random sample is the more representative of the population it is likely to be. Random sampling can be carried out by allocating a number to each member of the population and then either drawing numbered balls from a bag or using the random facility in a calculator.

Sampling techniques involve probability theory which is dealt with in Chapter 13.

12.3   Data presentation

Once a sample set of data has been obtained, it must then be organised and presented in such a way as to make its distinctive features more apparent.

Consider a survey carried out on rough engine castings produced by a new casting method during one working day. A sample of 66 castings was chosen at random during the day and the masses of these castings, in kilograms, were as shown in Table 12.1.

Table 12.1

51.4	56.4	54.4	54.5	56.0	52.1
52.8	53.7	50.9	54.1	54.2	57.2
52.1	55.3	53.1	57.2	54.7	52.9
56.5	55.6	52.8	54.5	56.0	54.5
57.3	54.8	56.1	51.8	53.2	58.4
54.6	55.3	55.3	54.0	56.9	55.1
54.1	53.9	49.6	50.5	52.8	55.4
53.0	55.4	54.0	50.2	56.4	54.0
56.2	55.0	53.5	57.0	55.5	50.5
55.1	52.7	55.9	51.6	56.1	55.0
53.9	55.8	54.2	56.1	52.0	53.8

12.4   Frequency distribution

The data in Table 12.1 can now be arranged in a more convenient form by tabulating the information in groups or classes. The size of each class (called the **class interval**) should be one that emphasises any pattern contained in the data. As a rough guide, between 8 and 15 class intervals should be used to cover the range of the data. In this example a class interval of 1 kg is chosen. Therefore masses between 49.5 and 50.4 inclusive are considered as 50 kg and so on. A **frequency distribution table** can now be compiled showing the number of castings (that is the **frequency**) in each class (Table 12.2). The information from this table can be represented graphically in the following ways.

Table 12.2

Mass of casting x (kg)	50	51	52	53	54	55	56	57	58
Number of castings (frequency) f	2	4	5	8	13	15	12	6	1

(a) Bar chart

In this type of chart the variable x is usually represented on the horizontal axis and the frequencies are represented by the heights of vertical lines or **bars** (Fig. 12.1).

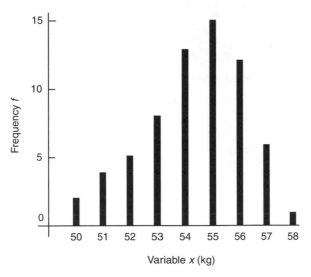

Fig. 12.1

(b) Histogram

This is similar to the bar chart except that the bars are widened to form rectangles. The width of each rectangle is equal to the class interval (Fig. 12.2). It will be noticed that the **area** of each rectangle is proportional to the class frequency. This concept of areas being proportional to frequencies will be seen to have applications in probability theory.

(c) Frequency polygon

If the tops of the bars on the bar chart or the mid-points of the tops of the rectangles on the histogram are joined together by straight lines a frequency polygon can be drawn (Fig. 12.3).

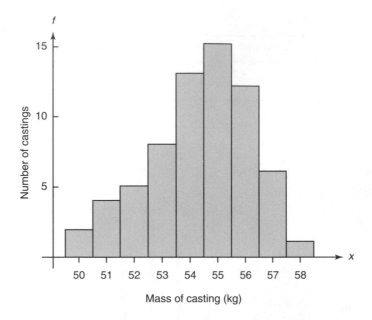

Fig. 12.2

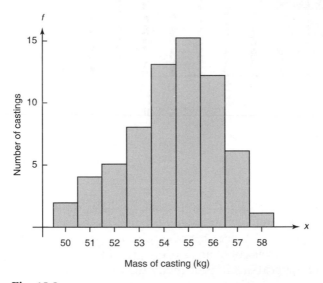

Fig. 12.3

(d) Frequency curve

If a large sample is considered and small class intervals are chosen there will be a large number of rectangles in the histogram. The frequency polygon will therefore consist of more points situated closer together. If this process is continued, using smaller and smaller class intervals, the frequency polygon will approximate more and more to a smooth curve known as the frequency curve (Fig. 12.4).

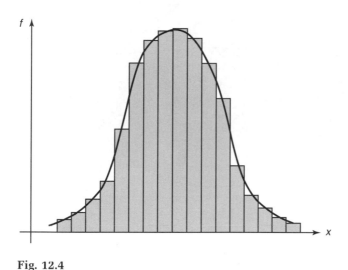

Fig. 12.4

For all the graphical representations discussed so far, the class frequency has been plotted in the vertical scale. It is sometimes convenient to plot **relative frequency** instead of class frequency, where

$$\text{relative frequency} = \frac{\text{class frequency}}{\text{total frequency of the sample}}$$

For example, the relative frequency of the 53 kg class is $\frac{8}{66}$ or 0.121.

Plotting relative frequency will not alter the pattern of these charts. It is usually used when the charts obtained from the sample are considered as direct representations of the population.

There are other methods of visual display that are suitable for certain types of data. One such method is the **pie chart**. The circle is divided into a number of sectors, the area of each sector being proportional to the quantity it represents.

The pie chart in Fig. 12.5 gives the results of a survey carried out to determine the pattern of the number of dependent children in families in Great Britain.

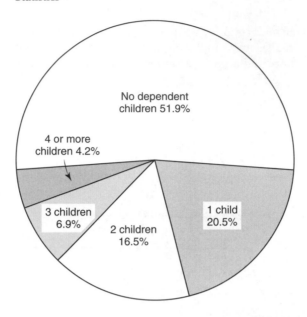

Fig. 12.5

EXERCISE 12.1

1. To test the pattern of results in an examination, a random sample was
    taken of fifty students. Their marks were as shown in Table 12.3.
    Compile a frequency distribution table using a class interval of 10
    marks. Hence draw a histogram, frequency polygon and bar chart.

    Table 12.3

94	50	65	65	72
78	65	60	74	63
60	72	74	68	35
63	87	64	55	53
64	70	81	21	51
65	61	72	57	37
30	38	25	21	11
58	26	34	33	67
54	56	18	75	69
64	31	51	12	61

2. In a technical college a student's day is divided up as follows:
    Mathematics 2 hours, Science $1\frac{1}{2}$ hours, Workshop Technology $2\frac{1}{4}$
    hours, General Studies $1\frac{3}{4}$ hours, Lunch, etc., $1\frac{1}{4}$ hours.
    Draw a pie chart to illustrate this pattern.

3. A random sample of 500 steel plates was chosen from a large production batch. The thicknesses were measured to the nearest tenth of a millimetre, giving the Table 12.4.

Table 12.4

Thickness (mm)	6.8	6.9	7.0	7.1	7.2	7.3	7.4	7.5	7.6
Number of plates	4	14	44	98	150	112	61	15	2

Use three graphical methods to present the distribution of thickness in the sample.

4. In a series of tests carried out in a laboratory to determine the resistances of 155 coils, the results in Table 12.5 were obtained.

Table 12.5

Resistance (ohms)	1.04	1.05	1.06	1.07	1.08	1.09	1.10
Number of coils	12	27	35	30	29	16	6

Assuming that this sample of 155 coils is considered to be representative of the entire week's production of a factory, represent the pattern of the week's production on a chart.

5. Table 12.6 gives the weekly wage of 100 of its employees.

Table 12.6

Weekly wages (£)	13–17	18–22	23–27	28–32	33–37	38–42	43–47
Frequency	6	17	24	21	14	10	8

Represent this distribution of wages in as many graphical ways as you can.

12.5 Numerical measures of a distribution

A frequency distribution can be very conveniently represented by two numerical quantities. One such number represents the central tendency or average value of the distribution. This is the value of the variable about which the variable seems to be located. The second number represents the dispersion or scatter of the variables about the average value.

It is much easier to compare samples if these two numbers are available for each sample.

12.6   Numerical measures of central tendency

(a) Mid-point of range

The range of a distribution is the difference between the largest and smallest values of the variable. The mid-point of the range can be a poor measure of central tendency since it depends only on the extreme values of the variable and therefore is not influenced by the form of the distribution.

(b) Mode

The mode is the most frequently occurring value of the variable. It is easily obtained by inspection of the graphs or the frequency table. For the engine casting survey, mode = 55 kg.

(c) Arithmetic mean

This is a good measure of average value since all the values of the variable are included in its derivation. With a small sample however, the mean can be distorted by one very high or very low value.

The mean value is determined by adding up all the values of the variable and dividing this total by the number of values. If $x_1, x_2, x_3 \ldots x_N$ are the N values then

$$\text{arithmetic mean} \quad \bar{x} = \frac{x_1 + x_2 + \cdots + x_N}{N}$$

$$\text{or} \quad \bar{x} = \frac{1}{N} \sum x$$

Example 12.1   Calculate the mean height of seven boys whose individual heights in metres are

1.26, 1.31, 1.22, 1.52, 1.18, 1.34, 1.25

$$\text{Mean height} = \frac{1.26 + 1.31 + 1.22 + 1.52 + 1.18 + 1.34 + 1.25}{7}$$

$$= \frac{9.08}{7}$$

$$= 1.30 \, \text{m}$$

For larger samples, where the information has been grouped as a frequency distribution table, the mean is calculated as follows:

If the values $x_1, x_2, \ldots, x_N$ occur with frequencies $f_1, f_2, \ldots, f_N$ then

$$\text{Arithmetic mean} = \frac{f_1 x_1 + f_2 x_2 + \cdots + f_N x_N}{f_1 + f_2 + \cdots + f_N}$$

where $f_1 + f_2 + \cdots + f_N = N$

or $\bar{x} = \dfrac{1}{N} \sum fx$

Example 12.2 The results in Table 12.7 were obtained when measuring the diameters of 50 washers. What is their mean diameter?

Table 12.7

Diameter mm	40	41	42	43	44
Number of washers	7	12	18	8	5

$$\text{Mean diameter} = \frac{7 \times 40 + 12 \times 41 + 18 \times 42 + 8 \times 43 + 5 \times 44}{50}$$

$$= 41.8\,\text{mm}$$

(d) Median

The median of a set of values is the value of the middle variable when the set is arranged in order of magnitude. For example, in the set (3, 8, 13, 19, 20) the median is 13.

For a set containing an even number of values there is no single middle value. In this case the median is considered to be the mean of the two middle values. For example, in the set (8, 9, 9, 12, 18, 20) the median is $(9 + 2)/2 = 10.5$. The significance of a median is that there is an equal number of values above and below the median value, that is, it divides the set into two halves. For a large distribution the median is best found by drawing a cumulative frequency curve as shown in Example 12.3.

Example 12.3 The times shown in Table 12.8 were taken by 170 trainees to learn to operate a particular machine.

Table 12.8

Time in hours	10–19	20–29	30–39	40–49	50–59	60–69	70–79	80–89
Number of trainees	3	4	20	48	57	20	13	5

Determine the median training time.

Table 12.9

Time (hours)	*Max* hours in each class	Frequency	Cumulative frequency
10–19	19	3	
20–29	29	4	3 + 4 = 7
30–39	39	20	7 + 20 = 27
40–49	49	48	27 + 48 = 75
50–59	59	57	75 + 57 = 132
60–69	69	20	132 + 20 = 152
70–79	79	13	152 + 13 = 165
80–89	89	5	165 + 5 = 170

Table 12.9 sets out the relevant information. The cumulative frequencies are now plotted against the **maximum** hours for each class (see Fig. 12.6). This curve is called an **ogive**.

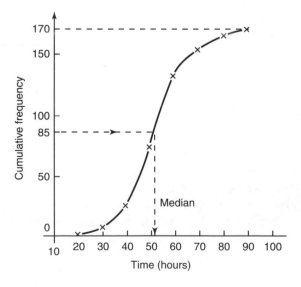

Fig. 12.6

Since there are 170 values the median will actually be between the 85th and 86th values. However, with larger samples there is little loss of accuracy incurred by considering the median as the value corresponding to the mid-point of the vertical scale for both odd and even sets. From graph median = 52 hours.

The median is not influenced by an extreme value in the same way as the mean. In such cases it can be a more appropriate measure of an average value. For example, the annual salaries in a small office are: 1 clerk at £5000, 1 typist at £7000, a personal assistant at £9000 and the manager at £19 000. The mean salary is £10 000; the median salary is £8000. It is seen that the mean value is distorted by the salary level of the manager. However, in engineering production where extreme values are rarely encountered the mean is a good measure of the average value.

All measures of central tendency can be useful, depending upon the information required from the survey.

If a distribution can be represented by a symmetrical histogram or frequency curve with a single peak, then the mean, mode and median have identical values. A distribution which is not symmetrical is called a **skewed** distribution.

EXERCISE 12.2

1. Eight rods were tested in a tensile machine and their failure loads were 15.8, 16.2, 16.0 15.7, 15.6, 15.3, 16.8, 17.0 kN. What is the mean failure load?
2. Determine the mean of the first twelve natural numbers.
3. Four men are 40 years of age, six men are 35 years of age, three men are 38 years of age and one man is 42 years of age. What is the average (i.e. mean) age of these men?
4. A factory producing boxes carried out a sample check on the area of cardboard being used to make their boxes. The results were as shown in Table 12.10.

Table 12.10

Area of box (m^2)	1.26	1.27	1.28	1.29	1.30
Number of boxes	3	8	16	12	5

What is the mean area of cardboard used to make one box?
5. The internal diameter of a pipe is measured at regular intervals along the pipe and the distribution shown in Table 12.11 was obtained.

Table 12.11

Internal diameter (m)	0.85	0.86	0.87	0.88	0.89	0.90	0.91
Number of readings	20	46	68	80	52	33	13

What is the mean internal diameter?

6.   From the results in questions 4 and 5 draw the ogives and hence determine the medians. What are the modal values in each case?

12.7   Numerical measures of dispersion

(a) Range

For the reasons previously mentioned, range is not always a good measure of dispersion. It is, however, a convenient and accurate measure of dispersion when used in the quality control of production processes, where items are being made to specified tolerance limits.

(b) Mean deviation from the mean or median

The deviation of each value of the variable from the mean or median is first obtained and then the average value of these deviations is calculated. It is not normally used in science and engineering.

(c) Standard deviation

The standard deviation is obtained by squaring the difference between each value of the variable and the mean, finding the average square value, and taking the square root of this average. It is the most commonly used measure of dispersion since it is based upon all the values and because of its significance in probability distributions (see Chapter 13).

For a sample of size N with values $x_1, x_2, \ldots, x_N$ occurring with frequencies $f_1, f_2, \ldots, f_N$ the standard deviation is determined from the formula

$$s = \sqrt{\left(\frac{\sum f(x - \bar{x})^2}{N} \right)}$$

The term **variance** is sometimes used when discussing dispersion, where variance $= s^2$. Standard deviation is to be preferred because it has the same units as the variable.

A low value of standard deviation indicates that the values of the variable tend to be closely packed about the mean. A high value of standard deviation indicates a much wider scatter of values.

Example 12.4 Calculate the standard deviation for the results given in Example 12.2 (Table 12.7).

Table 12.12

Diameter (mm) x	Frequency f	$x - \bar{x}$ $= x - 41.84$	$(x - \bar{x})^2$	$f(x - \bar{x})^2$
40	7	−1.84	3.40	23.80
41	12	−0.84	0.71	8.52
42	18	0.16	0.02	0.36
43	8	1.16	1.35	10.80
44	5	2.16	4.67	23.35
	$\sum f = N = 50$			$\sum f(x - \bar{x})^2 = 66.83$

Standard deviation $s = \sqrt{\left(\dfrac{\sum f(x - \bar{x})^2}{N}\right)}$

$$= \sqrt{\left(\frac{66.83}{50}\right)}$$

$$= 1.16 \, \text{mm}$$

(d) Interquartile range

This particular measure of dispersion can be obtained from the ogive. The vertical scale is converted to percentages (Fig. 12.7). The variable corresponding to the 50% cumulative frequency will be the median. The variables corresponding to the 25% and 75% values, labelled Q_1 and Q_2, are called the **lower** and **upper quartiles** respectively. The **interquartile range** is equal to $Q_2 - Q_1$ and it contains the middle 50% of the population. The **semi-interquartile range** is also used as a dispersion measure and it is equal to $\frac{1}{2}(Q_2 - Q_1)$

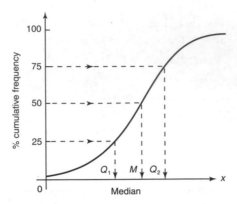

Fig. 12.7

Example 12.5   Find the semi-interquartile range for Example 12.3.

The ogive is reproduced in Fig. 12.8 with % cumulative frequency on the vertical scale.

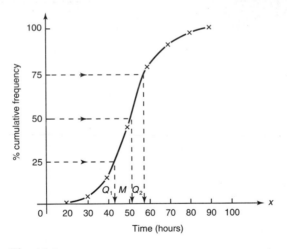

Fig. 12.8

From the curve $Q_1 = 43$, $Q_2 = 58$.

Therefore interquartile range $= Q_2 - Q_1$
$$= 58 - 43 = 15 \text{ hours}$$
Semi-interquartile range $= \tfrac{1}{2}(Q_2 - Q_1)$
$$= 7.5 \text{ hours}$$

EXERCISE 12.3

1. Calculate the mean and standard deviation of the set of numbers: 2, 4, 6, 8, 10, 12, 14.
2. 210 steel tubes were obtained from a factory and their lengths were checked. The results were as shown in Table 12.13.

Table 12.13

Length (m)	5.61	5.62	5.63	5.64	5.65	5.66	5.67
Number of bars	9	25	39	62	41	26	8

Calculate the mean length and the standard deviation.
3. The cross sectional areas of 300 brass bars were recorded. The results are presented in the frequency table (Table 12.14).

Table 12.14

Area (mm^2)	78.0	78.2	78.4	78.6	78.8	79.0
Number of bars	15	41	104	92	32	16

What is the mean cross-sectional area and the standard deviation?
4. The thickness of a concrete floor was tested with an ultrasonic tester and the 100 readings were accurate to the nearest 0.01 m (Table 12.15).

Table 12.15

Floor thickness (m)	0.31	0.32	0.33	0.34	0.35	0.36	0.37
Number of readings	5	14	18	31	15	12	5

Find the mean floor thickness and the standard deviation.
5. The heights of 320 men were measured to the nearest 30 mm (Table 12.16).

Table 12.16

Height (m)	1.62	1.65	1.68	1.71	1.74	1.77	1.80
Frequency	14	33	50	70	81	52	20

What is the mean height of this group and the standard deviation about this mean value?

6. Twenty concrete blocks were tested to failure in a crushing machine with the results shown in Table 12.17.

Table 12.17

Failure load (kN)	30.4	30.6	30.8	31.0	31.2	31.4
Frequency	1	4	7	5	2	1

What is the mean crushing strength and the standard deviation?

7. For questions 2 to 6 inclusive, prepare cumulative frequency tables and hence draw the ogives. From each ogive determine (a) the median, (b) the interquartile and semi-interquartile ranges.

8. For the distribution given in Table 12.18 plot the frequency polygon and determine the mean, mode, median and standard deviation.

Table 12.18

Centre of class	54	55	56	57	58	59	60	61	62
Frequency	1	3	8	14	27	45	32	18	5

9. The hourly wages received in a factory were determined for a random sample of its employees and Table 12.19 was compiled.

Table 12.19

Hourly rate (p)	60–64	65–69	70–74	75–79	80–84	85–89	90–94	95–99
Number of employees	18	33	71	94	86	62	46	22

Determine the median rate from the ogive and also find the semi-interquartile range.

10. The masses of a random sample of machine components taken from a production line are given in Table 12.20.

Find the mean mass and the median mass. Draw a histogram of the distribution.

Table 12.20

Mass (kg)	60	61	62	63	64	65	66
Frequency	2	9	13	15	11	8	3

12.8   Linear regression

The relationship between two variables x and y may be linear or non-linear, but in this section we shall deal only with linear or approximately linear relationships. Usually x is the independent variable and y the dependent variable. However, it is not always obvious which variable is the independent one.

Fig. 12.9 shows a set of data in which there is an approximately linear relationship between the x and y values. A line which seems to fit these points could be drawn directly by eye between the points.

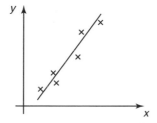

Fig. 12.9

A more accurate method of obtaining the 'best fit' line, however, is through the use of the 'method of least squares'. In this method the 'best fit' line is taken as that line which minimises the sum of the squares of the differences between each point and the line. The method is described below.

Note: Since the 'best fit' line will pass through the mean point $(\bar{x}, \bar{y})$ the sum of the differences will be zero, so that the minimum of the sum of squares of the differences has to be considered.

(a) Taking x as the independent variable

In Fig. 12.10(a) all the values of x are assumed to be correct and the differences or error (e) are considered to be in the values of y only. Let the 'best fit' straight line be $y = ax + b$, where a and b are constants to be calculated.

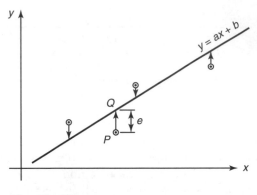

Fig. 12.10(a)

For any point such as $P(x_1, y_1)$

$$\text{error } e = y_P - y_Q$$
$$= y_1 - (ax_1 + b)$$

Therefore for n points

$$\sum_1^n e^2 = \sum_1^n (y - (ax + b))^2 \tag{i}$$

Since the position of the 'best fit' line is unknown at this stage a and b are treated as variables. The minimum value of expression (i) will occur when the partial differential coefficients are set to zero, from which the required values of a and b can be found.

Differentiating partially with respect to a, keeping b constant, for any given value of x,

$$\frac{\partial}{\partial a} \sum e^2 = 2 \sum -x(y - (ax + b)) = 0$$

which reduces to

$$\sum xy - a \sum x^2 - b \sum x = 0 \tag{ii}$$

Differentiating partially with respect to b, keeping a constant, for any given value of x,

$$\frac{\partial}{\partial b} \sum e^2 = 2 \sum (y - ax - b)(-1) = 0$$

which reduces to

$$\sum y - a \sum x - b \sum 1 = 0$$
$$\sum y - a \sum x - nb = 0 \tag{iii}$$

From the two equations (ii) and (iii) the values of a and b are calculated, and substituted into $y = ax + b$ to produce the 'best fit' straight line. The line is called the linear regression of **y on x**. It is used to predict values of y for given values of x.

(b) Taking y as the independent variable

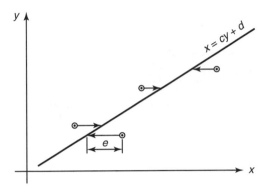

Fig. 12.10(b)

In Fig. 12.10(b) all the y values are assumed to be correct and differences or errors (e) are considered to be in the x-values only. The line of regression is now taken as $x = cy + d$ and the constants c and d found as above, using equations (iv) and (v):

$$\sum xy - c \sum y^2 - d \sum y = 0 \tag{iv}$$

and

$$\sum x - c \sum y - nd = 0 \tag{v}$$

The line $x = cy + d$ is the linear regression of **x on y**, and can be used to predict values of x for given values of y.

It is seen that there are two regression lines for a set of data, except where the points lie exactly on straight line. Both regression lines pass through the mean point $(\bar{x}, \bar{y})$. The more scattered the sample the greater the angle between the two regression lines. Example 12.6 shows the determination of the two regression lines.

Example 12.6 Values of temperature and voltage were recorded in an electrical control system and gave the results in Table 12.21.

Table 12.21

Temperature x (°C)	5.0	9.0	15.5	20.0	24.0	30.0
Voltage y (V)	2.6	3.2	4.4	8.0	8.6	8.8

Find the linear regression equations of (a) y on x and (b) x on y, and hence draw the regression graphs.

Table 12.22

x	y	x^2	y^2	xy
5.0	2.6	25.0	6.76	13.0
9.0	3.2	81.0	10.24	28.8
15.5	4.4	240.25	19.36	68.2
20.0	8.0	400.0	64.0	160.0
24.0	8.6	576.0	73.96	206.4
30.0	8.8	900.0	77.44	264.0
103.5	35.6	2222.25	251.76	740.4

Case 1: y on x, that is, x is the independent value

Substitute values from Table 12.22 into the equations:

$$\sum xy - a \sum x^2 - b \sum x = 0$$
$$\sum y - a \sum x - nb = 0$$

that is:

$$740.4 - 2222.25a - 103.5b = 0$$
$$35.6 - 103.5a - 6b = 0$$

which gives

$$a = 0.29, \qquad b = 0.95$$

The line of regression, y on x, is

$$y = 0.289x + 0.946$$

Case 2: x on y is the independent variable

Substitute values from the above table into the equations:

$$\sum xy - c \sum y^2 - d \sum_n y = 0$$
$$\sum x - c \sum y - nd = 0$$

that is

$740.4 - 251.76c - 35.6d = 0$

$103.5 - 35.6c - 6d = 0$

Solving these two simultaneous equations gives

$c = 3.116, \qquad d = -1.238$

The line of regression is

$x = 3.12y - 1.24$

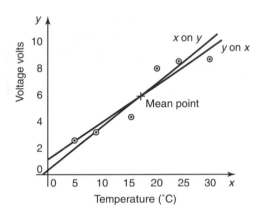

Fig. 12.11

The two lines of regression are shown in Fig. 12.11

The mean value $\qquad \bar{x} = \dfrac{\sum x}{n} = \dfrac{103.5}{6} = 17.3$

The mean value $\qquad \bar{y} = \dfrac{\sum y}{n} = \dfrac{35.6}{6} = 5.9$

Both regression lines are seen to pass through the point $(\bar{x}, \bar{y})$.

12.9   Linear correlation

When there is a 'cause and effect' relationship between two variables such as x and y a correlation exists between them. The relationship may be linear or non-linear. In this section data will be examined only for linear correlation.

Consider the five sets of data shown in Fig. 12.12.

In Fig. 12.12(a) all the points lie exactly on a straight line and this is an example of perfect linear correlation between x and y. The correlation is positive since both x and y are increasing together.

Fig. 12.12(a)

Fig. 12.12(b)

Fig. 12.12(c)

Fig. 12.12(d)

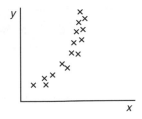

Fig. 12.12(e)

In Fig. 12.12(b) there is no pattern to the points and this is an example of no correlation.

In Fig. 12.12(c) although the points are scattered they show a good linear pattern, that is they show a positive linear correlation.

In Fig. 12.12(d) the points show some negative linear correlation.

In Fig. 12.12(e) the points show a good non-linear correlation but a poor linear correlation.

Correlation is expressed in terms of a **coefficient of linear correlation** r which can vary from -1 to $+1$, that is,

$$-1 \quad\longrightarrow\quad 0 \quad\longrightarrow\quad +1$$

-1	0	$+1$
perfect	no	perfect
negative	correlation	positive
correlation		correlation

The closer the value of r is to 0 the less the level of linear correlation between the two quantities. The value of r can be calculated from the following formulae.

(a) Using the regression equations

If $\;y = ax + b\;$ and $\;x = cy + d$
then

$$r = \sqrt{ac}$$

(b) Using the product moment formula

$$r = \frac{\sum XY}{\sqrt{\sum X^2 \sum Y^2}}$$

where $X = x - \bar{x}$ and $Y = y - \bar{y}$

Example 12.7 The stress in a structure was measured at two different positions and gave the results shown in Table 12.23.

Table 12.23

Stress x at position 1	10	11	12	13	14	15
stress y at position 2	35	37	36	33	25	15

Calculate the coefficient of linear correlation between the stresses at the two positions.

Table 12.24

	x	y	$X = (x - \bar{x})$	$Y = (y - \bar{y})$	X^2	Y^2	XY
	10	35	−2.5	4.8	6.25	23.04	−12
	11	37	−1.5	6.8	2.25	5.063	−10.2
	12	36	−0.5	5.8	0.25	33.64	−2.9
	13	33	0.5	2.8	0.25	7.84	1.4
	14	25	1.5	−5.2	2.25	27.04	−7.8
	15	15	2.5	−15.2	6.25	231.0	−38.0
Sum	75	181			17.5	368.84	−69.5

$$\bar{x} = \frac{75}{6} = 125$$

$$\bar{y} = \frac{181}{6} = 30.2$$

$$r = \frac{\sum XY}{\sqrt{\sum X^2 \sum Y^2}}$$

$$= \frac{-69.5}{\sqrt{17.5 \times 368.84}}$$

$$= -0.87$$

Since the stress in position 2 is decreasing as the stress at position 1 is increasing this results in r being negative. The value of r indicates a good negative linear correlation between the stresses but the graph in Fig. 12.13 is not very straight, showing that values of r need to be interpreted with caution.

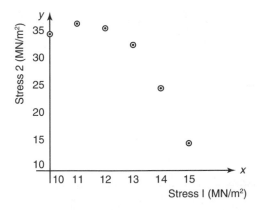

Fig. 12.13

EXERCISE 12.4

1. Table 12.25 shows a set of readings taken in an electric circuit.

    Table 12.25

Frequency f (Hz)	25	50	75	100	125	150
Current i (A)	1.6	2.8	4.8	5.2	5.8	7.6

    Find the linear regression formulae (a) for f on i and (b) for i on f. Use these formulae to find the coefficient of linear correlation. Draw the scatter diagram showing the regression lines.

2. The temperatures in two different positions of a hydraulic circuit are recorded in Table 12.26.

    Table 12.26

Temperature θ ($^\circ C$)	5.5	6.5	6.5	7.5	8.5	8.5	9.5	10
Temperature θ_2 ($^\circ C$)	20	20.5	21	22	22	24	23	25

    Find the regression fomula of (i) θ_1 on θ_2, (b) θ_2 on θ_1, and hence find the coefficient of linear correlation. Use the formula to find (i) θ_2 if $\theta_1 = 12°C$ (ii) θ_1 if $\theta_2 = 21.5°C$. Draw the scatter diagram showing the regression lines.

3. Use the figures given in Tables 12.27 and 12.28 to find the coefficients of linear correlation between the quantities x and y in each case. Draw

the scatter diagrams and use the information to comment upon the correlations in each case.

Table 12.27

(a)	x	7.8	8.3	10.0	10.5	10.8	11.5	13.5
	y	120	100	160	150	156	150	230

Table 12.28

(b)	x	198	204	350	410	450
	y	30	39	48	46	100

4. The marks obtained by 10 students in mathematics and engineering science are shown in Table 12.29.

Table 12.29

Mathematics	66	70	73	75	80	80	82	87	90	91
Engineering science	35	40	37	44	50	56	42	49	55	70

Draw the scatter diagram. Find the coefficient of linear correlation and comment on the correlation between the two sets of marks.

MISCELLANEOUS EXERCISE 12

1. A machine is set to produce a dimension of 27.5 mm. On measurement of 100 specimens, the results in Table 12.30 were obtained.

Table 12.30

Length of specimen (mm)	27.1	27.2	27.3	27.4	27.5	27.6	27.7	27.8	27.9
Number		1	3	8	18	32	20	9	5	4

Calculate the mean length and the standard deviation from the mean.

2. The amplification factor of a sample of 100 amplifier units
    is given in Table 12.31.

Table 12.31

Amplification factor	30	35	40	45	50
Frequency	16	25	36	17	6

    Find the mean value and the standard deviation of the
amplification factors of this sample

3. Table 12.32 shows the frequency distribution of overtime
    hours worked by employees in a factory. Find by a
    graphical method the median and quartile numbers of
    overtime hours worked per employee.

Table 12.32

No. of overtime hours	Under 10	10–20	20–30	30–40	40–50
No. of employees	41	96	143	85	36

No. of overtime hours	50–60	60–70	70–80	80–90	90–100
No. of employees	11	7	4	3	1

4. Table 12.33 shows the average speed, in km/h, of a sample
    of 35 journeys made by dump trucks on a motorway
    construction site.

Table 12.33

28.8	32.7	29.8	28.4	27.3	31.5	25.5
30.2	24.1	30.4	33.5	33.6	34.3	35.4
33.1	33.7	32.6	29.3	30.7	29.0	30.8
31.7	31.3	26.2	32.4	25.0	34.9	37.8
27.7	28.1	31.2	26.7	32.3	29.7	31.0

    Separate these values into seven suitable class intervals
and construct a histogram. Find the median value of this
sample.

5. Table 12.34 shows the range, in kilometres, of a series of
    36 test firings of a certain type of ground-to-ground
    missile.

Table 12.34

95.2	91.4	89.4	91.8	93.2	90.8
92.7	84.2	90.5	85.7	91.6	83.8
88.2	80.4	87.9	88.4	82.8	86.5
88.6	86.7	82.3	90.3	92.6	89.7
91.7	93.5	87.2	88.5	94.7	91.8
89.3	86.4	90.2	84.3	87.1	85.2

Separate these values into eight suitable class intervals and construct a histogram.

6. The working lifetime of 500 components was noted and the results were as shown in Table 12.35.

Table 12.35

| Time in hours | 300–399 | 400–499 | 500–599 | 600–699 |
| No. of components | 50 | 26 | 79 | 157 |

| Time in hours | 700–799 | 800 = 899 | 900–999 | 100–1099 |
| No. of components | 89 | 57 | 32 | 10 |

Construct a histogram, a frequency polygon, a cumulative frequency curve. Find the percentage of components with a working lifetime of more than 600 hours.

7. The time required to complete a similar job by each of 100 employees was recorded in Table 12.36.

Table 12.36

Time taken (min)	8	9	10	11	12	13	14	15
No. of employees	8	10	12	20	18	14	10	8

Construct a frequency polygon. Determine the mean, median and standard deviation for this set of values.

8. The number of matches in a box should be 35. Samples taken from a packing line give the numbers in Table 12.37.

Table 12.37

Matches per box	30	32	34	36	38	40
No. of boxes	12	20	30	60	60	15

Draw a histogram of this data. Determine the mean and standard deviation.

9. Find the mean and standard deviation of the set of numbers: 2, 5, 8, 10, 11, 14, 16, 22.

10. A sample of elderly people had their eyes tested for sensitivity to red light. The results are tabulated as Table 12.38. Represent this data in the form of a histogram and calculate, (a) the mean sensitivity index, and (b) the standard deviation.

Table 12.38

Sensitivity index	13	14	15	16	17	18
Frequency	2	4	9	8	1	1

11. Sixty students in a number of physics classes determined the value of g and the results were as shown in Table 12.39.

Table 12.39

9.76	9.71	9.75	9.81	9.87	9.90	9.78	9.66	9.74	9.84	9.86	9.72
9.61	9.56	9.82	9.79	9.82	9.70	9.89	9.84	9.75	9.76	9.83	9.77
9.87	9.77	9.92	9.84	9.69	10.04	9.81	9.66	9.80	9.98	9.76	9.82
9.73	9.63	9.70	9.86	9.74	9.77	9.67	9.88	9.76	9.81	9.69	9.79
9.79	9.97	9.84	9.72	9.80	9.94	9.81	9.74	9.83	9.77	9.93	9.85

Compile a frequency table in groups (9.55 to less than 9.60, 9.60 to less than 9.65, etc.) and draw the corresponding histogram. Determine the lower quartile and median values of the distribution.

12. A sample of bricks from a kiln was tested for transverse strength and the results were tabulated in Table 12.40.

Table 12.40

Transverse strength N/mm^2 (mid-point of class interval)	3.4	4.8	6.2	7.6	9.0	10.4
Number of bricks	11	51	115	82	33	8

Calculate the mean strength and the standard deviation from the mean. Draw a histogram to illustrate the data and use it to estimate the percentage of the sample whose strengths lie between the mean plus or minus twice the standard deviation.

13. Calculate the linear correlation between the two quantities shown Table 12.41.

Table 12.41

Load (kN)	33	27	14	8	7
Deflection (mm)	5.1	5.5	6.7	8.3	8.9

14. Table 12.42 shows readings taken of the temperature of an electronic component and the time taken for it to 'cut out'.

Table 12.42

Temperature θ (°C)	25	35	35	50	60	70	75
Time t (min)	29	30	31	26	27	27	25.5

Find both lines of linear regression and draw these lines on a scatter diagram. Calculate the coefficient of linear correlation. Is there a good correlation between the two quantities?

Chapter 13

Probability and sampling

13.1   Definition of probability

In everyday speech the word 'probability' implies how likely an event is of either taking place or not taking place. A more rigorous definition is required in mathematics.

A study of probability is essential to deal with such problems as the likelihood of an aircraft tyre failing on landing or the likelihood of a population having the same characteristics as a sample chosen from it, etc. Consider the result of throwing an unbiased, i.e. perfectly balanced, die (singular of dice) a number of times. It could well be of the form

3 4 1 1 5 2 1 6 2 4 6 6 2 2 1 5....

This sequence is a random sequence since the terms are patternless and unpredictable.

Examples of random number sequences that are produced deliberately are bingo, premium bond and national lottery numbers. For the above sequence the proportion

$$\frac{\text{total number of sixes}}{\text{total number of throws}} \text{ (called the relative frequency)}$$

can be calculated after 100, 200, 300 throws, etc. If the relative frequency is now plotted against the total number of throws it will be found that the points fluctuate erratically in the early stages but eventually settle down to a limiting value of $\frac{1}{6}$ after a large number of throws.

This limiting value of relative frequency is the mathematical probability of obtaining a six on any throw of the die.

Probability can be more easily evaluated by the following alternative definition.

If an event can occur x times and cannot occur y times then the probability of the event occurring is given by the ratio $\dfrac{x}{x+y}$.

In some situations it is not possible to find the value of y and the relative frequency method must be used instead.

Example 13.1   If a fair (i.e. unbiased) coin is tossed twice, what is the probability of obtaining one head only?
The possible results are *HH HT TH TT*.

Therefore probability of one head $= \dfrac{x}{x+y}$

$$= \tfrac{2}{4} = \tfrac{1}{2}$$

13.2   Scale of probability

Since probability has been defined on a mathematical basis, a probability scale can be used. If an event A is certain to occur then the probability, written $P(A)$, is equal to 1 or $P(A) = 1$, since $y = 0$ in the ratio $x/(x+y)$. If an event A cannot possibly occur then $P(A) = 0$, since $x = 0$. In general, the probability of any event A occurring must be in the range $0 \leqslant P(A) \leqslant 1$.

13.3   Compound probability

If more than one event is involved then probabilities must be added or multiplied together to give a compound probability. The following rules apply:

RULE I

If two events A and B are *mutually exclusive* (that is, either event can occur without the other event occurring), then the probability of *either A or B* occurring is given by the sum of the individual probabilities; that is,

$$P(A \text{ or } B) = P(A) + P(B).$$

Example 13.2   What is the probability of throwing either a one *or* a six with a single throw of an unbiased die?

$P(\text{one } or \text{ six}) = P(\text{one}) + P(\text{six})$

$$= \tfrac{1}{6} + \tfrac{1}{6}$$

$$= \tfrac{1}{3}$$

Example 13.3   What is the probability of selecting an ace *or* a king from a well-shuffled pack of 52 cards?

$$P(\text{ace}) = \tfrac{4}{52} \quad \text{since there are four aces in the pack}$$

$$= \tfrac{1}{13}$$

$$P(\text{king}) = \tfrac{1}{13}$$

Therefore $P(\text{ace } or \text{ king}) = \tfrac{1}{13} + \tfrac{1}{13}$

$$= \tfrac{2}{13}$$

RULE II

The probability of two events *A and B* occurring is given by the product of their individual probabilities. The events may be independent or dependent.

For independent events $P(A \text{ and } B) = P(A) \times P(B)$
For dependent events $P(A \text{ and } B) = P(A) \times P(B/A)$

Independent events are those where the result of event A has no effect on the result of event B. For example, the result of the second toss of a coin is not influenced by the result of the first toss.

   Dependent events are those where the result of the first event A affects the probability of the second event B occurring. This is written as $P(B/A)$. For example, the selection of an ace from a pack of cards will affect the probability of selecting, at random, a second ace if the first ace is not replaced.

Example 13.4   What is the probability *of* selecting at random, an ace and then a king of any suit from a well-shuffled pack (i) with replacement of the first selected card before selecting the second card, (ii) without replacement?

(i)   *With replacement* – therefore the events are independent.

$$P(\text{ace}) = \tfrac{4}{52} = \tfrac{1}{13}$$

$$P(\text{king}) \quad = \tfrac{1}{13}$$

$$P(\text{ace and king}) = P(\text{ace}) \times P(\text{king})$$

$$= \tfrac{1}{13} \times \tfrac{1}{13}$$

$$= \tfrac{1}{169}$$

(ii) *Without replacement* – the events are now dependent since the probability of selecting the second card is affected by removing the first card.

$$P(A) = P(\text{ace}) = \tfrac{4}{52} = \tfrac{1}{13}$$

$$P(B/A) = P(\text{king}) = \tfrac{4}{51}$$

Therefore $P(\text{ace and king}) = P(A) \times P(B/A)$

$$= \tfrac{1}{13} \times \tfrac{4}{51}$$

$$= \tfrac{4}{663}$$

(a) Tree diagrams

Diagrams of the type shown in Fig. 13.1 are useful for determining compound probabilities. Consider three tosses of a fair coin. The eight possible results are shown in the tree diagram (Fig 13.1), where $\tfrac{1}{2}$ represents the probability of getting a particular result. The probability of obtaining any one of these results can be obtained by using Rule II; that is, by multiplying the probabilities in each branch leading to that result. For example, $P(HHT) = \tfrac{1}{2} \times \tfrac{1}{2} \times \tfrac{1}{2} = \tfrac{1}{8}$.

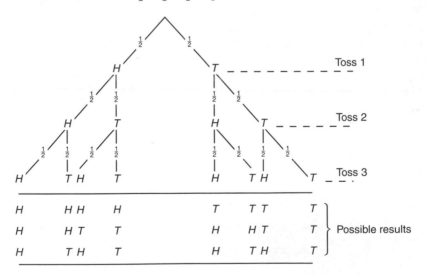

Fig. 13.1

It will be noticed that if the order in which the heads and tails appear is not important then some of these eight results are equivalent.

The probabilities of obtaining a given number of heads and tails, irrespective of order, can be found by applying the formula $x/(x+y)$ or applying the probability rules to the tree diagram.

Probability of 3 heads $= \dfrac{x}{x+y} = \dfrac{1}{1+7}$ (or $\frac{1}{2} \times \frac{1}{2} \times \frac{1}{2}$ using Rule 1.)

i.e. $\qquad P(3H) = \frac{1}{8}$

Probability of

$\qquad$ 2 heads, and 1 tail $= \dfrac{3}{3+5} \qquad\qquad$ (or $\frac{1}{8} + \frac{1}{8} + \frac{1}{8}$ using Rule 2.)

$\qquad\qquad P(2H, 1T) = \frac{3}{8}$

Similarly, $\quad P(2T, 1H) = \frac{3}{8}$

$\qquad\qquad P(3T) \qquad = \frac{1}{8}$

By addition total probability $= \frac{1}{8} + \frac{3}{8} + \frac{3}{8} + \frac{1}{8} = 1$ that is, the probability of any one of the results occurring is a certainty.

Note there are two branches from each junction; one branch for getting a head and the other branch for not getting a head (that is, getting a tail). For any tree diagram the sum of the probabilities of the two branches at any junction is always 1.

EXERCISE 13.1

1. What is the probability of throwing a one and then a six with two throws of an unbiased die?

2. What is the probability of throwing not less than a four with one throw of a die?

3. What is the probability of selecting at random (a) an even number, (b) a number not less than six from the prime digits 1, 2, ..., 9?

4. An urn contains 6 black balls and 4 white balls. What is the probability of selecting (a) 2 white balls, (b) 2 black balls, (c) 1 black ball and 1 white ball if there is (i) replacement after each selection, (ii) no replacement?

5. What is the probability of selecting, at random, a hand consisting of 2 aces and 1 king from a well-shuffled pack?

6. A bag contains 6 red balls and 4 black balls. Given that three balls are randomly withdrawn from the bag find the probability that (a) all 3 balls are black, (b) 1 black and 2 red balls are chosen, (c) at least one ball is black.

7. Given that two dice are thrown simultaneously determine the probability of obtaining a total of (a) 4, (b) less than 4, (c) at least 10.

8. An unbiased die is thrown three times. Determine the probability of obtaining (a) 3 sixes, (b) 2 sixes, (c) 1 six, (d) no sixes. Use the tree diagram method.
9. A packet of 48 matches contains 4 used matches. Find the probability of selecting with replacement (a) 2 used matches, (b) 2 good matches in two random selections.
10. A box contains 10 microscopic slides, 4 of which are of unhealthy cells. Find the probability of selecting with replacement (a) 3 unhealthy cells, (b) 1 unhealthy cell, (c) all healthy cells, in three selections from this box.

13.4   Probability distributions

It was shown in Section 12.4 that the distribution pattern of the frequencies with which all the events occur in a trial can be represented graphically (histogram, frequency polygon, etc.) and mathematically (mean and standard deviation).

Since probability has been defined as limiting relative frequency, then probability distributions can also be similarly represented.

The three theoretical distributions discussed below can be arrived at by probabilistic reasoning.

(a) Binomial distribution

For discrete events, such as the number of heads obtained when a coin is tossed several times or the number of faulty bolts in a sample, the probabilities can be calculated using a tree diagram (Section 13.3). Tree diagrams, however, are very cumbersome if there are more than say 5 tosses of a coin or more than 5 objects in a random sample. A more analytical approach is therefore required, and it is found that the binomial theorem will give the probabilities of discrete events, provided

(i)   There are only two possible outcomes in each trial, that is, a success or a failure;
(ii) The probability of a success or failure in each trial is always constant.

Let n = number of trials   (e.g. the throws of a die)

p = probability of a success in one trial

(e.g. of getting a six in one throw)

q = probability of a failure in one trial

(e.g. of not getting a six)

Therefore

$$p + q = 1$$

The probabilities of obtaining 0, 1, 2, 3, ... successes in n trials are given by the successive terms of the binomial expansion of $(q + p)^n$

$$(q+p)^n = q^n + nq^{n-1}p + \frac{n(n-1)}{1.2}q^{n-2}p^2 + \cdots + p^n \qquad \text{(i)}$$

$$1 = P(0) + P(1) \quad + P(2) \qquad\qquad + \cdots + P(n)$$

Consider the tree diagram problem in Section 13.4 where a coin is tossed 3 times.

$p =$ probability of a success, that is, a head obtained in 1 throw $= \frac{1}{2}$

$q =$ probability of failure, that is, a tail obtained in one throw $= \frac{1}{2}$

If the number of throws is 3 then $n = 3$
Substituting these values into equation (i) above gives

$$\left(\frac{1}{2}+\frac{1}{2}\right)^3 = \left(\frac{1}{2}\right)^3 + 3\left(\frac{1}{2}\right)^2\frac{1}{2} + \frac{3(3-1)}{1.2}\left(\frac{1}{2}\right)\left(\frac{1}{2}\right)^2 + \left(\frac{1}{2}\right)^3 \qquad \text{(ii)}$$

$$1 \quad = \quad \frac{1}{8} \quad + \quad \frac{3}{8} \quad + \quad \frac{3}{8} \quad + \quad \frac{1}{8}$$

$$\text{P(0 head)} \quad \text{P(1 head)} \qquad \text{P(2 heads)} \qquad\quad \text{P(3 heads)}$$

which are the results obtained with the tree diagram.

Note: If any term is examined in equation (i), say the third term, the coefficient $n(n-1)/(1.2)$ is the number of ways of obtaining two successes. In the second term in (ii), for example, the coefficient 3 is the number of ways of getting two heads, that is HHT, HTH, THH.

The probability of getting 2 heads

$$P(2 \text{ heads}) = 3 \times P(\text{HHT}) = 3 \times \tfrac{1}{2} \times \tfrac{1}{2} \times \tfrac{1}{2}$$

This statement uses the two rules of probability and applies to all terms of the binomial theorem in equation (i).

Example 13.5   What is the probability of getting 0, 1, 2, 3 or 4 sixes in four throws of an unbiased die?

Probability p of a six in one throw $= \frac{1}{6}$, therefore $q = \frac{5}{6}$.
Number of trials $n = 4$

$$\left(\frac{5}{6}+\frac{1}{6}\right)^4 = \left(\frac{5}{6}\right)^4 \qquad + 4\left(\frac{1}{6}\right)\left(\frac{5}{6}\right)^3 + \frac{4.3}{1.2}\left(\frac{1}{6}\right)^2\left(\frac{5}{6}\right)^2$$

$$1 = 0.482 \qquad\qquad + 0.386 \qquad\qquad + 0.116$$
$$P(\text{no sixes}) \qquad P(1 \text{ six}) \qquad\qquad P(2 \text{ sixes})$$

$$+ \frac{4.3.2}{1.2.3}\left(\frac{1}{6}\right)^3\left(\frac{5}{6}\right) + \left(\frac{1}{6}\right)^4$$

$$+ 0.015 \qquad\qquad\qquad + 0.001$$
$$P(3 \text{ sixes}) \qquad\qquad\qquad P(4 \text{ sixes})$$

It may be helpful to compare the binomial method with the tree diagram method for the above example.

Example 13.6   Over a long period of time it is known that 5% of the total production of transistors are below standard. If a random sample of 6 transistors is chosen, what is the probability of getting (a) 6 good transistors, (b) 1 bad transistor (c) at least 2 bad transistors?

Probability of 1 bad transistor in one choice $p = \frac{1}{20}$ therefore $q = \frac{19}{20}$ (a 'success' is getting 1 bad transistor). Also $n = 6$

$$\left(\frac{19}{20}+\frac{1}{20}\right)^6 = \left(\frac{19}{20}\right)^6 + 6\left(\frac{1}{20}\right)\left(\frac{19}{20}\right)^5 + \frac{6.5}{1.2}\left(\frac{1}{20}\right)^2\left(\frac{19}{20}\right)^4 + \cdots + \left(\frac{1}{20}\right)^6$$

$$1 = 0.735 + 0.232 + \qquad 0.031 + \cdots$$

$$1 P\left\{\begin{matrix}0 \text{ bad}\\6 \text{ good}\end{matrix}\right\} + P\left\{\begin{matrix}1 \text{ bad}\\5 \text{ good}\end{matrix}\right\} + P\left\{\begin{matrix}2 \text{ bad}\\4 \text{ good}\end{matrix}\right\} + \cdots + P\left\{\begin{matrix}6 \text{ bad}\\0 \text{ good}\end{matrix}\right\}$$

(a) Probability of 6 good
   transistors $= 0.735$

(b) Probability of 1 bad
   transistor $= 0.232$

(c) Probability of at least
   2 bad transistors $= P(2 \text{ bad}) + P(3 \text{ bad}) + P(4 \text{ bad})$
   $\qquad\qquad\qquad\qquad\qquad + P(5 \text{ bad}) + P(6 \text{ bad})$
   $\qquad\qquad\qquad\quad = 1 - [P(6 \text{ good}) + P(5 \text{ good})]$
   $\qquad\qquad\qquad\quad = 1 - [0.735 + 0.232]$
   $\qquad\qquad\qquad\quad = 1 - 0.967$
   $\qquad\qquad\qquad\quad = 0.033$

The probabilities given by the binomial expansion can be represented graphically. This has been done in Fig 13.2 by drawing a histogram for the expansion in Example 13.5.

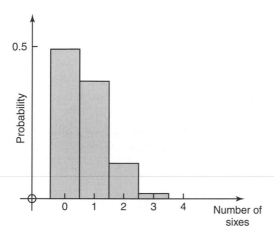

Fig. 13.2

Even though the vertical scale is one of probability, because the widths of the rectangles for these discrete results are made equal, the probabilities are also represented by the *areas* of the rectangles. This point will need to be remembered when dealing with normal distributions.

The binomial distribution may be represented mathematically by two figures:

1. mean value (or expectancy), $\bar{x} = np$;
2. standard deviation, $s = \sqrt{(npq)}$.

For Example 13.6, $p = \dfrac{1}{20}$ (i.e. of getting a bad transistor)

$$q = \frac{19}{20}, \quad n = 6$$

Mean value $\bar{x} = np = 6 \times \dfrac{1}{20} = 0.3$

Standard deviation $s = \sqrt{(npq)} = \sqrt{\left(6 \times \dfrac{1}{20} \times \dfrac{19}{20}\right)} = 0.534$

The mean value suggests that over a period of time one would reasonably expect an average of 0.3 bad transistors in each sample.

If the average number of bad transistors in each sample is found to differ significantly from this mean value it suggests either that the production is still continuing at a 5% failure rate but the sampling technique is biased in some way or that the production failure rate has changed from this 5% value. Further investigation would be desirable, such as immediately checking a second sample.

It will be seen later in this chapter that a 'significant' difference from this expectancy value can be expressed in terms of the standard deviation value.

EXERCISE 13.2

1. An unbiased coin is tossed 6 times. What is the probability of obtaining (a) 0, 1, 2, 3, 4, 5 or 6 heads, (b) not more than 3 heads? Represent the results for (a) on a histogram.

2. Four per cent of the components produced by a milling machine are below standard. If a sample of five components is chosen at random what are the chances that the sample contains (a) 2 below standard, (b) none below standard?

3. An inspector selects a sample of 10 articles at random from a large batch which contain 10% defectives. This batch will be rejected if he finds more than 2 defectives in the sample. What is the probability that the batch will be rejected?

4. An examination paper consists of 10 questions, and the answer to each question must be selected from 4 alternatives. If a student guesses the answer to each question what is the probability that he will gain 5 or more questions correct?

5. An unbiased coin is tossed 500 times. What is the mean number of heads expected and the standard deviation of the number of heads?

6. The cut-off valve in a chemical plant is found to operate on average twice in every five days. What is the probability that in any 5 successive days it will operate (a) once, (b) every day, (c) not at all?

7. An unbiased die is thrown 600 times. How many times would you expect to obtain (a) a six, (b) a result greater than 3?

8. Ten per cent of articles in a factory are defective. The articles are packed in cartons of 100 articles and the inspection process involves taking samples of 8 from each carton. A carton is rejected if the sample contains more than one defective. How many cartons are likely to be rejected from a consignment of 30 000 articles?

(b) Normal distribution

As the number of trials n increases the frequency polygon for a binomial distribution will contain an increasing number of points. When n is very large the polygon will approach a curve.

A binomial distribution is not symmetrical unless $p = q = \frac{1}{2}$. However, if n is very large then this limiting frequency curve will always be approximately symmetrical irrespective of the values of p and q. It will be appreciated that, strictly speaking, a continuous curve cannot be drawn for a discrete variable, such as the number of heads, but it can be drawn for a continuous variable such as the lengths of bars from a production line. This symmetrical bell-shaped curve (Fig. 13.3a) is known as the **normal** or **Gaussian** curve.

If a continuous variable is subjected to small random disturbances then it can have a range of values. The frequency or probability of each of these values occurring forms a normal distribution. Examples of this are the heights of all adults in this country or volumes of liquid being automatically measured into jars at a factory over a long period of time.

The normal curve has the equation

$$y = \frac{1}{s\sqrt{(2\pi)}} e^{-(x - \bar{x})^2 / 2s^2}$$

where x is the value of the variable, $\bar{x} = $ mean value and $s = $ standard deviation, and the curve is symmetrical about the mean value $\bar{x}$. The curve can be standardised by changing the horizontal axis to X where $X = (x - \bar{x})/s$ is the number of standard deviations as shown in Fig. 13.3(b). Since X has no units the properties of the standardised curve can be used for any variable that has a normal distribution.

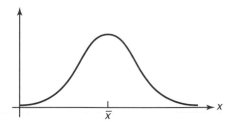

Fig. 13.3(a)

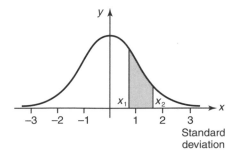

Fig. 13.3(b)

The equation now becomes

$$y = \frac{1}{s\sqrt{(2\pi)}} e^{-X^2/2}$$

and this curve has the following important property. It can be shown by integration that the total area under this curve between $X = -\infty$ and $X = +\infty$ is equal to unity. This represents the probability that the variable is in the range $-\infty$ to $+\infty$. Since a frequency curve is the limiting case of a histogram then the area between any two values X_1 and X_2 (shaded in Fig 13.3b) represents the probability of the variable occurring between these two values.

(c) Normal probability tables

Statistical tables are available giving the areas under the normal curve to the *left* of selected *positive* values of X. In other words it gives the probability of the variable occurring in the range $-\infty$ to X or $P(< X)$. For example, if $X = 1.2$, then from tables $P(> X) = 0.8849$ (this is the shaded area in Fig. 13.4).

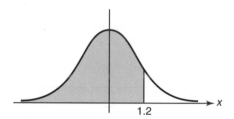

1.2

Fig. 13.4

Probabilities for negative X values can be obtained using the symmetry of the curve; that is,

$$P(< -X) = 1 - P(< X)$$

For $X = -1.2$

$$P(< -1.2) = 1 - P(1.2)$$

$$= 1 - 0.8849 = 0.115\,1$$

This is represented by the shaded area in Fig. 13.5.

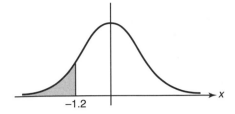

Fig. 13.5

(d) Confidence limits

For a normal distribution:
95% of the values of the variable lie in the range

$$-1.96 < X < 1.96$$

99% of the values of the variable lie in the range

$$-2.58 < X < 2.58$$

In other words, there is a 95% or 99% probability, that, if any object is chosen at random from a normal distribution, the value of the variable for the object will lie in ranges $\bar{x} \pm 1.96s$ or $\bar{x} \pm 2.58s$ respectively (since $X = (x - \bar{x})/s$). These limiting values of the variable are called the 95% *confidence limits* and the 99% *confidence limits*.

It is worth nothing that virtually all (99.7%) of all the values in a normal distribution are within 3 standard deviations either side of the mean.

Example 13.7 A machine produces components with a mean diameter of 20 mm and a standard deviation of 0.2 mm. Assuming a normal distribution, what proportion of the components have a diameter between 19.7 mm and 20.4 mm?

$$\bar{x} = 20 \, \text{mm}, \qquad s = 0.2 \, \text{mm}$$

For 19.7 mm, $X_1 = \dfrac{x - \bar{x}}{s} = \dfrac{19.7 - 20}{0.2} = -1.5$

For 20.4 mm, $X_2 = \dfrac{20.4 - 20}{0.2} = 2.0$

The area required is shaded in Fig. 13.6.

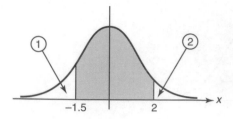

Fig. 13.6

Now unshaded area (1) $= P(< -15) = 1 - P(<1.5)$

$$= 1 - 0.933\,2 \quad \text{(from probability tables)}$$

$$= 0.066\,8$$

Unshaded area (2) $= P(>2)$ $= 1 - P(<2)$

$$= 1 - 0.977\,2 \quad \text{(from tables)}$$

$$= 0.022\,8$$

Therefore

proportion between 19.7 and 20.4 $= 1 - (0.066\,8 + 0.022\,8)$

$$= 0.910\,4 \text{ or } 91\%$$

Example 13.8   The production of capacitors is normally distributed about a mean value of $60\,\mu\text{F}$ with a standard deviation of $0.4\,\mu\text{F}$. If 2% of these capacitors are being rejected what are the design limits, assuming they are equispaced about the mean? Give the 99% confidence limits for these capacitors.

Let the design limits be x_1 and $x_2\,\mu\text{F}$. Since the acceptable range is symmetrically disposed about the mean, 1% of the capacitors

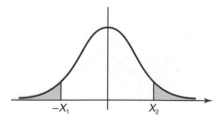

Fig. 13.7

are rejected for being too low and 1% are rejected for being too high (shaded areas of Fig. 13.7).

Therefore $|X_1| = |X_2|$ and $P(>X_2) = 0.01$ Therefore $P(<X_2) = 0.99$.

From tables $X_2 = 2.327$. But

$$X_2 = \frac{x_2 - 60}{0.4} \qquad \text{since } X = \frac{x - \bar{x}}{s}$$

$$2.327 = \frac{x_2 - 60}{0.4}$$

$$x_2 = 2.327 \times 0.4 + 60$$

$$= 60.93 \, \mu F$$

Similarly

$$x_1 = 60 - 0.93$$

$$= 59.07 \, \mu F$$

Therefore design limits are $60 \pm 0.93 \, \mu F$

99% confidence limits are $\bar{x} \pm 2.58s = 60 \pm 2.58 \times 0.4$

$$= 60 \pm 1.03 \, \mu F$$

That is, 99% of the capacitors will be in this range.

(e) Normal distribution test

In Section 12.6(d) it was seen that plotting cumulative frequency against the variable x for a set of data which appears to be normally distributed produces an ogive. If the same data is plotted on normal probability paper the points will lie on, or close to, a straight line if the data has a normal distribution. This is achieved by creating a non-linear vertical scale (see Fig. 13.8). In Fig. 13.3(b) it was seen that the probability of a value x in between two values x_1 and x_2 is the shaded area under the curve.

The vertical scale on normal probability paper on which the percentage cumulative frequency (CF) is plotted is non-linear because the distances between the percentage values are made proportional to the areas under the normal curve This non-linear scale has the effect of distorting the ogive into a straight line.

Once it is verified that that the data has a normal distribution the mean and the standard deviation can be found from the graph as follows:

1. The mean $\bar{x}$ occurs at 50% CF.
2. The standard deviation s occurs at 84.13% CF.
    Therefore $s = x_s - \bar{x}$ where x_s is the x value at 84.13% CF.

Example 13.9 A sample of 150 resistors was checked and gave the results shown in Table 13.1.

Table 13.1

Resistance (Ω)	100	101	102	103	104	105	106
Number	6	18	34	52	25	12	3

Use normal probability paper to check if these results have a normal distribution, and if so use the graph to find the mean and standard deviation.

The cumulative frequency is shown in Table 13.2 A graph of % cumulative frequency is plotted on probability paper as shown in Fig. 13.8. Since the graph is approximately straight the above results may be taken as normally distributed.

Table 13.2

Maximum resistance in class (Ω)	Cumulative frequency	% cumulative frequency
100.5	6	4
101.5	24	16
102.5	58	38.7
103.5	110	78.3
104.5	135	90
105.5	147	98
106.5	150	100

From the graph

the mean value at 50% CF $\bar{x} = 102.8\,\Omega$

the value of x_s at 84.13% CF $= 103.8\,\Omega$

the standard deviation: $s = 103.8 - 102.8 = 1.0\,\Omega$

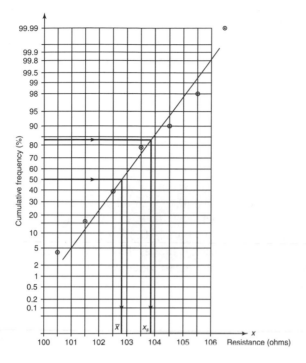

Fig. 13.8

EXERCISE 13.3

1. The mean mass of 600 students at a technical college is 70 kg with a standard deviation of 5 kg. Assuming that the mass is normally distributed, determine the number of students with a mass (a) more than 81 kg, (b) less than 64 kg, (c) between 67 and 75 kg.

2. The pass mark in an examination is 40. From previous examinations it is known that the marks are normally distributed with a mean of 48 and a standard deviation of 15. If 1000 students try this examination how many students should pass?

3. Washers are packed into boxes and it was found that the mean number in each packet was 250 washers with a standard deviation of 10. Assuming a normal distribution, what proportion of packets will contain less than 246 washers? Determine the 95% and 99% confidence limits.

4. The life of an electric light bulb is normally distributed about a mean value of 2500 hours with a standard deviation of 70 hours. Determine, for a batch of 6000 bulbs,
    (a) how many bulbs last between 2350 and 2600 hours?
    (b) how many bulbs will last more than 2650 hours?
    (c) for what length of time will 90% of the bulbs last?

5. If a set of measurements of the diameter of a human cell are normally distributed, what percentage of these cells differ from the mean by (a) more than $\frac{1}{4}$ of the standard deviation, (b) less than $\frac{1}{2}$ of the standard deviation?

6. A die is thrown 180 times. Assuming that the probability of the number of sixes is normally distributed, what is the probability that
    (a)   40 or more sixes will appear?
    (b)   the number of sixes will be between 20 and 40?
    (c)   more than 20 sixes will appear?
    (*Hint*: For a binomial distribution $\bar{x} = np$ and $s = \sqrt{(npq)}$).

7. Metal blocks are being produced with a mean length of 41.31 mm and a standard deviation of 0.03 mm. Assuming a normal distribution, what percentage of the blocks are likely to be rejected if the specification dimensions are 41.30 ± 0.03 mm?

8. Use probability paper to check if the results in Tables 13.3, 13.4 form a normal distribution. Find the mean value and the standard deviation from the graphs where possible.

Table 13.3

Failure load (kN)	40	41	42	43	44	45	46	47	48	49
Number of bolts	5	9	18	22	37	25	17	8	6	3

Table 13.4

Mass of wastings (kg)	50	51	52	53	54	55	56	57
Number of castings	15	37	97	135	113	48	23	7

(f) Poisson distribution

If the probability p of an event occurring is very small (that is, it is a rare event) and the number of trials n is large, then the terms of the binomial expansion are not easily evaluated.

The French mathematician Poisson showed that an easier method, involving the exponential function, can be used to give a close approximation to these terms, provided n is large and p is small (as a rough guide $n > 50$ and $\bar{x} = np < 5$).

The Poisson distribution will also be a model for rare events where the mean or expectancy value $\bar{x}$ is known but n and p are not known. This situation arises with time-based problems such as aircraft crashes, goals in a football match, or production flaws in manufacturing.

The probabilities are given by the terms in the series

$$e^{-\bar{x}}\left[1+\frac{\bar{x}}{1!}+\frac{(\bar{x})^2}{2!}+\frac{(\bar{x})^3}{3!}\cdots\right]$$

$$1 = P(0) + P(1) + P(2) + P(3) + \cdots$$

Since the distribution is for rare events the probability values in the series will diminish rapidly. The Poisson distribution is therefore skewed as shown in Fig. 13.9. A Poisson distribution has a mean $\bar{x}$ and a standard deviation $\sqrt{\bar{x}}$.

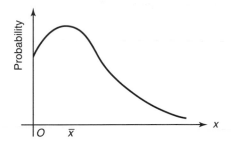

Fig. 13.9

Example 13.10   A machine is known to produce 1% defective components. The components are packed in packets containing 100 components. What is the probability that each packet will contain (a) 0, 1 or 2 defectives, (b) more than 2 defectives?

Draw the defective probability distribution for this machine. Mean value or expectancy of a defective in each box:

$$\bar{x} = np = 100 \times 0.01 = 1$$

(a)   Probability of 0 defectives in the box:

$$P(0) = e^{-\bar{x}} \quad = e^{-1} \quad = 0.3679$$

Probability of 1 defective in the box:

$$P(1) = \bar{x}e^{-\bar{x}} \quad = 1e^{-1} \quad = 0.3679$$

Probability of 2 defectives in the box

$$P(2) = \frac{\bar{x}^2}{2!}e^{-\bar{x}} = \frac{1^2}{2}e^{-1} = 0.1840$$

(b)   Probability of more than two defectives

$$P(>2) = 1 - P(\leqslant 2) \text{ where } \leqslant \text{ means less than or equal to}$$

$$= 1 - [P(0) + P(1) + P(2)]$$

$$= 1 - [0.367\,9 + 0.367\,9 + 0.184\,0]$$

$$= 1 - 0.9198$$

$$= 0.0802$$

In the same way as (a), $P(3) = 0.0613$, $P(4) = 0.0153$, $P(5) = 0.0031$. Plotting these values the probability distribution is shown in Fig. 13.10.

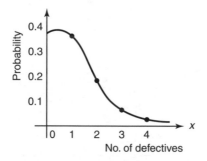

Fig. 13.5

Note: The binomial theorem could have been used since n, p and q are known.

Example 13.13   A garage workshop has one expensive tool which is used on average 1.6 times per 8-hour day. Each time it is used for 4 hours. How many days during a working period of 60 days is the tool (a) not required at all, (b) required once, (c) required twice, (d) required more than twice?

Expectancy $\bar{x} = 1.6$.

(a)   Probability of not being required

$$P(0) = e^{-\bar{x}} = e^{-1.6} = 0.2019$$

Therefore in 60 days the tool is not required for $60 \times 0.2019 = 12$ days.

(b) Probability of being required once

$$P(1) = \bar{x}e^{-\bar{x}} = 1.6e^{1.6} = 0.3230$$

Therefore in 60 days the tool is required once for 60 × 0.3230 = 19 days.

(c) Probability of being required twice

$$P(2) = \frac{\bar{x}^2}{2!}e^{-\bar{x}} = \frac{1.6^2}{2}e^{-1.6} = 0.2584$$

Therefore in 60 days the tool is required twice for 60 × 0.2584 = 16 days.

(d) The tool is required more than twice on 60 − (12 + 19 + 16) = 13 days.
For one quarter of the days a second tool would be useful to avoid waiting time. It is therefore worth considering buying another tool, even though it is expensive.

EXERCISE 13.4

Assume Poisson distributions in all questions.

1. A factory produces electrical resistors which are found to contain 1% defectives. What is the probability of obtaining (a) no defective, (b) 1 defective, (c) at least 2 defectives in a box of 50 resistors?
2. If the probability that a child is allergic to a particular drug is 0.003, determine the probability that if 500 children are injected with this drug (a) 4 children or (b) more than 1 child, will react unfavourably.
3. For many weeks the average number of goals scored by a team in each of their matches has been 2. What is the probability that they will score (a) 1 goal, (b) 2 goals, (c) 3 goals, (d) more than 3 goals in their next match?
4. The average number of telephone calls passing through a particular switchboard is 3 calls per minute. What is the probability that during any one minute there will be (a) 0, 1, 2 calls, (b) less than 2 calls, (c) more than 5 calls?
5. On average 0.8 lorries in a fleet of 40 break down in each week. What is the probability that in a given week (a) no lorries will break down, (b) at least one lorry will break down, (c) more than 2 lorries will break down?
6. Five per cent of the bolts produced by a machine are defective because they are oversize. Given that 50 boxes each containing 20 bolts are chosen at random, find the number of boxes which should contain 0, 1, 2, 3, 4, 5 defective bolts.

7. Samples of 50 are taken from a continuous process and successive samples gave the following number of defects:

    2,   0,   1,   0,   1,   2,   3,   2,   1,   1

    What are the probabilities of getting a particular sample with
    (a)   no defectives;
    (b)   1 defective;
    (c)   2 defectives;
    (d)   3 or more defectives;
    (e)   less than 3 defectives?
8. The number of painting defects on coated steel sheets are shown by the Table 13.5.

Table 13.5

Number of defects per sheet	0	1	2	3	4	5
Number of sheets	109	92	37	9	2	0

Show that these results have a Poisson distribution.

13.5   Sampling theory

The purpose of sampling is to obtain information about a large set of data by examining a smaller subset of the same data. This subset is called the **sample** and the total set the **parent population**.

The two values usually obtained from the sample are the mean value $\bar{x}$ and the standard deviation s. From these values an estimate is made of the mean value $\bar{x}_p$ and the standard deviation s_p of the parent population.

(a) Estimation of $\bar{x}_p$

If every possible sample of size n is taken from the parent population, with replacement, the mean values of these samples will themselves form a normal distribution provided

1. the population forms a normal distribution, or
2. for a non-normal population $n \geqslant 30$.

Fig. 13.11 shows that as one would expect the sample means are more closely packed around the parent population mean than the individual values.

Consider a sample of size n with values $x_1, x_2, x_3, ..., x_n$. The mean value of the sample is

$$\bar{x} = (x_1 + x_2 + x_3 + \cdots + x_n)/n$$

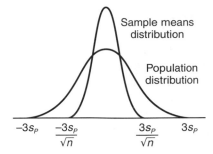

Fig. 13.11

This mean value will vary from sample to sample. The standard deviation of the sample means across all the samples is given by $s_p/\sqrt{n}$.

Thus assuming the means of the samples to be normally distributed, within a 95% confidence limit it is 95% certain that the population mean $\bar{x}_p$ is in the range ± 1.96 standard deviations about the mean. Therefore the estimate of $\bar{x}_p$ is

$$\bar{x}_p = \bar{x} \pm 1.96 \cdot \frac{s_p}{\sqrt{n}} \qquad \text{(i)}$$

(b) Estimation of s_p

It is necessary to know s_p and that condition (1) or (2) are satisfied in order to use equation (i).

For a small sample $(n < 30)$ from a normal distribution for which s_p is unknown the estimate for $\bar{x}_p$ is obtained using the sample standard deviation s and the small sample distribution table called the t-distribution. This approach is not considered here.

The estimate of the population standard deviation from the standard deviation of the sample is given by

$$s_p = s\sqrt{\frac{n-1}{n}}$$

For larger samples $(n \geq 30)$ $\quad \sqrt{\dfrac{n-1}{n}} \geq \sqrt{\dfrac{29}{30}} \simeq 1$

so that s_p may be taken as approximately equal to s.

13.6   Statistical quality control

To ensure that the articles produced by a continuous process are within the required limits it is necessary to take samples at regular intervals.

The information obtained from the samples can then be displayed on charts, called **control charts**, to decide if any variations between the samples are the result of chance fluctuations or due to changes in the production process, such as machine wear or operator error.

Limit lines are drawn on the control charts at certain prescribed probability values. Variations within these lines are considered to be due to chance fluctuations but if the points drift outside these lines the production process needs reassessing.

This control can be carried out by measuring the mean and dispersion values of the variables or by finding the number of defectives in each sample.

(a) Control by variable

For a process producing articles which are normally distributed about a mean value of x_p with a standard deviation s_p then for samples of size n

1. 95% of all sample mean values lie in the range $\bar{x}_p \pm 1.96 \dfrac{s_p}{\sqrt{n}}$

that is only 2.5%, or 1 in 40 samples taken at random should have a mean value below or above this range by chance. If this figure is exceeded it suggests a possible change in the mean value of the population and therefore the process needs increased monitoring. These limits are called **warning limits**.

2. 99.8% of all sample mean values lie in the range $\bar{x}_p \pm 3.09 \dfrac{s_p}{\sqrt{n}}$

that is only 0.1% or 1 in 1000 samples taken at random should have a mean value below or above this range by chance. Since this is a very small probability, if it is exceeded it indicates a change in the mean value of the population and immediate action is required to control the process. These limits are called **action limits**. A typical control chart is shown in Fig. 13.12.

The action and warning limits relate to the capability of the machine or process itself. These must be considered in conjunction with the upper and lower tolerance limits T_u and T_L specified for the articles. Since these tolerance values are for individual articles the width of the population curve $(x_p \pm 3.09s_p)$ should not be greater than AA (Fig. 13.13) if scrap articles are to be avoided. For this to be the case the range AB must not be less than

$$3.09s_p - 3.09\frac{s_p}{\sqrt{n}}$$

The control chart (Fig. 13.12) is showing a tendency to drift to the upper action limit. Further samples should be taken immediately to check if the last two values are occurring by chance or if the process is changing for some reason, such as machine wear etc.

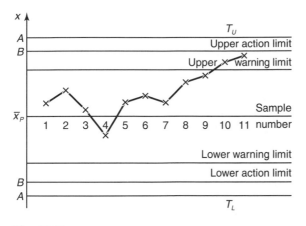

Fig. 13.12

Dispersion values are usually measured from standard deviation, but under workshop conditions this is more difficult. Consequently, keeping in mind that large fluctuations are unlikely in a controlled production process, the range of each sample is used instead in the control process. A range chart with suitable warning and action lines can be created in a similar way to Fig. 13.12 to monitor dispersion.

(b) Control by defective

For a process producing articles which are checked on a defective or good basis the probability distribution will be either binomial or Poisson.

A control chart can be created which records the number of defectives in each sample with warning and action lines drawn at suitably chosen probability levels.

Consider the monitoring of the process in Example 13.10

$$P(>2 \text{ defectives}) = 0.0802 = 8.02\%$$

$$P(>3 \text{ defectives}) = 1 - [0.3679 + 0.3679 + 0.1840 + 0613]$$

$$= 0.0189 = 1.89\%$$

The warning limit could be set at 2 defectives since, although there is a reasonable probability (8.02%) of getting 2 defectives in a box of 100 from a process operating at the accepted 1% defective rate, the probability of getting two successive boxes with more than two defectives in each box is $0.08 \times 0.08 = 0.0064$ (0.64%), which is very small.

The action limit could be set at 3 defectives since there is only a very small probability (1.89%) of getting more than 3 defectives in any box with a 1% defective rate. The control chart is shown in Fig. 13.13.

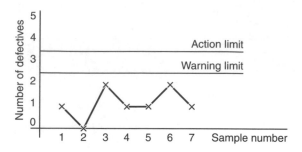

Fig. 13.13

Example 13.12   200 tablets are chosen at random from a wholesale chemist. Analysis of the tablets showed that the mean aspirin content in each tablet is 80.0 mg with a standard deviation of 1.80 mg. Calculate the 95% confidence limits for the mean aspirin content of the chemist's stock.

Since n is large $s_p = s = 1.80\,\text{mg}$

$$95\% \text{ confidence limits } \bar{x}_p = 80.0 \pm \frac{1.96s_p}{\sqrt{n}}$$

$$= 80.0 \pm \frac{1.96 \times 1.80}{\sqrt{200}}$$

$$= 80.0 \pm 0.25\,\text{mg}$$

or $\qquad\qquad \bar{x}_p = 79.75\,\text{mg to } 80.25\,\text{mg}$

EXERCISE 13.5

1. A sample of 50 nylon-reinforced driving belts are tested and the mean breaking stress is found to be $120\,\text{kN/m}^2$ with a standard deviation of $3.64\,\text{kN/m}^2$. Between what limits is it 99% certain that the mean breaking stress of all the belts produced by the factory will lie?

2. Resistors produced by a factory over a period of 12 months are known to have a nominal value of 500 ohms with a standard deviation of 25 ohms. What must be the size of the samples to be 95% confident that the sample mean is in the range 494 to 506 ohms?

3. A normal population has a mean of 21.2 and a standard deviation of 0.8. If a random sample of 200 is chosen from this population what is the probability that the sample mean will be within 0.1 of the population mean?

4. A certain type of electric light bulb is known to be normally distributed with a mean of 1500 hours and a standard deviation of 100 hours. What is the probability that
    (a) a single bulb chosen at random will have a life greater than 1570 hours?
    (b) a single bulb chosen from a random sample of 10 bulbs will have a mean life greater than 1570 hours?

5. A random sample of 50 shafts has a mean mass of 28.2 kg and a standard deviation of 1.7 kg. Assuming the population has a normal distribution calculate the 95% confidence estimate of the population mean.

6. Resistors are being produced with a normal distribution and a known standard deviation of $0.17\,\Omega$. If a random sample of 5 resistors had values of 60.2, 60.4, 61.0, 60.6, 60.3 ohms give an estimate of the production mean at the 99% confidence level. Do the results agree with a population standard deviation of $0.17\,\Omega$?

MISCELLANEOUS EXERCISE 13

1. (a) Twenty per cent of the items produced by a factory are defective. If 4 items are chosen at random what is the probability that (i) 2 are defective, (ii) 3 are defective, (iii) at least one is defective? (Assume a binomial distribution.)
    (b) Articles are being manufactured with a dimension whose mean is 4.522 cm with standard deviation 0.006 in. Tolerance limits are 4.509 cm and 4.531 cm. Find what percentage of the articles will fail to meet the tolerances if the distribution is normal.

2. A computer has 5 terminals; the probabilities of each being engaged are respectively $\frac{3}{4}, \frac{1}{5}, \frac{2}{3}, \frac{2}{5}, \frac{1}{3}$. What is the probability that at any given time:
    (a) all are engaged;
    (b) at least one is engaged;
    (c) one and one only is not engaged?

3. (a) Twenty samples, each of 20 items, contained defectives as shown in Table 13.6. Calculate the mean number of defectives and thus obtain a suitable Poisson distribution for the same conditions. Comment on the

Table 13.6

1	3	1	0	4	1	1	2	3	2
2	1	0	1	2	0	1	0	2	1

closeness of fit between the actual and the theoretical results.

(b) By random sampling of goods supplied by a manufacturer, it is found that a particular consignment contains 5% defectives. Given that a random sample of 10 is taken from this consignment, use the binomial distribution to determine the probability that this sample will contain
(i)   no defectives,
(ii) two defectives,
(iii) more than two defectives.

4. The yields from certain seed potatoes are known to be normally distributed with mean 1 256 g and standard deviation 28 g. Calculate:
(a) the percentage of the seed with an expected yield greater than 1 300.8 g,
(b) the value above which 90.3% of the yields would lie,
(c) the percentage of the seed producing yields in the range $1\,256 \pm 56$ g.

5. (a) Name two measures of position and two measures of dispersion for a set of data.
(b) A scientist thinks that the percentage impurity of batches of a chemical is normally distributed. Assuming that there is a large number of batches being produced, explain briefly how he could test this idea.
(c) A man throws four darts at a target. Each time he throws there is a probability of $\frac{1}{3}$ that he will hit the target. The number of darts out of four that actually hit the target will follow a binomial distribution. Calculate the probability that at least two darts hit the target.

6. (a) In the manufacture of screws by a certain process it was found that 5% of the screws were rejected because they failed to satisfy tolerance requirements. What was the probability that a sample of 8 screws contained (i) exactly one, (ii) more than one reject?
(b) Assuming that the chance of being killed in a colliery accident during a year is $\frac{1}{1000}$, use the Poisson distribution to calculate the probability that in a mine employing 400 miners there will be at least two fatal accidents in a year.

7. The ball bearings produced by a machine have a mean mass of 0.150 kg and a standard deviation of 0.010 kg. What is the probability that, if a random sample of 80 bearings is

chosen, the sample mean is not greater than 0.152 kg? Assume a normal distribution.

8.   Use (a) the binomial distribution, (b) the Poisson distribution to find the probabilities of a box of 100 bolts containing 0, 1 or 2 defective bolts if the bolts have an average failure rate of 1%.

9.   A builder has acquired a large quantity of nuts and bolts of a certain standard size, but some of them are faulty. Given that 4% of the nuts are faulty and 5% of the bolts are faulty, find the probability that if any bolt is selected and paired with any nut chosen at random,
(a)   only the nut will be faulty,
(b)   only the bolt will be faulty,
(c)   both of them will be faulty,
(d)   at least one of them will be faulty,
(e)   neither of them will be faulty.

10.   A container holds 20 articles, four of which are rejects. Determine the probability of randomly selecting two rejects in a sample of five articles. If the same container holds a large number of articles, 20% of which are rejects, find the probability of selecting the same sample as before.

11.   The number of defects in a process was recorded and the results are shown in Table 13.7. check how close this production process is to a Poisson distribution.

Table 13.7

Number of defects	0	1	2	3	4	5
Number of samples	12	29	29	16	10	4

Chapter 14

Complex numbers

14.1 Definition of a complex number

Consider the quadratic equation $x^2 - 4x + 13 = 0$. As shown in Section 1.1(b), the equation can be solved using the formula

$$x = \frac{-b \pm \sqrt{(b^2 - 4ac)}}{2a}$$

$$= \frac{4 \pm \sqrt{(-36)}}{2}$$

$$= 2 \pm 3\sqrt{-1}$$

It is not possible to evaluate $\sqrt{-1}$, but if $\sqrt{-1}$ is written as j then the solutions may be expressed as $2 \pm$ j3. $2 +$ j3 and $2 -$ j3 are known as **complex numbers**. Complex numbers have important applications in vector calculations and alternating current theory.

A complex number z is composed of a **real** part 'a' and an **imaginary** part 'jb', where a and b are ordinary numbers.
Thus

$$z = a + jb$$

In mathematics i is used instead of j, but in electrical work j is used to avoid confusion with the symbol used for current. Two complex numbers are said to be **conjugate** if the only difference between them is the sign of the imaginary part; that is, $a + jb$ is the conjugate of $a - jb$, and vice versa. The conjugate of z is written as $\bar{z}$.

14.2 Operations on complex numbers

The usual arithmetic operations of addition, subtraction, multiplication and division can be performed on complex numbers.

(a) Addition and subtraction

The real and imaginary parts of the complex numbers are collected together independently.

Example 14.1   Add the complex numbers $z_1 = 3 + j4$, $z_2 = -2 + j5$.

$$z_1 + z_2 = (3 + j4) + (-2 + j5)$$
$$= (3 - 2) + j(4 + 5)$$
$$= 1 + j9$$

Example 14.2   Subtract the numbers in Example 14.1.

$$z_1 - z_2 = (3 + j4) - (-2 + j5)$$
$$= 3 + j4 + 2 - j5$$
$$= (3 + 2) + j(4 - 5)$$
$$= 5 - j$$

(b) Multiplication

The usual methods for multiplying brackets are applicable remembering that $j^2 = (\sqrt{-1})^2 = -1$.

Example 14.3   If $z_1 = 3 + j2$ and $z_2 = 5 - j3$ evaluate $z_1 z_2$.

$$z_1 z_2 = (3 + j2)(5 - j3)$$
$$= 15 + j10 - j9 - 6j^2$$
$$= 15 + j + 6$$
$$= 21 + j$$

(c) Division

Division of complex numbers is carried out by converting the complex number in the denominator into an ordinary number. This can be done by multiplying both the numerator and the denominator by the conjugate of the denominator.

Example 14.4 Evaluate the $\dfrac{z_1}{z_2}$ for the numbers in Example 14.3.

$$\frac{z_1}{z_2} = \frac{3+j2}{5-j3}$$

$$= \frac{(3+j2)\,(5+j3)}{(5-j3)\,(5+j3)}$$

$$= \frac{15+j10+j9-6}{25+9}$$

$$= \frac{9}{34} + j\frac{19}{34}$$

$$= 0.27 + j0.56$$

EXERCISE 14.1

1.   Solve the following equations:
     (a) $x^2 + 6x + 13 = 0$
     (b) $3x^2 - 2x + 5 = 0$

2.   Simplify (a) j^3, (b) j^9, (c) $\dfrac{2}{j^2}$, (d) $\dfrac{2}{j^5}$

3.   Given that $z_1 = 3+j2$, $z_2 = -2+j4$, $z_3 = -6-j$ and $z_4 = j3$ evaluate the following, leaving your answers in the form $a+jb$:

   (a) $-z_1$ (b) $z_1 + z_2$ (c) $z_2 - z_3$ (d) $z_1 z_2$

   (e) z_3^2 (f) $\bar{z}_2$ (g) $z_1 z_4$ (h) $\dfrac{z_1}{z_3}$

   (i) $\dfrac{z_1 z_3}{z_2}$ (j) $\dfrac{1}{z_1^2}$ (k) $z_1^2 - z_4^2$ (l) z_1^3

   (m) $\dfrac{1}{\bar{z}_2}$ (n) $\dfrac{1}{z_4}$ (o) $\dfrac{1}{z_4^2}$

14.3   The Argand diagram

A complex number can be represented graphically by a point on an Argand diagram. In an Argand diagram the real part of the complex number is represented on the horizontal axis and the imaginary part is represented on the vertical axis. Cartesian or polar coordinate systems can be used.

(a) Cartesian representation $z = x + jy$

The complex number z is shown in Fig. 14.1. z is represented by the point P where the Cartesian coordinates of P are (x, jy). The addition and

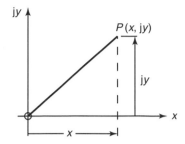

Fig. 14.1

subtraction of complex numbers can be represented graphically. Consider the complex numbers $z_1 = x_1 + jy_1$ and $z_2 = x_2 + jy_2$.

(i) Addition: $z_1 + z_2 = (x_1 + x_2) + j(y_1 + y_2)$

In Fig. 14.2 P_1 and P_2 represent the complex numbers z_2 and z_2. $P_1 R$ is drawn parallel and equal in length to OP_2. It is seen that the horizontal and vertical coordinates of R are $(x_1 + x_2)$ and $j(y_1 + y_2)$. Thus the point R represents the complex number $z_1 + z_2$. Now if these lines are considered as vectors then

$$\overline{OP}_1 + \overline{P_1 R} = \overline{OR}$$

Therefore, since

$$\overline{P_1 R} = \overline{OP}_2$$
$$\overline{OP}_1 + \overline{OP}_2 = \overline{OR}$$

It is seen that complex number addition and vector addition are comparable operations. This is one of the important aspects of complex numbers.

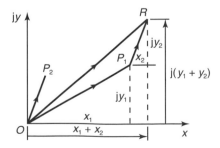

Fig. 14.2

(ii) Subtraction: $z_1 - z_2 = (x_1 - x_2) + j(y_1 - y_2)$

As before, P_1 and P_2 represent the complex numbers z_1 and z_2. P_1R is drawn parallel and equal in length to OP_2 but is now drawn in the opposite sense to OP_2. From the diagram it is seen that the horizontal and vertical coordinates of R are $(x_1 - x_2)$ and $j(y_1 - y_2)$. Point R therefore represents the complex number $z_1 - z_2$.

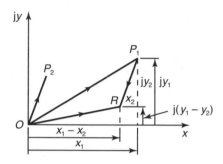

Fig. 14.3

In vector notation

$$\overline{OR} = \overline{OP}_1 + \overline{P_1R}$$

But,

$$\overline{P_1R} = -\overline{RP}_1 = -\overline{OP}_2$$

Therefore

$$\overline{OR} = \overline{OP}_1 - \overline{OP}_2$$

Again it is seen that complex number subtraction and vector subtraction are comparable operations.

EXERCISE 14.2

1. Represent the following numbers on an Argand diagram.
    (a) $2 + j8$,   (b) $3 - j5$,   (c) $-8 + j2$,   (d) $-5 - j6$.
2. Plot the following complex numbers on an Argand diagram:

    $z_1 = 5 + j8$, $z_2 = 7 + j$

    Use your diagram to obtain the values of
    (a) $z_1 + z_2$ (b) $z_1 - z_2$ (c) $\bar{z}_1$ (d) $\bar{z}_2$ (e) $\bar{z}_1 - z_2$
    (f) $\bar{z}_1 + \bar{z}_2$
    Check your answers by calculation.

3.   Represent the vector $6 + j3$ by a line on the Argand diagram. What is the effect on this line of multiplying the vector by j, j^2, j^3, j^4 respectively?

(b) Polar representation

The polar form is convenient for multiplication and division of complex numbers. This form also enables roots and powers to be calculated.

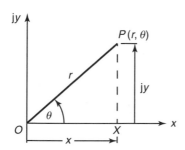

Fig. 14.4

In Fig. 14.4 (r, θ) are the polar coordinates of P, where θ is measured anticlockwise from the positive x-axis. Then in triangle OPX

$$x = r \cos \theta \qquad (i)$$

$$y = r \sin \theta \qquad (ii)$$

But $\quad z = x + jy$

$$= r \cos \theta + jr \sin \theta$$

$$= r(\cos \theta + j \sin \theta)$$

In shortened form this last result is written $r \angle \theta$, where r is called the **modulus** of z and θ is called the **argument** of z.

$$r = \sqrt{(x^2 + y^2)} \qquad (iii)$$

$$\theta = \tan^{-1}\left(\frac{y}{x}\right) \qquad (iv)$$

Equations (i) to (iv) can be used to convert complex numbers from one coordinate system to another.

In complex numbers the **principal value** of θ is always used, so that θ always lies in the range $-\pi \leqslant \theta \leqslant \pi$. Therefore for complex numbers represented by points in the third and fourth quadrants of the Argand diagram θ will be measured in a clockwise direction from the $+x$-axis. In this case, such angles will be negative.

Example 14.5   Convert

(a) $z = 13 \angle 60°$ into its Cartesian form
(b) $z = -3 - j4$ into its polar form

(a) $z = 13 \angle 60°$

This complex number is shown in Fig. 14.5(a)
The modulus $r = 13$, the argument $\theta = 60°$
From equations (i) and (ii)

$x = r \cos \theta = 13 \times 0.5 = 6.5$

$y = r \sin \theta = 13 \times 0.866 = 11.3$

In Cartesian form

$z = 6.5 + j11.3$

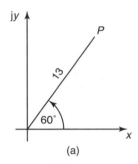

Fig. 14.5

(b) $z = -3 - j4$

The point representing z is in the third quadrant since $x = -3$
and $y = -4$ (Fig. 14.5(b)).
From equations (iii) and (iv)

Modulus $r = \sqrt{\{(-3)^2 + (-4)^2\}} = 5$

Argument $\theta = \tan^{-1}\left(\dfrac{-4}{-3}\right)$ $= \tan^{-1} 1.333\,3$

From the calculator $\angle POQ = 53.13°$
Therefore the principal value of $\theta = -(180 - 53.13) = -126.9°$

In polar form $z = 5\{\cos(-126.9°) + j \sin(-126.9°)\}$
that is $z = 5 \angle -126.9°$

(c) Multiplication and division in polar form

(i) Multiplication

$$z_1 z_2 = r_1 r_2 (\cos \theta_1 + j \sin \theta_1)(\cos \theta_2 + j \sin \theta_2)$$
$$= r_1 r_2 [(\cos \theta_1 \cos \theta_2 - \sin \theta_1 \sin \theta_2)$$
$$+ j(\sin \theta_1 \cos \theta_2 + \cos \theta_1 \sin \theta_2)]$$
$$= r_1 r_2 [\cos(\theta_1 + \theta_2) + j \sin(\theta_1 + \theta_2)]$$
$$= r_1 r_2 \underline{/(\theta_1 + \theta_2)}$$

Multiplication in polar form can be carried out by multiplying the modulii and adding the arguments.

(ii) Division

$$\frac{z_1}{z_2} = \frac{r_1(\cos \theta_1 + j \sin \theta_1)}{r_2(\cos \theta_2 + j \sin \theta_2)}$$

$$= \frac{r_1(\cos \theta_1 + j \sin \theta_1)}{r_2(\cos \theta_2 + j \sin \theta_2)} \frac{(\cos \theta_2 - j \sin \theta_2)}{(\cos \theta_2 - j \sin \theta_2)}$$

$$= \frac{r_1}{r_2} \frac{[(\cos \theta_1 \cos \theta_2 - j^2 \sin \theta_1 \sin \theta_2) + j(\sin \theta_1 \cos \theta_2 - \cos \theta_1 \sin \theta_2)]}{\cos^2 \theta_2 - j^2 \sin^2 \theta_2}$$

$$= \frac{r_1}{r_2} \frac{[\cos(\theta_1 - \theta_2) + j \sin(\theta_1 - \theta_2)]}{\cos^2 \theta_2 + \sin^2 \theta_2}$$

$$= \frac{r_1}{r_2} [\cos(\theta_1 - \theta_2) + j \sin(\theta_1 - \theta_2)]$$

$$= \frac{r_1}{r_2} \underline{/(\theta_1 - \theta_2)}$$

Division in polar form can be carried out by dividing the modulii and subtracting the arguments.

Example 14.6 Given $z_1 = 6\underline{/30°}$ and $z_2 = 4\underline{/-10°}$ evaluate $z_1 z_2$ and z_1/z_2

$$z_1 z_2 = 6\underline{/30°} \times 4\underline{/-10°}$$
$$= 24 \underline{/30° + (-10°)}$$
$$= 24 \underline{/20°}$$

$$\frac{z_1}{z_2} = \frac{6\,\underline{/30^\circ}}{4\,\underline{/-10^\circ}}$$

$$= 1.5\,\underline{/30^\circ - (-10^\circ)}$$

$$= 1.5\,\underline{/40^\circ}$$

14.4   De Moivre's theorem

De Moivre's theorem states that

$$(\cos\,\theta + j\,\sin\,\theta)^n = \cos\,n\theta + j\,\sin\,n\theta$$

The theorem can be proved as follows:
It has been shown in the previous section that

$$(\cos\,\theta_1 + j\,\sin\,\theta_1)(\cos\,\theta_2 + j\,\sin\,\theta_2)$$

$$= \cos(\theta_1 + \theta_2) + j\,\sin(\theta_1 + \theta_2) \quad \text{(i)}$$

If $\theta_1 = \theta_2 = \theta$ then

$$(\cos\,\theta + j\,\sin\,\theta)^2 = \cos\,2\theta + j\,\sin\,2\theta$$

Multiply both sides by $\cos\,\theta + j\,\sin\,\theta$

$$(\cos\,\theta + j\,\sin\,\theta)^3 = (\cos\,2\theta + j\,\sin\,2\theta)(\cos\,\theta + j\,\sin\,\theta)$$

$$= \cos(2\theta + \theta) + j\,\sin(2\theta + \theta) \quad \text{using (i)}$$

$$= \cos\,3\theta + j\,\sin\,3\theta$$

Similarly

$$(\cos\,\theta + j\,\sin\,\theta)^4 = \cos\,4\theta + j\,\sin\,4\theta$$

and so on. Thus it follows that the theorem is true for any positive integer n. Although the theorem has been examined for integral values of n, it will also apply if n is negative or fractional. The theorem is useful for determining roots and powers of complex numbers. The procedure is shown in Examples 14.7 and 14.8.

Example 14.7   If $z = 16(\cos\,60^\circ + j\,\sin\,6\theta)$ evaluate $\sqrt{z}$, giving the answer in Cartesian form.

$$\sqrt{z} = 16^{\frac{1}{2}}(\cos\,60^\circ + j\,\sin\,60^\circ)^{\frac{1}{2}}$$

$$= 4(\cos\,30^\circ + j\,\sin\,30^\circ), \text{ using De Moivre's theorem.}$$

$$= 4(0.866 + j0.5)$$

$$= 3.5 + j2$$

Example 14.8   Find the value of $(1 + j2)^6$

The complex number is first converted into polar form:

$$r = \sqrt{x^2 + y^2} = \sqrt{1^2 + 2^2} = 2.24$$

$$\theta = \tan^{-1}\left(\frac{y}{x}\right) = \tan^{-1} 2 = 63.4°$$

Therefore

$$(1 + j2)^6 = [r(\cos \theta + \sin \theta)]^6$$
$$= (2.24)^6[\cos 63.4 + \sin 63.4]^6$$

Applying De Moivre's theorem

$$(1 + j2)^6 = 126[\cos 6 \times 63.4 + \sin 6 \times 63.4]$$
$$= 126[\cos 20 + \sin 20]$$

14.5   Exponential form of the complex number

Consider the series of $e^{j\theta}$ and $e^{-j\theta}$.

$$e^{j\theta} = 1 + j\theta + \frac{(j\theta)^2}{2!} + \frac{(j\theta)^3}{3!} + \frac{(j\theta)^4}{4!} + \cdots$$

$$= 1 + j\theta - \frac{\theta^2}{2!} - \frac{j\theta^3}{3!} + \frac{\theta^4}{4!} + \cdots \tag{14.1}$$

since

$$j^2 = -1, \qquad j^3 = -1j = -j, \qquad j^4 = -1 \times -1 = 1$$

Similarly

$$e^{-j\theta} = 1 - j\theta + \frac{(-j\theta)^2}{2!} + \frac{(-j\theta)^3}{3!} + \frac{(-j\theta)^4}{4!} + \cdots$$

$$= 1 - j\theta - \frac{\theta^2}{2!} + \frac{j\theta^3}{3!} + \frac{\theta^4}{4!} + \cdots \tag{14.2}$$

Adding equations (14.1) and (14.2)

$$e^{j\theta} + e^{-j\theta} = 2 - 2\frac{\theta^2}{2!} + 2\frac{\theta^4}{4} + \cdots$$

$$= 2\left(1 - \frac{\theta^2}{2!} + \frac{\theta^2}{4!} + \cdots\right)$$

$$= 2 \cos \theta \qquad \text{from equation (10.2)} \tag{14.3}$$

Subtracting equations (14.1) and (14.2)

$$e^{j\theta} - e^{-j\theta} = 2j\theta - 2j\frac{\theta^3}{3!} + 2j\frac{\theta^5}{5!} + \cdots$$

$$= 2j\left(\theta - \frac{\theta^3}{3!} + \frac{\theta^5}{5!} + \cdots\right)$$

$$= 2j \sin \theta \qquad \text{from equation (10.1)} \tag{14.4}$$

From equations (14.3) and (14.4)
adding gives

$$e^{j\theta} = \cos \theta + j \sin \theta \tag{14.5}$$

subtracting gives

$$e^{-j\theta} = \cos -j \sin \theta \tag{14.6}$$

From equation (14.5) $z = re^{j\theta} = r(\cos \theta + j \sin \theta)$

Therefore it can be seen that the exponential form of a complex number is $z = re^{j\theta}$, where r is the modulus and θ is the argument, which in this form, must be in radians. It is a convenient form for multiplication and division, and for finding roots and powers. It also allows the logarithm of a complex number to be found.

Example 14.9   If $z_1 = 16e^{j0.6}$ and $z_2 = 2e^{j0.2}$ find the values of

(a) $z_1 z_2$ (b) $\dfrac{z_1}{z_2}$ (c) $\sqrt{z_1}$ (d) $\ln z_2$

(a) $z_1 z_2 = 16e^{j0.6} \times 2e^{j0.2} = 32e^{j0.8}$

(b) $\dfrac{z_1}{z_2} = \dfrac{16e^{j0.6}}{2e^{j0.2}} = 8e^{j0.4}$

(c) $\sqrt{z_1} = (z_1)^{0.5} = (16e^{j0.6})^{0.5} = 4e^{j0.3}$

(d) $\ln z_2 = \ln 2e^{j0.2} = \ln 2 + \ln e^{j0.2}$ using law I Section 1.6
$ = 0.64 + j0.2$ using property (e), Section 1.6.

EXERCISE 14.3

1. Express the following complex numbers in polar form:
    (a) $3 + j2$,   (b) $-2 + j4$,   (c) $-6 - j$,   (d) $j3$,
    (e) $1 - j$,   (f) $-1 + j$.

2. Express in Cartesian form:
   (a) $2(\cos 30° + j \sin 30°)$,   (b) $3(\cos \frac{1}{3}\pi - j \sin \frac{1}{3}\pi)$,
   (c) $5\underline{/120°}$ (d) $10\underline{/\frac{1}{4}\pi}$ (e) $2.2\underline{/-40°}$.

3. Represent each of the polar forms in Q1 and Q2 on an Argand diagram.

4. Find the modulus and argument of
   (a) $5(\cos 60° - j \sin 60°)$,   (b) $-5 - j12$.

5. Use the Argand diagram to evaluate
   (a) $4\underline{/60°} + 2\underline{/20°}$
   (b) $1.6\underline{/78°} - 1.2\underline{/10°}$
   (c) $19\underline{/140°} - 25\underline{/-110°}$

6. Simplify the following:
   (a) $2\underline{/62°} \times 7\underline{/39°}$
   (b) $2.9\underline{/176°} \times 3.1\underline{/34°}$
   (c) $12\underline{/29°} \div 10\underline{/18°}$
   (d) $17\underline{/96°} \div 4\underline{/140°}$
   (e) $(9\underline{/45°})^2$
   (f) $\dfrac{6.4\underline{/81°} \times 2.6\underline{/42°}}{2\underline{/33°}}$

7. By converting each number to polar form evaluate:
   (a) $(3 - j4)^5$,   (b) $\sqrt{(3 - j4)}$,   (c) $\dfrac{1}{\sqrt{(3 - j4)}}$
   (d) $\dfrac{\sqrt{(1 + j)}}{(1 + j2)}$,   (e) $\dfrac{1}{(1 + j2)^2}$

8. Show that $\dfrac{1}{cos\,\theta + j\,sin\,\theta} = \cos(-\theta) + j \sin(-\theta)$.

9. Change the following complex numbers into exponential form:
   (i) $z_1 = 2 - j$ (ii) $z_2 = 5\underline{/40°}$.
   (iii) Use the most appropriate form to find (a) $z_1 z_2$, (b) z_1^2, (c) z_1/z_2
   (d) $z_2^{0.6}$ (e) $\ln z_1$ (f) $\ln z_1^3$

10. Change the following numbers into Cartesian form:
    (a) $2e^j$ (b) $4e^{-j1.8}$

11. If $z = 3e^{j1.4}$ what is the effect on z if it is multiplied by $e^{j1.1}$? Show the effect on an Argand diagram.

14.6 Applications of complex numbers

There are many applications of complex numbers in science and engineering. This section shows how complex numbers are used in vector analysis and in alternating current theory.

(a) Vector analysis

Example 14.10   Three forces acting on a body at one point are 10 kN at 30°, 6 kN at 240° and 12 kN at 300°. All the angles are measured anticlockwise from the positive x-axis. Determine the magnitude and direction of the resultant force on the body.

The three forces are represented by complex numbers. The angles quoted above must first be converted into principal values; that is,

$$240° = -120°, \qquad 300° = -60°$$

Therefore the complex numbers representing the forces are

$$z_1 = 10\,\underline{/30°}, \quad z_2 = 6\,\underline{/-120°}, \quad z_3 = 12\,\underline{/-60°}$$

The resultant force $= z_1 + z_2 + z_3$

$$= 10[\cos 30° + j \sin 30°]$$
$$+ 6[\cos(-120°) + j \sin(-120°)]$$
$$+ 12[\cos(-60°) + j \sin(-60°)]$$
$$= 10\left(\frac{\sqrt{3}}{2} + j\frac{1}{2}\right) + 6\left(-\frac{1}{2} - j\frac{\sqrt{3}}{2}\right)$$
$$+ 12\left(\frac{1}{2} - j\frac{\sqrt{3}}{2}\right)$$
$$= 8.66 + j5 - 3 - j5.20 + 6 - j10.39$$
$$= 11.66 - j10.59 \qquad\qquad\qquad (i)$$

Modulus of resultant force $= \sqrt{(x^2 + y^2)}$
$$= \sqrt{(11.66^2 + 10.59^2)}$$
$$= 15.8 \text{ kN}$$

Argument of resultant force $= \tan^{-1}\dfrac{y}{x}$
$$= \tan^{-1}\left(-\frac{10.59}{11.66}\right)$$
$$= \tan^{-1}(-0.9082)$$
$$= -42.3° \text{ by reference to equation (i)}$$

Therefore the resultant force is 15.8 kN at 42.3° measured clockwise from the positive x-axis.

(b) Alternating current theory

In an alternating current circuit the current flowing through a pure resistance is in phase with the voltage applied across the resistance. For a pure capacitance the current leads the voltage across it by $90°$ and for a pure inductance the current lags the voltage across it by $90°$.

Consider the series circuit shown in Fig. 14.6. In a series circuit the same current flows through each element and therefore has the same phase in each element. The current I is taken as the reference vector and the phasor

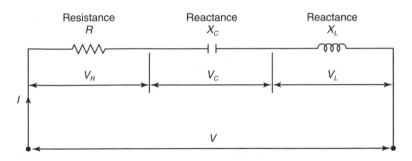

Fig. 14.6

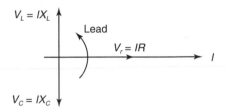

Fig. 14.7(a)

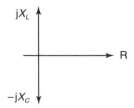

Fig. 14.7(b)

diagram for voltage is drawn as shown in Fig. 14.7(a). Each voltage is expressed as the product of current I and the corresponding impedance X_L, X_C or R. Fig.14.7(a) is redrawn in terms of impedance only, as shown in Fig. 14.7(b), since I is common to each voltage term.

Using the complex number approach, the vertical axis can be regarded as the imaginary axis, so that the two impedances become jX_L and jX_C. Fig. 14.7(b) can be used to determine the impedance of the complete series circuit as shown in Example 14.11.

Example 14.11   A 200 volt, 50 Hz alternating voltage is applied across a series circuit comprising a 50 ohm resistor, a coil of inductance 0.15 henry and a capacitor of capacitance 100 μF. Find the resultant current flowing through the circuit. The circuit diagram is shown in Fig. 14.8.

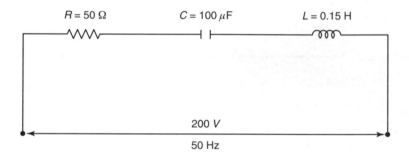

Fig. 14.8

From electrical theory

$X_L = 2\pi f L$, where f is the frequency,

$\qquad = 2\pi 50 \times 0.15$

$\qquad = 47.1 \text{ ohms}$

$X_C = \dfrac{10^6}{2\pi f C}$ Since C must be in Farads

$\qquad = \dfrac{10^6}{2\pi 50 \times 100}$

$\qquad = 31.8 \text{ ohm}$

Figure 14.9 shows the Argand diagram for this circuit, where Z is the impedance of the whole circuit.

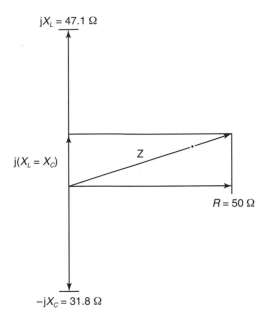

Fig. 14.9

From the diagram $Z = R + j(X_L - X_C)$

$$= 50 + j(47.1 - 31.8)$$

$$= 50 + j15.3$$

Modulus $\qquad z = \sqrt{(50^2 + 15.3^2)} = 52.3 \text{ ohm}$

Argument $\qquad \theta = \tan^{-1}\left(\dfrac{15.3}{50}\right)$

$$= \tan^{-1}(0.306) = 17°$$

The current in the circuit

$$I = \frac{V}{Z}$$

$$= \frac{200}{52.3} = 3.82 \text{ A}$$

Therefore the current is 3.82 A, lagging the voltage by 17°.

EXERCISE 14.4

1. Two voltages are represented by the complex numbers $18 + j10$ and $15 - j3$. Determine the resultant voltage.

2. Three forces of magnitude 5, 10 and 15 kN are inclined to the positive x-axis at angles of $30°$, $60°$ and $120°$ respectively, measured clockwise. Calculate the magnitude and direction of the resultant force by using complex numbers.

3. An electrical circuit has a 30 ohm resistor in series with a coil of inductance 0.16 henry. A 50 Hz, 240 V supply is applied across the circuit. Find the magnitude and phase angle (indicating lag or lead) of the current.

4. An alternating current circuit consists of a capacitor in series with a resistor. Given that the current flowing in the circuit is $(10 + j12)$ A when the voltage applied is $(16 - j)$ V determine (a) the current phase angle relative to the voltage, (b) the circuit impedance.

5. For a parallel circuit the total impedance Z is given by

$$\frac{1}{Z} = \frac{1}{Z_1} + \frac{1}{Z_2}$$

where Z_1 and Z_2 are the branch impedances.

A two branch circuit has the following details:
Branch 1   20 ohm resistor in series with 50 ohm inductive reactance.
Branch 2   50 ohm resistor in series with 80 ohm capacitive reactance.
Given that 200 V, 50 Hz is applied to the circuit determine (a) the impedance Z to the nearest whole number, (b) the supply current.

MISCELLANEOUS EXERCISE 14

1. Given $u = 2 + j$, $v = 1 - j2$ and $w = 3 - j4$, represent u, v and w on an Argand diagram. Given $z = vw$, find z in the form $a + jb$ and in polar form.

   Determine the argument of jz.

   Express $\dfrac{u}{v}$ in the form $a + jb$.

2. (a) Given $x + jy = (2 - j3)^2/(1 + j)$, find values for x and y.

   (b) P is the point $2 + j3$ on the Argand diagram, origin O. OP is rotated anticlockwise through $\frac{1}{4}\pi$ rad to OP_1 and then extended to Q such that $OQ : OP_1$ is $3 : 1$. Find the complex number in the form $a + jb$, which represents the point Q.

3. Express $10\underline{/21°}$ and $20\underline{/49°}$ in the form $a + jb$ and hence calculate r and θ given that

   $$10\underline{/21°} + 20\underline{/49°} = r\underline{/\theta}$$

   Illustrate the relationship on an Argand diagram.

4. (a) *ABCD* is a square on the Argand diagram with point
        A at the origin. If *B* is the point $-1 + j5$, and *C* and *D*
        lie in the first quadrant, find the complex numbers
        represented by the points *C* and *D*.

    (b) *ABCD* is a square on the Argand diagram with point
        A at the origin. *AD* is a line 4 units long inclined at
        $60°$ to the positive *x*-axis, and *C* and *D* lie in the
        second quadrant. Find the coordinates of *B* and *C* in
        the form $r \underline{/\theta}$.

5. (a) Find (i) the cube root of $\cos 30° - j \sin 30°$, (ii) the
        square root of $\cos \theta - j \sin \theta$, (iii) the square root of
        $\cos 6\theta + j \sin 6\theta$.

    (b) Solve the following equations, expressing the answers as
        complex numbers: (i) $x^2 + 1 = 0$, (ii) $x^2 + 3x + 4 = 0$.

6. (a) Simplify: (i) $j^2(2 + j7) + j(3 - j6) - 16 - j$

        (ii) $\dfrac{(4 - j2)(2 - j7)}{(4 + j6)}$

        (iii) $\dfrac{3 - j2}{1 + j} - \dfrac{1 - j3}{2 + j3}$

    (b) Rationalise:

        $$\frac{1}{\cos \theta + j \sin \theta}$$

    (c) Convert the following numbers to polar form and
        represent them on an Argand diagram:
        (i) $-2 + j4$,   (ii) $-3 - j4$

7. (a) Use De Moivre's theorem to evaluate $(5 + j12)^6$ in
        polar form.

    (b) Show that $(\cos x + j \sin x)^2 = \cos 2x + j \sin 2x$.

    (c) Express the following in polar form:
        (i) $(3 - j2)(2 + j7)$ (ii) $-1 + j\sqrt{3}$ (iii) $1 + j$

        (iv) $-j3$ (v) $\dfrac{1}{3 - j2}$.

8. (a) Calculate the resultant of the four voltages, $20 - j15$,
        $40 + j80$, $70 + j0$, $0 - j80$ volts. What is the phase angle
        of the resultant voltage relative to the $70 + j0$ voltage?

    (b) Find the impedance *Z* of a circuit if the voltage of
        $(100 + j60)$ volts produces a current of $(30 - j35)$ A.

    (c) Given that the voltage applied to a circuit is $(20 + j30)$
        volts and the current in the circuit is $(4 + j6)$ A, find
        the power dissipated in the circuit. (*Note*: Power in
        watts = volts × amperes).

Chapter 15

Binary arithmetic, sets, logic and Boolean algebra

15.1 Binary number system

The most commonly used number system is the **denary** system. This system contains the ten digits 0, 1, 2, 3, ..., 8, 9, and is said to have a base of ten. Consider the denary number 328. It is made up of three digits. The digit on the extreme right of the number indicates how many units there are in the number. The next digit indicates how many tens there are in the number, the next digit indicates how many hundreds, and so on; that is,

$$328 = 3 \text{ hundreds} \quad \text{plus} \quad 2 \text{ tens} \quad \text{plus} \quad 8 \text{ units}$$
$$= 3 \times 100 \quad + \quad 2 \times 10 \quad + \quad 8 \times 1$$
$$= 3 \times 10^2 \quad + \quad 2 \times 10^1 \quad + \quad 8 \times 10^0$$

It is seen that each digit, moving from right to left, indicates how many 10^0, 10^1, 10^2, etc., there are in the number.

For normal calculations denary numbers are satisfactory. However, the denary system is unsuitable for computer calculations because digital computers depend for their operation upon two state devices, such as switches, which are in either the OFF or ON position. The OFF position can represent the digit 0 and the ON position the digit 1. Digital computers therefore perform their calculations using a system containing these two digits only. It has a base of two and is called the **binary system**.

15.2 Binary counting

Counting in the denary system proceeds as follows,

0, 1, 2, 3, ... 8, 9,
10, 11, 12, 13, ... 18, 19,
20, 21, 22, , ...

It involves the repeated addition of 1. It is seen that

$$
\begin{array}{cccccc}
0 & 4 & 9 & 19 & 49 & 199 \\
\underline{+1} & \underline{+1} & \underline{+\ 1} & \underline{+\ 1} & \underline{+\ 1} & \underline{+\ \ 1} \\
1 & 5 & 10 & 20 & 50 & 200
\end{array}
$$

It is seen that when 1 is added to the highest digit 9, the column reverts back to 0 and a 1 (which represents ten) is 'carried over' to the next column on the left.

The same pattern is used in the binary system, except that the highest digit is now 1. If 1 is added to this highest digit 1, then the column reverts back to 0, and 1 (which now represents two in denary) is carried over to the next column on the left. Counting proceeds as follows, where the carry over numbers are shown in [ ].

$$
\begin{array}{cccccccc}
0 & 1 & 1\,0 & 1\,1 & 1\,0\,0 & 1\,0\,1 & 1\,1\,0 & 1\,1\,1 \\
\underline{+1} & \underline{+\ \ 1} & \underline{+\ \ 1} & \underline{+\quad 1} & \underline{+\quad 1} & \underline{+\quad 1} & \underline{+\quad 1} & \underline{+\qquad 1} \\
1 & 1\,0 & 1\,1 & 1\,0\,0 & 1\,0\,1 & 1\,1\,0 & 1\,1\,1 & 1\,0\,0\,0 \\
 & [1] & & [1][1] & & [1] & & [1][1][1]
\end{array}
$$

Table 15.1 shows the binary numbers and their denary equivalent:

Table 15.1

Binary number	0	1	10	11	100	101	110	111	1000	1001	1010	1011
Denary equivalent	0	1	2	3	4	5	6	7	8	9	10	11

Whereas each place in the denary number represents the numbers of 10^0, 10^1, 10^2, etc., each place in the binary system, by the same procedure, represents the numbers of 2^0, 2^1, 2^2, etc. For example, 10011 means

$$
1 \times 2^4 + 0 \times 2^3 + 0 \times 2^2 + 1 \times 2^1 + 1 \times 2^0
$$

that is

$$
1 \times 16 + 0 \times 8 + 0 \times 4 + 1 \times 2 + 1 \times 1
$$
$$
= \quad 16 \quad + \quad 0 \quad + \quad 0 \quad + \quad 2 \quad + \quad 1
$$
$$
= 19 \text{ in the denary system.}
$$

15.3 Number conversion

Since both number systems are used in practice, it is necessary to be able to convert numbers from one system to another.

(a) Binary to denary

Such a conversion is easily done, simply by writing down the place value of each digit. For example,

$$11011 \text{ (binary)} = 1 \times 2^4 + 1 \times 2^3 + 0 \times 2^2 + 1 \times 2^1 + 1 \times 2^0$$
$$= \quad 16 \quad + \quad 8 \quad + \quad 0 \quad + \quad 2 \quad + \quad 1$$
$$= \quad 27 \text{ (denary)}$$

(b) Denary to binary

The denary number is first written in terms of the sum of the powers of two, that is, 1, 2, 4, 8, 16, 32, 64, etc. The largest possible power of two is first subtracted, then the next possible power, and so on until the number is completely expressed as a combination of the numbers shown above. For example,

$$54 \text{ (denary)} = \quad 32 \quad + \quad 16 \qquad\qquad + \quad 4 \quad + \quad 2$$
$$= 1 \times 2^5 + 1 \times 2^4 + 0 \times 2^3 + 1 \times 2^2 + 1 \times 2^1 + 0 \times 2^0$$
$$= 110110 \text{ (binary)}$$

An easier method of conversion involves repeated division of the denary number by 2. The remainders form the binary number, as shown in Example 15.1.

Example 15.1   Convert the denary number 327 to binary form

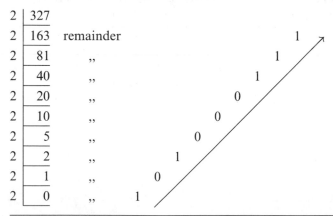

Therefore the denary number $327 = 101\,000\,111$.

EXERCISE 15.1

1.   Convert the following binary numbers to the denary system:
      (a) 110   (b) 1001   (c) 11 011   (d) 100 110

(e) 1 110 101   (f) 1 111 111   (g) 1 010 011 001

(h) 101 001 101 110.

2. Convert the following denary numbers to the binary system:

(a) 22   (b) 53   (c) 140   (d) 110   (e) 512   (f) 943

(g) 2017.

15.4   Operations on binary numbers

Binary numbers can be added, subtracted, multiplied and divided as shown below.

(a) Addition

The method of binary counting in Section 15.2 forms the basis of addition. The carry over numbers are shown in square brackets.

Example 15.2   Add 11 110 and 10 101.

```
    1 1 1 1 0      denary equivalent   30
+   1 0 1 0 1                         +21
    1 1 0 0 1 1                        51
       [1]
      [1]
    [1]
```

Example 15.3   Add 10 011, 1100, 10 111 and 1111.

```
    1 0 0 1 1      denary equivalent   19
    1 1 0 0                           12
    1 0 1 1 1                         23
+     1 1 1 1                        +15
    1 0 0 0 1 0 1                     69
         [1]
       [1 0]
        [1]
     [1 0]
      [1]
    [1]
```

In Example 15.3 it is seen that the carry-over numbers become quite complicated. This may be avoided by using a method that only adds two numbers at any time as shown below. For example, consider the addition of 1 011 + 101 + 1 001 + 110.

First of all the sum of 1 011 and 101 is determined; then 1 001 is added to the result; and finally 110 is added.

```
 1011
  101
10000     partial sum
 1001
11001     partial sum
  110
11111     final sum
```

(b) Subtraction

If a particular column cannot be subtracted because the top digit is 0 and the bottom digit is 1, then 10 is borrowed from the next column on the left (this is column (:) in the example below). The 10 must be 'paid back' by adding 1 to the bottom digit in column (:).

Example 15.4   Subtract the following:

(i)
```
        :  10
     1 1 0̸ 1 1
     1̸ 1 0 1
   1  0
     ───────
     1 1 1 0
     1   ← – – – – – – – –   'pay back' number
```

When 1 is paid back in column (:) the bottom line changes to 10, since $1 + 1 = 10$

(ii)
```
           :  10
     1 1 0 1 0 0̸ 1
     1̸ 1̸ 1̸ 1̸ 1 0
   1  0 0 0 0
     ─────────────
     1 0 1 0 1 1
     1
```

When 1 is paid back in column (:) the 1111 changes to 10 000.

(c) Multiplication

Multiplication in the binary system follows the same procedure as multiplication in the denary system, but is simpler since the multiplier is either 0 or 1.

Example 15.5   Multiply the two numbers 11011×1101

```
  11011    multiplicand
× 1101    multiplier
  11011  ⎫
  00000  ⎪
  11011  ⎬ partial products
  11011  ⎭
101011111
```

(d) Division

Division in the binary system is the same as division in the denary system.

Example 15.6   Divide 10 010 001 by 11 101

```
1 1 1 0 1 ) 1 0 0 1 0 0 0 1 ( 1 0 1
            1 1 1 0 1
              1 1 1 0 1
              1 1 1 0 1
              · · · · ·
```

(e) Fractional binary numbers

In a decimal denary number the places to the right of the fractional point represent values of 10^{-1}, 10^{-2}, 10^{-3} etc. (that is 0.1, 0.01, 0.001 etc.). Similarly in a fractional binary number, the places to the right of the point represent values of $2^{-1}, 2^{-2}, 2^{-3}$ etc. (That is 0.5, 0.25, 0.125 etc.) Therefore the number binary 10.101 is $2 + 1 \times 0.5 + 0 \times 0.25 + 1 \times 0.125 = 2 + 0.5 + 0 + 0.125 = 2.625$ in denary.

EXERCISE 15.2

All the numbers in this exercise are binary numbers. Check each answer by converting into the denary system.

1.   Evaluate
   (a) $110 + 101$
   (b) $1001 + 1111$
   (c) $10011 + 1101$
   (d) $11100 + 10111$
   (e) $111111 + 11111$
   (f) $1011 + 1101 + 1111$
   (g) $1 + 101 + 110 + 10101$
   (h) $1 + 11 + 111 + 1111 + 11111$.

2. Evaluate
   (a) $110 - 11$ (b) $1\,010 - 111$
   (c) $10\,100 - 1\,011$ (d) $10\,000 - 1\,111$
   (e) $1\,001\,101 - 10\,011$ (f) $110\,101 - 100\,111$
   (g) $1\,011 - 100 - 11$ (h) $1\,110\,101 - 101\,101$
                                             $-10\,001 - 11\,100.$

3. Find the value of
   (a) 101×11 (b) $1\,010 \times 110$
   (c) $10\,110 \times 1\,010$ (d) $110\,101 \times 1\,001$
   (e) $11\,011 \times 1\,000$ (f) $101\,011 \times 101\,011.$

4. Find the value of
   (a) $110 \div 11$ (b) $1010 \div 10$
   (c) $111\,000 \div 1\,000$ (d) $10\,010\,110 \div 1\,111$
   (e) $10\,000\,100 \div 1\,011$ (f) $1\,001\,000\,000 \div 11\,000$

15.5   Computer arithmetic

This section shows how the answers to the subtraction, multiplication and division of two binary numbers may be obtained using an addition process. In terms of computer hardware it means that a circuit designed for binary addition may also be used for the other arithmetic processes.

(a) Subtraction by complementing

The complement of any number can be found by subtracting each digit in the number from the highest digit in that system. For example.

(i)   In the denary system the complement of 647 is $999 - 647 = 352$.
(ii) In the binary system the complement of 1 011 is $1\,111 - 1\,011 = 0100$.

It will be noticed from this example that the complement of a binary number is obtained very simply by changing 0 to 1 and 1 to 0. Subtraction can be carried out by finding the complement of the number being subtracted and adding this to the larger number. The complement, however, must contain the same number of digits as this larger number.

Example 15.7   Evaluate $395 - 48$ by the complementation method.

$$
\begin{array}{r}
395 \\
- \ 48 \\
\hline
347
\end{array}
\qquad \text{or} \qquad
\begin{array}{r}
395 \\
+\,951 \\
\hline
1\,346 \\
\end{array}
\quad \text{(three digit complement of 48)}
$$

$$+ \quad \longrightarrow 1$$
$$347$$

It will be noticed that with the complementation method the 1 on the extreme left is transferred to the unit column, and added, to give the correct answer. This also applies to the binary system, as shown in Example 15.8.

Example 15.8   By the method of the complementation evaluate
(a) $11\,010 - 1\,011$,   (b) $1\,100 - 1\,001$.

(a)
```
   11 010              |11 010
 − 1 011     or      + |10 100      (5 digit complement
   1 111              1|01 110       of 1 011)
                   + \————→1
                      |————
                      | 1111
```

(b)
```
   1 100              |1 100
 −1 001      or      + |0 110      (4 digit complement
     11              1|0 010        of 1 001)
                  + \————→1
                     |————
                     |    11
```

(b) Multiplication by shifting

In Example 15.5 it can be seen that the partial products are repeats of the multiplicand when multiplying by 1, and zeros when multiplying by 0. In the usual way, any partial product is shifted one place to the left of the previous partial product. Multiplication in the binary system can therefore be regarded as the additions of the multiplicand with appropriate shifting.

In a computer each partial product is added on as it is obtained, as illustrated in Example 15.9.

Example 15.9   Using the shifting technique, multiply 11 011 by 1 101.

$$1\ 1\ 0\ 1\ 1 \times \overset{\vdots}{1}\ \overset{\vdots}{1}\ \overset{\cdot}{0}\ 1$$

× 1	Partial product 1	11 011	
× 0̇	Partial product 2	000 00∗	Shift one place
	Partial sum	11 011	
× 1̈	Partial product 3	1 101 1∗∗	Shift two places
	Partial sum	10 000 111	
× 1̈	Partial product 4	11 011 ∗∗∗	Shift three places
	Final sum	101 011 111	

(c) Division by repeated subtraction

Consider the denary division $116 \div 29$. This may be evaluated by repeated subtraction of the divisor 29.

$$
\begin{array}{ll}
116 & \\
-\ \ 29 & \text{Subtraction 1} \\
\overline{87} & \\
-\ \ 29 & \text{Subtraction 2} \\
\overline{58} & \\
-\ \ 29 & \text{Subtraction 3} \\
\overline{29} & \\
-\ \ 29 & \text{Subtraction 4} \\
\overline{\cdot\ \cdot} &
\end{array}
$$

that is, $116 \div 29 = 4$.

Example 15.10   Using repeated subtraction evaluate the binary division $1\,001\,011 \div 11\,001$.

$$
\begin{array}{ll}
1\,001\,011 & \\
-\ \ \ \ \ 11\,001 & \text{Subtraction 1} \\
\overline{110\,010} & \\
-\ \ \ \ \ 11\,001 & \text{Subtraction 10} \\
\overline{11\,001} & \\
-\ \ \ \ \ 11\,001 & \text{Subtraction 11} \\
\overline{\cdot\ \cdot\ \cdot\ \cdot\ \cdot} &
\end{array}
$$

The result is 11.
   The repeated subtraction process can be replaced by the repeated addition of the complement.

EXERCISE 15.3

Use the methods of complements, shifting and repeated subtraction to answer Questions 2, 3, and 4 in Exercise 15.2.

15.6   Sets

(a) Description of a set

Much of the work of scientific investigation involves the grouping of information and looking at the properties of the groups. A group in which the members have a common property is called a **set**. For example, a curve drawn on graph paper is in reality a set of points obeying a particular mathematical equation.

Other examples of sets are:

1.   the even numbers between 1 and 10;
2.   the days of a week;
3.   the proteins that constitute a living cell.

The members of a set are written between a pair of brackets, { }. Therefore we can write

$$\text{set A} = \{2, 4, 6, 8, 10\}$$

$$\text{set W} = \{\text{Sunday, Monday, } \ldots, \text{Saturday}\}$$

An empty set contains no members, and is given the symbol ϕ, thus

$$\phi = \{ \ \}$$

(b) Element of a set $\in$

Each member of a set is called an element. The statement 'is an element of a set' is given the symbol $\in$. In set A above, 4 is an element of the set, that is

$$4 \in A$$

(c) Diagrammatic representation of sets – Venn diagrams

Much of the introductory work on sets can be more easily understood by representing them diagrammatically. The elements of a set are shown within a boundary. Set A of Section 15.6(a) is shown in Fig. 15.1. Such a diagram is called a **Venn diagram**.

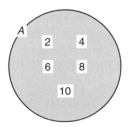

Fig. 15.1

(d) Intersection and union of two sets

Let two sets A and B contain common elements. Then a new set can be formed with common elements of a set A and set B. This new set is called the intersection of set A and set B, and is written $(A \cap B)$. The symbol $\cap$ represents intersection. The two sets

$$A = \{2, 4, 6, 8, 10\}$$
$$B = \{1, 2, 3, 4, 5\}$$

have common elements 2, 4 so that

$$(A \cap B) = \{2, 4\}$$

The intersection is easily represented on a Venn diagram by the overlapping region, which is common to both sets. In Fig. 15.2 the intersection $(A \cap B)$ is represented by the shaded area.

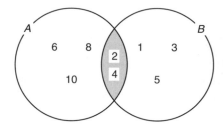

Fig. 15.2

All the elements in either set A or set B form another set, which is called the **union** of A and B, and written $(A \cup B)$. Union is represented by the symbol $\cup$. In the two sets A and B above, the union is

$$(A \cup B) = \{1, 2, 3, 4, 5, 6, 8, 10\}$$

The union of two sets $(A \cup B)$ is represented by the shaded region in Fig. 15.3.

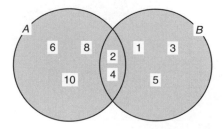

Fig. 15.3

(e) Empty set ϕ, universal set Ɛ and the complement set

An empty set contains no elements and on a Venn diagram will be a region having zero area.

Any set A can be considered as a subset of a much larger set, called a Universal set $\mathcal{E}$. On the Venn diagram in Fig. 15.4 the universal set is represented by the rectangle, and the set A by the shaded area.

The unshaded area in Fig. 15.4 represents the elements of the universal set $\mathcal{E}$ which are **NOT** in set A. This region represents the complement of set A and is written as $\bar{A}$. In sets the complement is a concept which is similar to the complement in the binary system.

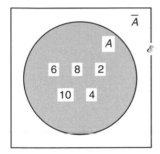

Fig. 15.4

EXERCISE 15.4

1.   Given that set $A = \{1, 2, 3, 4, 5, 6, 7, 8, 9, 10\}$
                set $B = \{1, 2, 4, 8\}$
                set $C = \{2, 4, 6, 8, 10\}$
                set $D = \{1, 3, 5, 7, 9\}$
     find $(B \cap C), (B \cap D), (B \cup C), (B \cap A), (C \cap D)$.
     Show that $D \cup (B \cap C) = (D \cup B) \cap (D \cup C)$.

2.   (a)   Find the set A whose elements are the coordinate pairs on the circle $x^2 + y^2 = 16$, for integer values of x between -4 and $+4$.
     (b)   Find the set B whose elements are the coordinate pairs of the line $y = -x + 4$, for integer values of x between -4 and $+4$.
     (c)   Find $(A \cap B)$. What does this set signify?

3.   Given that set $A = \{1, 2, 3, \ldots 16\}$
                set $B = \{1, 2, 4, 8, 16\}$
                set $C = \{1^2, 2^2, 3^3, 4^2\}$
     find

     $(A \cap B), (A \cap C), (B \cap C), (B \cup C)$

Show that

    (a) $C \cup (A \cap B) = (C \cup A) \cap (C \cup B)$

    (b) $A \cap (B \cup C) = (A \cap B) \cup (A \cap C)$

15.7   Algebra of sets

Following on from the definitions of sets, certain laws can be shown to be true. These laws form a basis of analytical work involving sets. The laws can be verified using Venn diagrams.

(a) Commutative laws

In ordinary algebra and arithmetic, the order of arithmetic or multiplication does not change the result; for example,

$$x + y = y + x$$

and

$$3 \times 4 = 4 \times 3$$

These two results obey the commutative law. Sets also obey the commutative law; that is,

$$(A \cup B) = (B \cup A) \tag{15.1}$$
$$(A \cap B) = (B \cap A) \tag{15.2}$$

(b) Associative laws

In algebra or arithmetic the addition or multiplication of three numbers is independent of the order in which the operation is carried out; that is,

$$(6 + 7) + 9 = 6 + (7 + 9)$$

and

$$l \times (m \times n) = (l \times m) \times n$$

These results illustrate the associative laws.

The associative laws are also obeyed by sets, so that,

$$(A \cup B) \cup C = A \cup (B \cup C) \tag{15.3}$$

and

$$(A \cap B) \cap C = A \cap (B \cap C) \tag{15.4}$$

The commutative and associative laws given by (15.1) to (15.4) can easily be verified using Venn diagrams, since the order in which the diagrams are constructed is immaterial.

(c) Distributive laws

In algebra we have the distributive property

$$a \times (b + c) = (a \times b) + (a \times c)$$

A similar property holds in sets, where

$$A \cap (B \cup C) = (A \cap B) \cup (A \cap C) \tag{15.5}$$

which forms one of the distributive laws.
The law may be verified using Fig. 15.5.

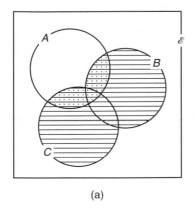   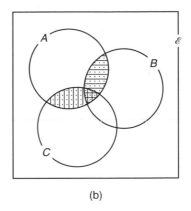

(a)   (b)

Fig. 15.5

In Fig. 15.5(a), $(B \cup C)$ is represented by the shading

$A \cap (B \cup C)$ is represented by the shading

In Fig. 15.5(b), $(A \cap B)$ is represented by the shading

$(A \cap C)$ is represented by the shading

$(A \cap B) \cup (A \cap C)$ is represented by the shading

The dot-shaded regions are identical in both figures, hence verifying the law.
The other distributive law which is valid in sets, but which does not have a counterpart in ordinary algebra or arithmetic, is

$$A \cup (B \cap C) = (A \cup B) \cap (A \cup C) \tag{15.6}$$

Law 15.6 can easily be verified using Venn diagrams, as shown in Fig. 15.6.

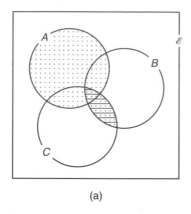 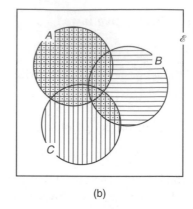

(a)   (b)

Fig. 15.6

In Fig. 15.6(a), $(B \cap C)$ is represented by the shading [shading]
                $A \cup (B \cap C)$ is represented by the shading [shading]
In Fig. 15.6(b), $(A \cup B)$ is represented by the shading [shading]
                $(A \cup C)$ is represented by the shading [shading]
$(A \cup B) \cap (A \cup C)$ is represented by the shading [shading]

The dot-shaded regions in both figures are identical thus verifying the law.

(d) De Morgan's or the complementation laws

These two laws apply to the complement of sets and may be stated as follows:

$$\bar{A} \cap \bar{B} = \overline{A \cup B} \qquad (15.7)$$

$$\bar{A} \cup \bar{B} = \overline{A \cap B} \qquad (15.8)$$

Again the two laws may be verified using Venn diagrams. Starting with law (15.7),

In Fig. 15.7(a), $\bar{A}$ is represented by the shading [shading]
                $\bar{B}$ is represented by the shading [shading]
                $\bar{A} \cap \bar{B}$ is represented by the shading [shading]
In Fig. 15.7(b), $A \cup B$ is represented by the shading [shading]
                $\overline{A \cup B}$ is represented by the shading [shading]

The dot-shaded and cross hatch regions in the figures are identical, thus proving the law.

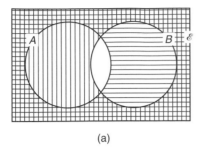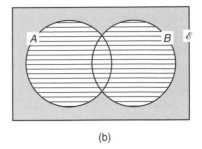

(a)   (b)

Fig. 15.7

Law (15.8) may be verified in a similar way using Figs. 15.8 and 15.7(a).

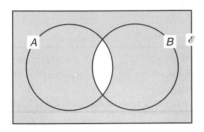

Fig. 15.8

(e) Intersection and union of a set with itself

Any set such as A in Fig. 15.1 overlaps completely with itself. Hence, it is obvious that

$$(A \cup A) = A \tag{15.9}$$

$$(A \cap A) = A \tag{15.10}$$

(f) Intersection and union of a set with a universal set $\mathcal{E}$ and empty set ϕ

By examining the Venn diagram in Fig. 15.4, the following laws may be verified:

$$A \cup \mathcal{E} = \mathcal{E} \tag{15.11}$$

$$A \cap \mathcal{E} = A \tag{15.12}$$

$$A \cup \bar{A} = \mathcal{E} \tag{15.13}$$

$$A \cap \bar{A} = \phi \tag{15.14}$$

With reference to this last law, the intersection of A and $\bar{A}$ occurs only over the boundary of A, which has no area. Such a region defines an empty set.

Using this boundary to represent an empty set it is possible to show that

$$A \cup \phi = A \tag{15.15}$$

$$A \cap \phi = \phi \tag{15.16}$$

The use of sets may be illustrated by the following example. The following faults had occurred in 100 fan heaters returned for repair:

Heating element and fan	6
Heating element and thermostat	10
Fan and thermostat	7
All three faults	4

Of those with single faults only there were three times as many with faulty thermostats as with faulty heaters. There were 29 with only faulty fans. Find the total number with faulty thermostats.

Draw a Venn diagram, as in Fig. 15.9, with three overlapping sets. where faulty heaters, fans and thermostats are represented by H, F, T.

It is seen that $H \cap F = 6$

         $H \cap T = 10$

         $F \cap T = 7$

     $H \cap F \cap T = 4$

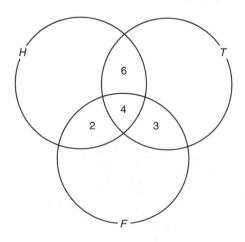

Fig. 15.9

Let the number of heater faults only be x so that number of thermostat faults only is $3x$. Hence from the Venn diagram

$$H \cup F \cup T = 100$$
$$3x + x + 29 + 2 + 6 + 3 + 4 = 100$$
$$4x + 44 = 100$$
$$x = 14$$

Therefore the number with thermostat faults only $= 3 \times 14 = 42$.
Total number with thermostat faults $= 42 + 6 + 4 + 3 = 55$.

EXERCISE 15.5

1. In a sample of 30 cars which had faulty door hinges and locks there were 18 with faulty hinges only, and 6 with faulty hinges and locks. Use the method of sets to find how many had faulty locks only. Illustrate the answer with a Venn diagram.

2. In a street of 27 houses a boy delivers 19 copies of newspaper A and 14 copies of newspaper B. Given that no house receives more than one copy of the same paper, find
    (a) the maximum number of houses to which the boy could deliver both papers;
    (b) the minimum number of houses to which the boy could deliver both papers.
    Illustrate both answers by means of Venn diagrams.

3. In a school where 16 study foreign languages, 26 study French, 18 study German, 11 study French and German, 10 study French and Latin, 6 study German and Latin, and 4 study French, German and Latin. Using Venn diagrams find:
    (a) the number who study French and German but not Latin;
    (b) the number who study Latin;
    (c) the number who study at least two languages.

4. In a sample of 50 faulty electrical machines,
    ● 8 had electrical faults only;
    ● 7 had mechanical faults only;
    ● 3 had assembly faults only;
    ● 11 had mechanical and electrical faults only;
    ● 8 had mechanical and assembly faults only;
    ● 4 had electrical and assembly faults only.
    Find how many had all three faults.

15.8 Boolean algebra

In the work on sets it was seen that an element x must be a member of a set

A or a member of the complement set $\bar{A}$. For a given universal set

$$x \in A \quad \text{or} \quad x \in \bar{A}$$

Sets may be considered as two state systems, consisting of sets A, B, C, etc., and their complements $\bar{A}$, $\bar{B}$, $\bar{C}$, etc. Certain laws were obtained for sets, and it is reasonable to expect that these laws will extend to any other two-state systems, such as **logic** (where statements are **true** or **false**) and **switching circuits** (where a switch is open or closed).

The algebra associated with such two-state systems is called **Boolean algebra**. The laws are the same as the laws of sets obtained in Section 15.7, except they are written in accordance with the following symbols:

a is equivalent to set A

$\bar{a}$ is equivalent to the complement set $\bar{A}$

$\times$ is equivalent to intersection $\cap$

$+$ is equivalent to union $\cup$

1 is equivalent to universal set $\mathscr{E}$

0 is equivalent to empty set ϕ

The laws of Boolean algebra are given in Table 15.2 together with the equivalent laws from sets.

The laws can be used to simplify Boolean expressions, as shown in Example 15.11 below. The $\times$ sign will be omitted from here on, so that, for instance, $a \times b$ will be written as ab.

Table 15.2

Laws of sets			Laws of Boolean algebra	
$(A \cup B)$	$= (B \cup A)$	1	$a + b$	$= b + a$
$(A \cap B)$	$= (B \cap A)$	2	$a \times b$	$= b \times a$
$(A \cup B) \cup C = A \cup (B \cup C)$		3	$(a + b) + c = a + (b + c)$	
$(A \cap B) \cap C = A \cap (B \cap C)$		4	$(a \times b) \times c = a \times (b \times c)$	
$A \cap (B \cup C) = (A \cap B) \cup (A \cap C)$		5	$a \times (b + c) = (a \times b) + (a \times c)$	
$A \cup (B \cap C) = (A \cup B) \cap (A \cup C)$		6	$a + (b \times c) = (a + b) \times (a + c)$	
$\bar{A} \cap \bar{B}$	$= \overline{A \cup B}$	7	$\bar{a} \times \bar{b}$	$= \overline{a + b}$
$\bar{A} \cup \bar{B}$	$= \overline{A \cap B}$	8	$\bar{a} + \bar{b}$	$= \overline{a \times b}$
$A \cup A$	$= A$	9	$a + a$	$= a$
$A \cap A$	$= A$	10	$a \times a$	$= a$
$A \cup \mathscr{E}$	$= \mathscr{E}$	11	$a + 1$	$= 1$
$A \cap \mathscr{E}$	$= A$	12	$a \times 1$	$= a$
$A \cup \bar{A}$	$= \mathscr{E}$	13	$a + \bar{a}$	$= 1$
$A \cap \bar{A}$	$= \phi$	14	$a \times \bar{a}$	$= 0$
$A \cup \phi$	$= A$	15	$a + 0$	$= a$
$A \cap \phi$	$= \phi$	16	$a \times 0$	$= 0$

Example 15.11   Simplify the following Boolean expressions:

(a) $Z = (a + bc)(\bar{a}\bar{b} + c)$

(b) $Z = \overline{a + \bar{b}\bar{c}}$

(a) $Z = (a + bc)(\bar{a}\bar{b} + c)$

$\quad Z = a(\bar{a}\bar{b} + c) + bc(\bar{a}\bar{b} + c)$ (using law 5)

$\quad = a\bar{a}\bar{b} + ac + \bar{a}\bar{b}bc + bcc$

$\quad = 0 \cdot \bar{b} + ac + \bar{a} \cdot 0 \cdot c + bc$ (using laws 10/14)

$\quad = 0 + ac + 0 + bc$ (using law 16)

$\quad = ac + bc$ (using law 15)

$\quad Z = c(a + b)$ (using law 5)

(b) $Z = \overline{a + \bar{b}\bar{c}}$

$\quad = \bar{a}(\overline{\bar{b}\bar{c}})$ (using law 7)

$\quad = \bar{a}(\overline{\bar{b}} + \overline{\bar{c}})$ (using law 8)

$\quad = \bar{a}(b + c)$

EXERCISE 15.6

Simplify the following expressions using the laws of Boolean algebra

1. $\bar{a}\bar{b}(a + b)$
2. $a + b\bar{c} + \bar{b}\bar{c}$
3. $(a + bc)[a + b\bar{c}]$
4. $(a + b)(a + \bar{b})(\bar{a} + b)$
5. $(a + ac)(b + bc)(c + ca)$
6. $(a + b)(a + c)(a + d)$
7. $(a + b)(a + \bar{b})$
8. $(a + bc)(a + b + c)$
9. $(a + b)(a + b^2 + b)$
10. $(a + bc)(a + \bar{b} + \bar{c})$
11. $a^4 + a^3 + a^2 + a$
12. $\overline{(\bar{a} + b)} + \bar{b}$
13. $\overline{(\bar{a}b + c)}$
14. $\overline{(\bar{a} + \bar{b})} + \overline{(a\bar{b})}$
15. $\overline{(\bar{y} + yz)} + \overline{(\bar{y} + yz)}$

15.9   Switching circuits

As already pointed out, Boolean algebra is readily applicable to circuits containing two-state switches. Switches can be represented by Boolean symbols a, b, c. These symbols are equal to 1 when the switch is closed and equal to 0 when the switch is open.

$\qquad\qquad a \qquad\qquad\qquad b$

Fig. 15.10

Consider two switches a and b in series, as in Fig. 15.10. A signal will be transmitted if switch a AND switch b are closed. The connection AND implies that two conditions must be realised, so that it is equivalent to the intersection $\cap$ in sets and hence to $\times$ in Boolean algebra. If Z is the Boolean function for the transmission of a signal, then

$$Z = a \times b = ab.$$

For three switches in series the current will flow if

$$Z = abc$$

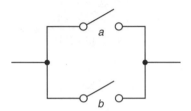

Fig. 15.11

If two switches a and b are connected in parallel, as in Fig. 15.11 a signal will be transmitted through the circuit if switch a is closed OR switch b is closed. The connection OR implies that one or other of the conditions must be realised, so that it is equivalent to union $\cup$ in sets, and hence to $+$ in Boolean algebra. If Z is the Boolean function for the transmission of a signal, then

$$Z = a + b$$

It is possible to represent complicated switching circuits by a Boolean expression. The laws of Boolean algebra can then be applied to simplify the expression to obtain a simpler circuit, as shown in Example 15.12 below. In Fig. 15.12(a) the labels b and $\bar{b}$ in the same circuit mean that if one switch is open the other switch will be closed and vice versa.

Example 15.12   Write down the Boolean function for the circuit in Fig 15.12(a). Simplify the expression, and draw the simplified circuit.

The Boolean expression for circuit PQ is

$$a\bar{b} + c$$

The expression for the upper branch of circuit RS is

$$(a\bar{b} + c)b$$

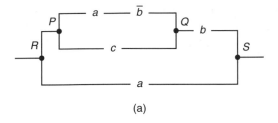

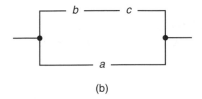

(b)

Fig. 15.12

so that the Boolean expression for the complete circuit is

$Z = a + b(a\bar{b} + c)$

The expression may be simplified as follows:

$Z = a + ab\bar{b} + bc$	(using law 5)
$= a + a.0 + bc$	(using law 14)
$Z = a + bc$	(using law 16)

The circuit represented by this Boolean function is shown in Fig. 15.12(b).

EXERCISE 15.7

In Exercise 15.6 draw the switching circuits for Questions 2 and 4–11. Draw the circuit represented by the simplified Boolean expression, in each case.

15.10 Logic and truth tables

(a) Logic

A logical sequence of steps is required for most operations that we are required to perform, such as starting a car, controlling a set of automatic traffic lights, using a lathe, etc.

The use of logic involves statements and making decisions about whether or not the statements are **true** or **false**. Logic is therefore a two-state system to which Boolean algebra may be applied. A statement can be represented by a symbol such as a, where $a = 1$ if the statement is true and $a = 0$ if the statement is false.

Consider the sentence

'A car is to be started if the doors are closed AND the seat belts are fastened.'

Let Z represent the statement: 'A car is to be started'
      a represent the statement: 'the doors are closed'
      b represent the statement: 'the seat belts are fastened'
      $\times$ represent the connection: AND

The sentence can be written in symbolic form as

$$Z = a \times b$$

Consider the sentence
'Insurance premiums on cars are high if the owner is young OR has a record of accidents.'

      Z represents the statement: 'Insurance premiums are high'
      a represents the statement: 'the owner is young'
      b represents the statement: 'has a record of accidents'
      $+$ represents the connection: OR

The sentence can be written in symbolic form as

$$Z = a + b$$

For any Boolean expression a truth table can be set up which lists all possible combinations of true and false values for the function. This is shown for two Boolean functions $Z = ab$ and $Z = a + b$ in Tables 15.3 and 15.4.

Table 15.3

a	b	$Z = a + b$
0	0	0
0	1	1
1	0	1
1	1	1

Truth tables are useful to check if two expressions in Boolean algebra are equivalent. The two expressions are equivalent if identical inputs to either

Table 15.4

a	b	Z = ab
0	0	0
0	1	0
1	0	0
1	1	1

expression produce identical outputs. Example 15.13 below illustrates the procedure.

Example 15.13 Show that $ab + ac = a(b + c)$.

Table 15.5

(1) a	(2) b	(3) c	(4) ab	(5) ac	(6) ab + ac	(7) b + c	(8) a(b + c)
0	0	0	0	0	0	0	0
0	0	1	0	0	0	1	0
0	1	0	0	0	0	1	0
0	1	1	0	0	0	1	0
1	0	0	0	0	0	0	0
1	0	1	0	1	1	1	1
1	1	0	1	0	1	1	1
1	1	1	1	1	1	1	1

All possible combinations of inputs a, b, c are considered in columns (1), (2) and (3). Column (4) is obtained from columns (1) and (2) using the AND statement. Column (6) is obtained from columns (4) and (5) using the OR statement, and so on. Since columns (6) and (8) are identical the relationship $ab + bc = a(b + c)$ is verified.

15.11 Gates

A piece of equipment which provides an output which is a logical connection of the inputs is called a **logic gate**. Many types of gates are available, such as electronic, pneumatic, fluidic, electromechanical, etc. Circuits can be designed using these logic gates to provide the desired Boolean functions as outputs. In other words, circuits can be designed to carry out specific logical processes. An example of such a circuit is the addition section of a digital computer.

An AND gate is one which gives an output $Z = ab$ for two inputs a, b and an OR gate which gives an output $Z = a + b$ for two inputs a and b. Other gates are those which represent the logic connections NOT. NOT AND (NAND), NOT OR (NOR). These gates are discussed below.

(a) AND gate

The AND gate gives all output signal with a truth value 1, if, and only if, the input signals have truth values of 1. Two switches connected in series form an AND gate, as in Fig. 15.10. The output is given by $Z = ab$. A signal is transmitted (that is, output is 1) if and only if the switches A AND B are closed (that is, $a = 1$ AND $b = 1$). The AND gate is represented diagrammatically in Fig. 15.13(a). The truth table (Fig. 15.13(b)) shows the output Z of an AND gate for all combinations of a and b. It shows that the output Z is zero if $a = 0$ or $b = 0$.

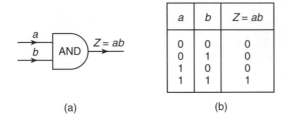

a	b	Z = ab
0	0	0
0	1	0
1	0	0
1	1	1

(a)   (b)

Fig. 15.13

For three inputs a, b, c output $Z = abc$.

(b) OR gate

The OR gate gives an output signal of 1 if either one OR other of the inputs is 1. Two switches connected in parallel form an OR gate (Fig. 15.11). The

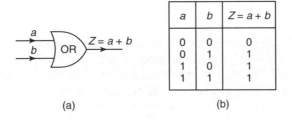

a	b	Z = a + b
0	0	0
0	1	1
1	0	1
1	1	1

(a)   (b)

Fig. 15.14

output is given by

$$Z = a + b$$

A signal is transmitted (that is $Z = 1$) if either switch is closed, (that is, if $a = 1$ OR $b = 1$). The OR gate is shown diagrammatically in Fig. 15.14(a). The truth table in Fig. 15.14(b) shows the output Z for all combinations of inputs a and b.

(c) NOT gate

This gate provides the condition that there will NOT be an output when there is an input signal, that is

$$Z = 0 \quad \text{when } a = 1$$

and

$$Z = 1 \quad \text{when } a = 0$$

It provides an output Z which is an inverted input. It is equivalent to the complement in Boolean algebra; that is,

$$Z = \bar{a}$$

The NOT gate is represented in Fig. 15.15.

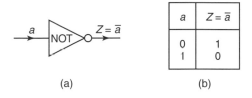

a	$Z = \bar{a}$
0	1
1	0

(a)   (b)

Fig. 15.15

(d) NAND gate

NAND is the gate which is NOT AND. The output from a NAND gate is a the inverted output of an AND gate. Therefore the output from a NAND gate is the complement of the output of an AND gate; that is,

$$Z = \overline{ab}$$

The NAND gate and its truth table is shown in Fig. 15.16.

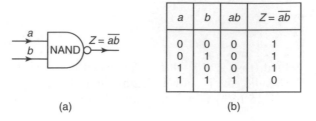

a	b	ab	$Z = \overline{ab}$
0	0	0	1
0	1	0	1
1	0	0	1
1	1	1	0

(a)   (b)

Fig. 15.16

(e) NOR gate

NOR is the gate which is NOT OR. The output from a NOR gate is the inverted output of an OR gate. Therefore the output from a NOR gate is the complement of the output of the OR gate, that is

$$Z = \overline{a + b}$$

The NOR gate and its truth table is shown in Fig 15.17

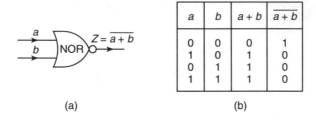

a	b	a + b	$\overline{a + b}$
0	0	0	1
1	0	1	0
0	1	1	0
1	1	1	0

(a)   (b)

Fig. 15.17

Example 15.14 What is the logic system required to control automatically a hydraulic press for pressing a bearing on a shaft? The press will not operate until the old assembly has been removed

AND the new shaft assembly inserted
AND the safety guard closed
AND the start button pressed
OR
   a special lock is operated by the foreman to override the safety guard condition
AND the old shaft assembly removed
AND the start button pressed.

Let 'the old assembly removed' $= a$
'the new assembly inserted' $= b$
'the safety guard closed' $= c$
'the start button pressed' $= e$
'the special lock operated' $= d$

The logic function for starting the press is

$S = abce + ade$

$\quad = ae(bc + d)$

The logic circuit for this function is shown in Fig. 15.18.

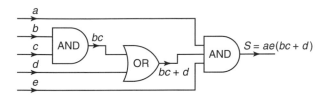

Fig. 15.18

Example 15.15   Obtain the Boolean function for the logic circuit shown in Fig. 15.19. Simplify the expression and draw the equivalent logic circuit.

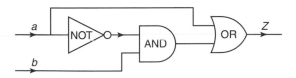

Fig. 15.19

The output from the NOT gate is $\bar{a}$
The input into the AND gate is $\bar{a}$ and b
The output from the AND gate is $\bar{a}b$
The input into the OR gate is $\bar{a}b$ and a
The output from the OR gate is

$Z = a + \bar{a}b$

$(a + \bar{a})(a + b)$ using laws 6 and 14

$Z = a + b$ using law 1.3

Such an output may be obtained from an OR gate as shown in Fig. 15.20.

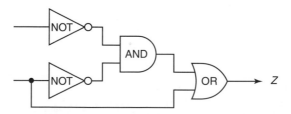

Fig. 15.20

Example 15.16 Obtain the Boolean output formula for the truth table in Table 15.6 in as simple a form as possible. Design a circuit to provide this output.

Table 15.6

a	b	Z
0	0	1
0	1	1
1	0	0
1	1	1

The technique is to obtain the Boolean formula for each of the lines with an output of 1, that is,

$$Z = \bar{a}\bar{b} \quad \text{OR} \quad \bar{a}b \quad \text{OR} \quad ab$$
$$= \bar{a}\bar{b} + \bar{a}b + ab$$
$$= \bar{a}\bar{b} + b(\bar{a} + a)$$
$$= \bar{a}\bar{b} + b1$$
$$= \bar{a}\bar{b} + b$$

The logic circuit is shown in Fig. 15.21.

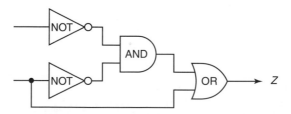

Fig. 15.21

15.12 NAND logic

All logic functions may be obtained using NAND gates only in a circuit. Although this may result in more gates being used the overall cost can be less because a single type of gate is being used. Consider the following functions using NAND gates only in Fig. 15.22.

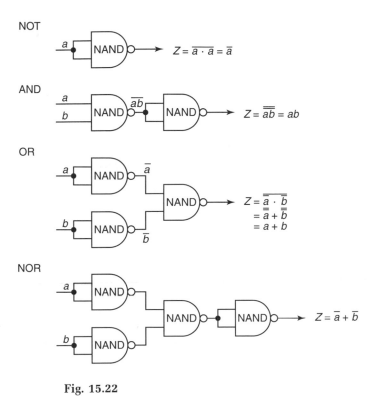

NOT

$Z = \overline{a \cdot a} = \overline{a}$

AND

$Z = \overline{\overline{ab}} = ab$

OR

$Z = \overline{\overline{a} \cdot \overline{b}}$
$= \overline{\overline{a}} + \overline{\overline{b}}$
$= a + b$

NOR

$Z = \overline{a} + \overline{b}$

Fig. 15.22

Circuits may be changed over to NAND gates by replacing each of the other gates with the equivalent group of NAND gates from Fig. 15.22. Alternatively it may be possible to use fewer NAND gates by modifying the logic function for the complete circuit. Consider the modification of the circuit shown in Fig. 15.18.

$$Z = abce + ade$$

First the OR operation must be eliminated using a double complement,

$$Z = \overline{\overline{(abce + ade)}}$$

Use De Morgan's law for the lower complement

$$Z = \overline{(\overline{abce} \cdot \overline{ade})}$$

The function can now be obtained using NAND gates as shown in Fig. 15.23.

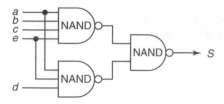

Fig. 15.23

Note: Using the other method would have required 7 NAND gates. All circuit modifications can be checked using truth tables.

EXERCISE 15.8

1. Use truth tables to check the validity of the following Boolean equations. Draw the logic circuits to represent each side of the equation:
   (a) $a + ab = a$
   (b) $a(a + b) = a$
   (c) $a(\bar{a} + b) = ab$
   (d) $a + bc = (a + b)(a + c)$
   (e) $\overline{a + b} = \bar{a}\bar{b}$
   (f) $\overline{ab} = \bar{a} + \bar{b}$

2. Obtain the Boolean expression for each of the circuits in Fig. 15.24. By simplifying these expressions obtain a simplified equivalent circuit in each case. Replace the simplified circuits with NAND gates only and check the NAND logic circuits using truth tables.

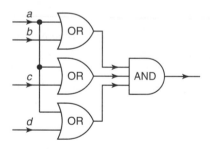

Fig. 15.24(a)

3. Using NAND gates only provide the logic circuit with an output of $\overline{a + b + ab}$ and show that the same output can be provided using one AND, one OR and one NAND gate. Check both circuits using truth tables.

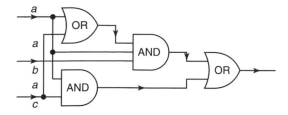

Fig. 15.24(b)

4. Use four NAND gates to provide the logic expression $a + b + c$. and use a truth table to show how the logic circuit operates.
5. Find the logic output for the circuit shown in Fig. 15.25. Set up a truth table showing how the circuit operates.

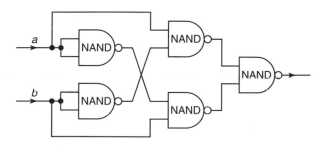

Fig. 15.25

6. For the circuit shown in Fig. 15.26, prove that the output $S = \bar{a}b + a\bar{b}$ and the output $C = ab$. By means of truth tables show that the outputs S and C are the sum and carry-over digits respectively, obtained when two binary digits are added together.

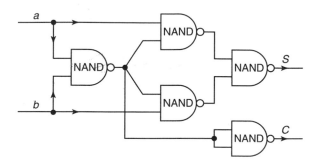

Fig. 15.26

7. Obtain the Boolean expression in as simple form as possible for the circuit operating according to the truth table in Table 15.7.

Table 15.7

a	b	Z
0	0	1
0	1	0
1	0	1
1	1	0

Design this circuit using NAND gates only.

8. Design a circuit to operate as shown in Table 15.8 using gates of your choice.

Table 15.8

a	b	c	Z
0	0	0	0
0	0	1	1
0	1	0	0
1	0	0	0
0	1	1	1
1	0	1	0
1	1	0	0
1	1	1	1

9. An exclusive OR circuit operates as shown in Table 15.9. Design a circuit using NOR gates only.

Table 15.9

a	b	Z
0	0	0
0	1	1
1	0	1
1	1	0

Chapter 16

Determinants and matrices

16.1 Definition of a determinant

In mathematics methods are developed to solve problems as routine processes by formulating sets of rules, such as, for example, the laws of indices for dealing with numbers with powers, or the laws of Boolean algebra for designing logic circuits. Such is the case with determinants and matrices, which are systems for handling numbers in applications such as solving simultaneous equations, etc. These systems are suitable for use with computers to solve problems.

A determinant is an array of numbers, called elements, arranged in columns and rows. These numbers multiply out according to given rules to produce a single final answer. A determinant, therefore, represents an arithmetic or algebraic number. Two examples of determinants are

$$\det A = \begin{vmatrix} 2 & 4 \\ 1 & 3 \end{vmatrix}, \quad \det B = \begin{vmatrix} 1 & 3 & 9 \\ 2 & 6 & 3 \\ 8 & 7 & 4 \end{vmatrix}$$

det A is a 2×2 determinant, that is, it has 2 rows and 2 columns, det B is a 3×3 determinant.

Note that determinants are always square, that is, they always have the same number of rows and columns.

Determinants multiply out as follows:

$$\det A = 2 \times 3 - 1 \times 4 = 2$$

$$\det B = 1 \times \begin{vmatrix} 6 & 3 \\ 7 & 4 \end{vmatrix} - 3 \times \begin{vmatrix} 2 & 3 \\ 8 & 4 \end{vmatrix} + 9 \times \begin{vmatrix} 2 & 6 \\ 8 & 7 \end{vmatrix} \tag{16.1}$$

$$= 1(6 \times 4 - 3 \times 7) - 3(2 \times 4 - 3 \times 8) + 9(2 \times 7 - 8 \times 6)$$

$$= 3 + 48 - 306$$

$$= -255$$

It is seen in (16.1) that each element of the top row is multiplied by the 2×2 determinant formed from the remaining rows and columns not containing the element, with the added requirement that the middle element of the top row changes sign. These 2×2 determinants are called **minors** of the corresponding element. The general rule is

For a 2×2 determinant

$$\begin{vmatrix} a_{11} & a_{12} \\ a_{21} & a_{22} \end{vmatrix} = a_{11}a_{22} - a_{21}a_{12}$$

For a 3×3 determinant

$$\begin{vmatrix} a_{11} & a_{12} & a_{13} \\ a_{21} & a_{22} & a_{23} \\ a_{31} & a_{32} & a_{33} \end{vmatrix} = a_{11} \begin{vmatrix} a_{22} & a_{23} \\ a_{32} & a_{33} \end{vmatrix} - a_{12} \begin{vmatrix} a_{21} & a_{23} \\ a_{31} & a_{33} \end{vmatrix} + a_{13} \begin{vmatrix} a_{21} & a_{22} \\ a_{31} & a_{32} \end{vmatrix}$$

$$= a_{11}(a_{22}a_{33} - a_{32}a_{23}) - a_{12}(a_{21}a_{33} - a_{31}a_{23}) + a_{13}(a_{21}a_{32} - a_{31}a_{22})$$

It is seen that each element is designated a precise label in the determinant, thus making determinants suitable for computer evaluation.

EXERCISE 16.1

Evaluate the following determinants:

1. $\begin{vmatrix} 3 & 7 \\ 1 & 2 \end{vmatrix}$

2. $\begin{vmatrix} -6 & -7 \\ 4 & 3 \end{vmatrix}$

3. $\begin{vmatrix} x & 2x \\ y & -3y \end{vmatrix}$

4. $\begin{vmatrix} 0 & 7 \\ 3 & 6 \end{vmatrix}$

5. $\begin{vmatrix} 1 & 0 \\ 0 & 1 \end{vmatrix}$

6. $\begin{vmatrix} 3 & 1 & 1 \\ 2 & -1 & 2 \\ -1 & 2 & 0 \end{vmatrix}$

7. $\begin{vmatrix} -1 & 2 & 0 \\ -3 & 1 & 1 \\ -2 & 2 & 5 \end{vmatrix}$

8. $\begin{vmatrix} 1 & 0 & 0 \\ 0 & 1 & 0 \\ 0 & 0 & 1 \end{vmatrix}$

9. $\begin{vmatrix} x & 2x & -x \\ 3 & 4 & 1 \\ 2 & 0 & -3 \end{vmatrix}$

16.2 Cofactors

The cofactor of any element is the minor determinant formed from the rows and columns **not** containing that element, together with the insertion of the appropriate sign given below for 2×2, 3×3 and 4×4 determinants.

$$\begin{vmatrix} + & - \\ - & + \end{vmatrix} \qquad \begin{vmatrix} + & - & + \\ - & + & - \\ + & - & + \end{vmatrix} \qquad \begin{vmatrix} + & - & + & - \\ - & + & - & + \\ + & - & + & - \\ - & + & - & + \end{vmatrix}$$

Thus the cofactor of elements 2 and 7 in det B, Section 16.1, are, respectively,

$$-\begin{vmatrix} 3 & 9 \\ 7 & 4 \end{vmatrix}, \qquad -\begin{vmatrix} 1 & 9 \\ 2 & 3 \end{vmatrix}$$

Example 16.1   What are the cofactors of each element in the first row of the following determinant?

$$\begin{vmatrix} 3 & 1 & 2 \\ 4 & 9 & 3 \\ 1 & 2 & 1 \end{vmatrix}$$

Cofactor of element $3 = \begin{vmatrix} 9 & 3 \\ 2 & 1 \end{vmatrix} = 9 - 6 = 3$

Cofactor of element $1 = -\begin{vmatrix} 4 & 3 \\ 1 & 1 \end{vmatrix} = -4 + 3 = -1$

Cofactor of element $2 = \begin{vmatrix} 4 & 9 \\ 1 & 2 \end{vmatrix} = 8 - 9 = -1$

EXERCISE 16.2

In the following determinants find the cofactors of the ringed elements:

1. $\begin{vmatrix} ④ & ① & 2 \\ 3 & 6 & ⑦ \\ ⓪ & 1 & 2 \end{vmatrix}$ 2. $\begin{vmatrix} 3 & 1 & ① \\ 2 & ⓪ & 6 \\ 3 & 2 & 1 \end{vmatrix}$ 3. $\begin{vmatrix} x & ②y & z \\ 4x & -2y & -z \\ -x & y & -2z \end{vmatrix}$

16.3   Properties of determinants

Determinants have the advantage that rows and columns may be changed in certain ways so as to alter the value of the determinant in a prescribed way or to leave it unchanged. These processes can result in reduced computation, especially if some of the elements are reduced to zero in the process. These alterations are described in this section.

(a) The value of a determinant is unchanged if rows become columns and columns rows

Therefore

$$\begin{vmatrix} -1 & 3 \\ 2 & 6 \end{vmatrix} = \begin{vmatrix} -1 & 2 \\ 3 & 6 \end{vmatrix} \quad \text{and} \quad \begin{vmatrix} 1 & 2 & 4 \\ 3 & 0 & 6 \\ 5 & 2 & 1 \end{vmatrix} = \begin{vmatrix} 1 & 3 & 5 \\ 2 & 0 & 2 \\ 4 & 6 & 1 \end{vmatrix}.$$

(b) To multiply a determinant by a number each element of one row or one column only is multiplied by that number

Therefore if

$$\det C = \begin{vmatrix} 1 & 2 & 4 \\ 3 & 0 & 6 \\ 5 & 2 & 1 \end{vmatrix} \quad \text{then} \quad 3\det C = \begin{vmatrix} 3 & 6 & 12 \\ 3 & 0 & 6 \\ 5 & 2 & 1 \end{vmatrix} \quad \text{or} \quad \begin{vmatrix} 3 & 2 & 4 \\ 9 & 0 & 6 \\ 15 & 2 & 1 \end{vmatrix}$$

Each element of the first row is multiplied by 3, or each element of the first column is multiplied by 3.

In reverse, a common factor in any row or column can be factorised outside the determinant. as shown in Example 16.2.

Example 16.2 Factorise the following determinant

$$\begin{vmatrix} 1 & 3 & 7 \\ 2 & 28 & -21 \\ -20 & 5 & 105 \end{vmatrix}$$

The third row has a common factor of 5 and the third column a common factor of 7. The determinant factorises to

$$5\begin{vmatrix} 1 & 3 & 7 \\ 2 & 28 & -21 \\ -4 & 1 & 21 \end{vmatrix} = 5 \times 7 \begin{vmatrix} 1 & 3 & 1 \\ 2 & 28 & -3 \\ -4 & 1 & 3 \end{vmatrix}$$

(c) Interchanging two rows or two columns causes the determinant to be multiplied by (−1)

For example,

$$\begin{vmatrix} 1 & 2 & 3 \\ 4 & 5 & 6 \\ 3 & 3 & 2 \end{vmatrix} = 3$$

$$\begin{vmatrix} 4 & 5 & 6 \\ 1 & 2 & 3 \\ 3 & 3 & 2 \end{vmatrix} = -3$$

(d) It two rows or two columns are identical the value of the determinant is 0

For example,

$$A = \begin{vmatrix} 3 & 3 & 1 \\ 1 & 1 & 2 \\ 2 & 2 & 1 \end{vmatrix} = 3\begin{vmatrix} 1 & 2 \\ 2 & 1 \end{vmatrix} - 3\begin{vmatrix} 1 & 2 \\ 2 & 1 \end{vmatrix} + 1\begin{vmatrix} 1 & 1 \\ 2 & 2 \end{vmatrix}$$

$$= 3(1 - 4) - 3(1 - 4) + 1(2 - 2)$$

$$= 0$$

(e) The value of a determinant is unaltered if a multiple of one row (or column) is added or subtracted from another row (or column)

For example,

$$\det A = \begin{vmatrix} 4 & 2 & 2 \\ 2 & 0 & 1 \\ -1 & 1 & 1 \end{vmatrix} = 4 \times \begin{vmatrix} 0 & 1 \\ 1 & 1 \end{vmatrix} - 2 \times \begin{vmatrix} 2 & 1 \\ -1 & 1 \end{vmatrix} + 2 \times \begin{vmatrix} 2 & 0 \\ -1 & 1 \end{vmatrix}$$

$$= -4 - 6 + 4 = -6$$

Multiply row 2 by 2 and subtract from row 1 gives

$$\begin{vmatrix} 0 & 2 & 0 \\ 2 & 0 & 1 \\ -1 & 1 & 1 \end{vmatrix} = 0 \times \begin{vmatrix} 0 & 1 \\ 1 & 1 \end{vmatrix} - 2 \times \begin{vmatrix} 2 & 1 \\ -1 & 1 \end{vmatrix} + 0 \times \begin{vmatrix} 2 & 0 \\ -1 & 1 \end{vmatrix}$$

$$= 0 - 2 \times 3 + 0 \qquad = -6$$

Using this property it is seen that the computation can be much simplified by modifying some elements to zero in the process.

Example 16.4 Evaluate the following determinant

$$\det A = \begin{vmatrix} 4 & 10 & 8 \\ 3 & 5 & 4 \\ 1 & 2 & 7 \end{vmatrix}$$

Multiply row 2 by 2 and subtract from row 1

$$\det A = \begin{vmatrix} -2 & 0 & 0 \\ 3 & 5 & 4 \\ 1 & 2 & 7 \end{vmatrix} = -2 \begin{vmatrix} 5 & 4 \\ 2 & 7 \end{vmatrix} = -2(35 - 8) = -54$$

16.4 Factorising a determinant

As already stated in Section 16.1 a common factor in any row or column can be factorised outside the bracket. Consider the following determinant

$$\begin{vmatrix} 1 & a & a^2 \\ 1 & b & b^2 \\ 1 & c & c^2 \end{vmatrix}$$

Using property (e) the value of the determinant is unaltered if row 2 is subtracted from row 1

$$\begin{vmatrix} 0 & a-b & a^2-b^2 \\ 1 & b & b^2 \\ 1 & c & c^2 \end{vmatrix} = \begin{vmatrix} 0 & a-b & (a-b)\cdot(a+b) \\ 1 & b & b^2 \\ 1 & c & c^2 \end{vmatrix}$$

The first row has now a comon factor $(a-b)$. The determinant reduces to

$$(a-b)\begin{vmatrix} 0 & 1 & a+b \\ 1 & b & b^2 \\ 1 & c & c^2 \end{vmatrix}$$

Similarly subtracting row 3 from row 2 and factorising as above produces

$$(a-b)(b-c)\begin{vmatrix} 0 & 1 & a+b \\ 0 & 1 & b+c \\ 1 & c & c^2 \end{vmatrix}$$

Further subtracting row 2 from row 1 gives

$$(a-b)(b-c)\begin{vmatrix} 0 & 0 & a-c \\ 0 & 1 & b+c \\ 1 & c & c^2 \end{vmatrix}$$

$$= (a-b)(b-c)(a-c)\begin{vmatrix} 0 & 0 & 1 \\ 0 & 1 & b+c \\ 1 & c & c^2 \end{vmatrix}$$

$$= (a-b)(b-c)(a-c)(-1)$$

$$= (a-b)(b-c)(c-a)$$

EXERCISE 16.3

1.   Show that

$$\begin{vmatrix} 0 & x & y \\ z & 0 & z \\ y & x & 0 \end{vmatrix} = 2xyz$$

2.   Factorise

(a) $\begin{vmatrix} a & a^2 & a^3 \\ b & b^2 & b^3 \\ c & c^2 & c^3 \end{vmatrix}$ (b) $\begin{vmatrix} 1 & a & a^2 \\ -1 & -b & -b^2 \\ 1 & c & c^2 \end{vmatrix}$

3.   Evaluate

(a) $\begin{vmatrix} 4 & 6 & 8 \\ -3 & 12 & 15 \\ 2 & 1 & 3 \end{vmatrix}$ (b) $\begin{vmatrix} 2 & 4 & 16 \\ 3 & 9 & 81 \\ 5 & 25 & 125 \end{vmatrix}$ (c) $\begin{vmatrix} x & a & b \\ a & x & b \\ a & b & x \end{vmatrix}$

16.5   Solution of simultaneous equations by Cramer's method

Determinants may be used to solve simultaneous equations using Cramer's method. If the three equations are written as

$$a_{11}x + a_{12}y + a_{13}z = c_1$$
$$a_{21}x + a_{22}y + a_{23}z = c_2$$
$$a_{31}x + a_{32}y + a_{33}z = c_3$$

Let D be the determinant comprising the coefficients of x, y and z, and D_x, D_y D_z respectively be the determinants obtained from D by replacing its first, second, and third columns by the column (c_1, c_2, c_3), as shown below:

$$\det D = \begin{vmatrix} a_{11} & a_{12} & a_{13} \\ a_{21} & a_{22} & a_{23} \\ a_{31} & a_{32} & a_{33} \end{vmatrix} \quad \text{and} \quad D_x = \begin{vmatrix} c_1 & a_{12} & a_{13} \\ c_2 & a_{22} & a_{23} \\ c_3 & a_{32} & a_{33} \end{vmatrix}$$

$$D_y = \begin{vmatrix} a_{11} & c_1 & a_{13} \\ a_{21} & c_2 & a_{23} \\ a_{31} & c_3 & a_{33} \end{vmatrix} \quad\quad D_z = \begin{vmatrix} a_{11} & a_{12} & c_1 \\ a_{21} & a_{22} & c_2 \\ a_{31} & a_{32} & c_3 \end{vmatrix}$$

Cramer's solution states that

$$x = \frac{D_x}{D} \quad y = \frac{D_y}{D} \quad z = \frac{D_z}{D}$$

Because D is in the denominator, if $D = 0$ the equations cannot be solved and they are said to be **singular**.

Example 16.5 Solve the equations $2x + 3y = 4$

$x - y = -3$

Using the above definitions

$$\det D = \begin{vmatrix} 2 & 3 \\ 1 & -1 \end{vmatrix} = -2 - 3 \quad = -5$$

$$\det D_x = \begin{vmatrix} 4 & 3 \\ -3 & -1 \end{vmatrix} = -4 - (-9) = 5$$

$$\det D_y = \begin{vmatrix} 2 & 4 \\ 1 & -3 \end{vmatrix} = -6 - 4 \quad = -10$$

$$x = \frac{D_x}{D} = \frac{5}{-5} = -1$$

$$y = \frac{D_y}{D} = \frac{-10}{-5} = 2$$

EXERCISE 16.4

Solve the simultaneous equtions using Cramer's method.

1. $3x + 2y = 12$
$5x - 3y = 1$

2. $2u + 5v = 4$
$7u + 4v = 14$

3. $3p - 4q = 1$
$6p - 6q = 5$

4. $x + y + z = 6$
$2x + 3y + z = 11$
$3x + 2y + 2z = 13$

5. $2x + 3y + 4z = 2$
$4y + 3z = -1$
$8x + 5z = 1$

6. $2x + y - z = 1$
$3x - y + 2z = 10$
$4x - 2y - 3z = 9$

16.6 Matrices

A matrix is an array of numbers which does not reduce to a single figure through internal multiplication. For example, let the forces along the x, y, and z direction acting on particle A be 5 N, 7 N, 10 N, and on particle B be 3 N, 5 N, 12 N. These forces can be recorded as an array:

$$\begin{pmatrix} 5 & 7 & 10 \\ 3 & 5 & 12 \end{pmatrix}$$

It is meaningless to evaluate these numbers any further, but there are advantages to their representation as an array, with rules developed to use such arrays to solve engineering and science problems using routine processes. Such arrays are called matrices.

Whereas determinants are always square, with the number of rows equal to the number of columns, matrices need not be.

(a) Addition and subtraction

Two matrices can be added or subtracted if both have the same order, that is, the number of rows is the same in both matrices and the number of columns is the same in both. Addition and subtraction is carried out by adding or subtracting the corresponding elements. This is illustrated in Example 16.6.

Example 16.6   Two matrices A and B are shown below. Find (a) $A + B$, (b) $B - A$

$$A = \begin{pmatrix} 2 & 4 & 7 \\ -6 & 3 & -1 \end{pmatrix} \qquad B = \begin{pmatrix} -2 & 0 & 1 \\ 3 & 4 & 7 \end{pmatrix}$$

(a) $A + B = \begin{pmatrix} 2-2 & 4+0 & 7+1 \\ -6+3 & 3+4 & -1+7 \end{pmatrix} = \begin{pmatrix} 0 & 4 & 8 \\ -3 & 7 & 6 \end{pmatrix}$

(b) $B - A = \begin{bmatrix} -2-2 & 0-4 & 1-7 \\ 3-(-6) & 4-3 & 7-(-1) \end{bmatrix} = \begin{pmatrix} -4 & -4 & -6 \\ 9 & 1 & 8 \end{pmatrix}$

(b) Multiplication or division of a matrix by a scalar

To multiply or divide a matrix by a scalar number each element is multiplied or divided by that number. (Compare with a determinant where only 1 row or 1 column is multiplied by the number.) This is shown in Example 16.7

Example 16.7   Given

$$A = \begin{pmatrix} 3 & 0 \\ 1 & 2 \\ 6 & 5 \end{pmatrix} \text{ find (a) } 3A \text{ (b) } xA$$

(a) $3A = 3 \times \begin{pmatrix} 3 & 0 \\ 1 & 2 \\ 6 & 5 \end{pmatrix} = \begin{pmatrix} 9 & 0 \\ 3 & 6 \\ 18 & 15 \end{pmatrix}$

(b) $xA = x \begin{pmatrix} 3 & 0 \\ 1 & 2 \\ 6 & 5 \end{pmatrix} = \begin{pmatrix} 3x & 0 \\ x & 2x \\ 6x & 5x \end{pmatrix}$

(c) Multiplication of two matrices

Two matrices A and B are multiplied together by multiplying the rows of A by the columns of B term by term and adding the products. **Multiplication can only take place if the number of columns in the first matrix is equal to the number of rows in the second matrix.**

This can be illustrated by an example. In Fig. 16.1 a rod 2 m long is shown. Two different sets of forces can act perpendicular to the rod, in two different combinations of distances from P. These forces can take on combinations of distances from end P, of (10 cm, 70 cm, 120 cm), and (15 cm, 90 cm, 110 cm). Each set of forces with each combination of distances will create a moment of forces about P, which is obtained by multiplication.

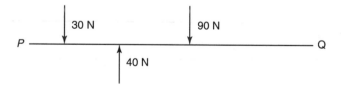

Fig. 16.1(a)

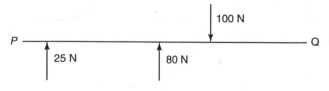

Fig. 16.1(b)

First of all the various combinations are written as matrices.

Matrix of forces 2×3 Matrix of distances from P 3×2

$$\begin{pmatrix} 30 & -40 & 90 \\ -25 & -80 & 100 \end{pmatrix} \quad \begin{pmatrix} 10 & 15 \\ 70 & 90 \\ 120 & 110 \end{pmatrix}$$

With the forces in Fig. 16.1(a) and the first combination of distances, the sum of moments of forces about P is $30 \times 10 + (-40) \times 70 + 90 \times 120 = 8300$. The first row in the first matrix is multiplied by the first column in the second matrix. This is repeated for the other rows and columns: the second row in the first matrix is multiplied by the first column of the second matrix to give $(-25) \times 10 + (-80) \times 70 + 100 \times 120 = 6150$. The complete multiplication is

$$\begin{pmatrix} 30 & -40 & 90 \\ -25 & -80 & 100 \end{pmatrix} \begin{pmatrix} 10 & 15 \\ 70 & 90 \\ 120 & 110 \end{pmatrix} = \begin{pmatrix} 8300 & 6750 \\ 6150 & 3425 \end{pmatrix}$$

The matrix formed from the multiplication is an array of the various combinations of the moments of the forces. This can be extended to any number of forces and any combination of distances. The advantage of so doing is that as the size of the matrix gets larger manual calculation becomes cumbersome. However, in this form computer programs will evaluate the products quite easily.

Multiplication can be summarised as follows:

---

1. The condition for multiplication is
    Number of columns in the first matrix = Number of rows in the second matrix

    In the above example we have $(2 \times 3) \times (3 \times 2) = 2 \times 2$

2. The multiplication is carried out as
    Rows of the first matrix by columns of the second matrix.

It is seen therefore that if two matrices $A \times B$ can be multiplied because they meet the above condition, multiplication $B \times A$ will not meet the above condition, unless both are square matrices. Even when this is the case $AB \neq BA$ except under certain circumstances.

EXERCISE 16.4

Multiply the following matrices:

1. $\begin{pmatrix} 3 & 1 & -2 \\ 4 & -1 & 2 \end{pmatrix} \cdot \begin{pmatrix} 4 & 1 \\ -1 & 0 \\ 2 & -2 \end{pmatrix}$

2. $\begin{pmatrix} 1 & 2 & 1 \\ -2 & 3 & 5 \\ -1 & 4 & -2 \end{pmatrix} \cdot \begin{pmatrix} 2 & -1 \\ 4 & 3 \\ -3 & -2 \end{pmatrix}$

3. $\begin{pmatrix} 1 & 2 & -1 & -4 \\ 0 & 3 & 7 & 1 \end{pmatrix} \begin{pmatrix} 3 & -1 \\ 2 & 0 \\ 4 & -1 \\ 3 & 7 \end{pmatrix}$

4. $\begin{pmatrix} 2 & -3 & 4 \\ -1 & 4 & -2 \\ 1 & 0 & 5 \end{pmatrix} \begin{pmatrix} 3 & 0 & -1 \\ 2 & 5 & -3 \\ 1 & 2 & -3 \end{pmatrix}$

16.7   Special matrices

(a) Square matrix

In a square matrix the number of rows = number of columns, for example

$$\begin{pmatrix} 3 & 2 & 6 \\ 1 & 0 & 3 \\ 1 & 5 & 1 \end{pmatrix}$$

(b) Diagonal matrix

This is a square matrix where all the elements are zero except along the diagonal, for example

$$\begin{pmatrix} 4 & 0 & 0 \\ 0 & 3 & 0 \\ 0 & 0 & 7 \end{pmatrix}$$

(c) Unit matrix

A unit matrix is a diagonal matrix in which all the elements are 1, that is

$$\begin{pmatrix} 1 & 0 & 0 \\ 0 & 1 & 0 \\ 0 & 0 & 1 \end{pmatrix}$$

Multiplying a matrix A by a unit matrix I leaves A unchanged. Consider the matrix A multplied by a unit matrix I

$$AI = \begin{pmatrix} 4 & 3 & 0 \\ 0 & 1 & 5 \\ 1 & -2 & 1 \end{pmatrix} \cdot \begin{pmatrix} 1 & 0 & 0 \\ 0 & 1 & 0 \\ 0 & 0 & 1 \end{pmatrix} = \begin{pmatrix} 4 & 3 & 0 \\ 0 & 1 & 5 \\ 1 & -2 & 1 \end{pmatrix} = A$$

This result is also true if the order of multiplication is changed, that is

$$IA = A$$

(d) Transpose matrix A^T of a matrix A

The transpose of a matrix is obtained by writing the rows as columns in order, as shown with the example

$$A = \begin{pmatrix} 4 & 3 & 0 \\ 0 & 1 & 5 \\ 1 & -2 & 1 \end{pmatrix} \qquad A^T = \begin{pmatrix} 4 & 0 & 1 \\ 3 & 1 & -2 \\ 0 & 5 & 1 \end{pmatrix}$$

(e) Adjoint matrix adj A of a square matrix A

This is obtained first by finding the matrix of cofactors, and then transposing this matrix of cofactors. The cofactor of each element is the determinant obtained from rows and columns not containing that element, as described in Section 16.2, including the appropriate sign from

$$\begin{pmatrix} + & - & + \\ - & + & - \\ + & - & + \end{pmatrix}$$

Let

$$A = \begin{pmatrix} 1 & 3 & 2 \\ -1 & 4 & 2 \\ 2 & 3 & -2 \end{pmatrix}$$

The matrix of cofactors is

$$B = \begin{pmatrix} \begin{vmatrix} 4 & 2 \\ 3 & -2 \end{vmatrix} & -\begin{vmatrix} -1 & 2 \\ 2 & -2 \end{vmatrix} & \begin{vmatrix} -1 & 4 \\ 2 & 3 \end{vmatrix} \\ -\begin{vmatrix} 3 & 2 \\ 3 \cdot & -2 \end{vmatrix} & \begin{vmatrix} 1 & 2 \\ 2 & -2 \end{vmatrix} & -\begin{vmatrix} 1 & 3 \\ 2 & 3 \end{vmatrix} \\ \begin{vmatrix} 3 & 2 \\ 4 & 2 \end{vmatrix} & -\begin{vmatrix} 1 & 2 \\ -1 & 2 \end{vmatrix} & \begin{vmatrix} 1 & 3 \\ -1 & 4 \end{vmatrix} \end{pmatrix} = \begin{pmatrix} -14 & 2 & -11 \\ 12 & -6 & 3 \\ -2 & -4 & 7 \end{pmatrix}$$

Then adj A is the transpose of matrix B:

$$\begin{pmatrix} -14 & 12 & -2 \\ 2 & -6 & -4 \\ -11 & 3 & 7 \end{pmatrix}$$

(e) Inverse matrix A^{-1} of a square matrix A

The inverse of a square matrix A is $A^{-1} = 1/A$ that is, $AA^{-1} = 1$. The inverse of a matrix A is adj A divided by the determinant of A,

$$A^{-1} = \frac{1}{|A|} \cdot \text{adj } A$$

The inverse matrix exists provided det $A \neq 0$. If det $A = 0$ the matrix is said to be singular.

Using

$$\text{matrix } A = \begin{pmatrix} 1 & 3 & 2 \\ -1 & 4 & 2 \\ 2 & 3 & -2 \end{pmatrix}$$

its adjoint has been found in (v) above.

$$\text{adj } A = \begin{pmatrix} -14 & 12 & -2 \\ 2 & -6 & -4 \\ -11 & 3 & 7 \end{pmatrix}$$

$$\det A = \begin{vmatrix} 1 & 3 & 2 \\ -1 & 4 & 2 \\ 2 & 3 & -2 \end{vmatrix} = -30$$

Therefore, the inverse matrix

$$A^{-1} = -\frac{1}{30} \cdot \begin{pmatrix} -14 & 12 & -2 \\ 2 & -6 & -4 \\ -11 & 3 & 7 \end{pmatrix}$$

It can be seen that this is the case by calculating AA^{-1}, that is,

$$AA^{-1} = \begin{pmatrix} 1 & 3 & 2 \\ -1 & 4 & 2 \\ 2 & 3 & -2 \end{pmatrix} \times -\frac{1}{30} \cdot \begin{pmatrix} -14 & 12 & -2 \\ 2 & -6 & -4 \\ -11 & 3 & 7 \end{pmatrix}$$

$$= -\frac{1}{30} \begin{pmatrix} -30 & 0 & 0 \\ 0 & -30 & 0 \\ 0 & 0 & -30 \end{pmatrix}$$

$$= \begin{pmatrix} 1 & 0 & 0 \\ 0 & 1 & 0 \\ 0 & 0 & 1 \end{pmatrix}$$

For a 2×2 matrix obtaining the adjoint and inverse matrices is much simpler. Let

$$B = \begin{pmatrix} 2 & 4 \\ 3 & -5 \end{pmatrix}$$

The cofactors are the elements diagonally opposite, with the appropriate signs from

$$\begin{pmatrix} + & - \\ - & + \end{pmatrix}$$

which is

$$\begin{pmatrix} -5 & -3 \\ -4 & 2 \end{pmatrix}$$

$$\text{Adj } B = \begin{pmatrix} -5 & -4 \\ -3 & 2 \end{pmatrix}$$

Det B $-5 \times 2 - -3 \times -4 = -22$

Therefore

$$B^{-1} = \frac{1}{22} \begin{pmatrix} -5 & 4 \\ -3 & 2 \end{pmatrix}$$

Checking

$$BB^{-1} = \begin{pmatrix} 2 & 4 \\ 3 & -5 \end{pmatrix} \times \frac{1}{22} \begin{pmatrix} -5 & -4 \\ -3 & 2 \end{pmatrix} = \frac{1}{22} \begin{pmatrix} -22 & 0 \\ 0 & -22 \end{pmatrix} = I$$

EXERCISE 16.6

Determine the inverse matrix of each of the following:

1. $\begin{pmatrix} 3 & 1 \\ -2 & -5 \end{pmatrix}$
2. $\begin{pmatrix} -4 & 2 \\ -3 & -2 \end{pmatrix}$

3. $\begin{pmatrix} 2 & -1 & 3 \\ 0 & -2 & 4 \\ 1 & 5 & -2 \end{pmatrix}$ 4. $\begin{pmatrix} -3 & 3 & -1 \\ -1 & 4 & -2 \\ -3 & 0 & -1 \end{pmatrix}$

16.8   Solution of linear simultaneous equations

Matrix theory can be used to solve linear simultaneous equations of any number of unknowns. The three equations, in three unknowns will be solved using this matrix method.

$$2x - 3y + 4z = 13$$
$$3x + y - 2z = -2$$
$$-x + 2y + z = -1$$

We can write the equations in a matrix form

$$\begin{pmatrix} 2 & -3 & 4 \\ 3 & 1 & -2 \\ -1 & 2 & 1 \end{pmatrix} \begin{pmatrix} x \\ y \\ z \end{pmatrix} = \begin{pmatrix} 13 \\ -2 \\ -1 \end{pmatrix}$$

In matrix form this equation is $AX = B$
Multiply both sides by A^{-1}:

$$A^{-1}AX = A^{-1}B$$
$$X = A^{-1}B$$

Therefore to find x, y, z the inverse matrix is first determined.

$$A = \begin{pmatrix} 2 & -3 & 4 \\ 3 & 1 & -2 \\ -1 & 2 & 1 \end{pmatrix}$$

$$\text{Matrix of cofactors} = \begin{pmatrix} 5 & -1 & 7 \\ 11 & 6 & -1 \\ 2 & 16 & 11 \end{pmatrix}$$

$$\text{Adj } A = \begin{pmatrix} 5 & 11 & 2 \\ -1 & 6 & 16 \\ 7 & -1 & 11 \end{pmatrix}$$

$$\text{Det } A = 2 \begin{vmatrix} 1 & -2 \\ 2 & 1 \end{vmatrix} + 3 \begin{vmatrix} 3 & -2 \\ -1 & 1 \end{vmatrix} + 4 \begin{vmatrix} 3 & 1 \\ -1 & 2 \end{vmatrix}$$
$$= 2(1 + 4) + 3(3 - 2) + 4(6 + 1)$$
$$= 41$$

$$A^{-1} = \frac{1}{41} \cdot \begin{pmatrix} 5 & 11 & 2 \\ -1 & 6 & 16 \\ 7 & -1 & 11 \end{pmatrix}$$

Check

$$AA^{-1} = \begin{pmatrix} 2 & -3 & 4 \\ 3 & 1 & -2 \\ -1 & 2 & 1 \end{pmatrix} \cdot \frac{1}{41} \cdot \begin{pmatrix} 5 & 11 & 2 \\ -1 & 6 & 16 \\ 7 & -1 & 11 \end{pmatrix}$$

$$= \frac{1}{41} \begin{pmatrix} 41 & 0 & 0 \\ 0 & 41 & 0 \\ 0 & 0 & 41 \end{pmatrix}$$

$$= \begin{pmatrix} 1 & 0 & 0 \\ 0 & 1 & 0 \\ 0 & 0 & 1 \end{pmatrix}$$

Therefore equation (16.2) becomes

$$\begin{pmatrix} x \\ y \\ z \end{pmatrix} = \frac{1}{41} \cdot \begin{pmatrix} 5 & 11 & 2 \\ -1 & 6 & 16 \\ 7 & -1 & 11 \end{pmatrix} \cdot \begin{pmatrix} 13 \\ -2 \\ -1 \end{pmatrix}$$

$$= \frac{1}{41} \begin{pmatrix} 65 - 22 - 2 \\ -13 - 12 - 16 \\ 91 + 2 - 11 \end{pmatrix}$$

$$= \frac{1}{41} \begin{pmatrix} 41 \\ -41 \\ 82 \end{pmatrix}$$

$$\begin{pmatrix} x \\ y \\ z \end{pmatrix} = \begin{pmatrix} 1 \\ -2 \\ 2 \end{pmatrix}$$

Therefore

$$x = 1$$
$$y = -1$$
$$z = 2$$

Example 16.8   Solve the simultaneous equations:

$5x - 2y = 13$

$2x + y = 7$

In matrix form

$$\begin{pmatrix} 5 & -2 \\ 2 & 1 \end{pmatrix} \begin{pmatrix} x \\ y \end{pmatrix} = \begin{pmatrix} 13 \\ 7 \end{pmatrix}$$

$$AX = B$$

$$X = A^{-1}B$$

Matrix of cofactors of $A = \begin{pmatrix} 1 & -2 \\ 2 & 5 \end{pmatrix}$

$$\text{Adj } A = \begin{pmatrix} 1 & 2 \\ -2 & 5 \end{pmatrix}$$

$$\det A = 5 - (-2) = 9$$

$$A^{-1} = 1/9 \begin{pmatrix} 1 & 2 \\ -2 & 5 \end{pmatrix}$$

$$X = \begin{pmatrix} x \\ y \end{pmatrix} = 1/9 \begin{pmatrix} 1 & 2 \\ -2 & 5 \end{pmatrix} \begin{pmatrix} 13 \\ 7 \end{pmatrix} = 1/9 \begin{pmatrix} 27 \\ 9 \end{pmatrix} = \begin{pmatrix} 3 \\ 1 \end{pmatrix}$$

$x = 3, \quad y = 1$

EXERCISE 16.7

1.   Solve the simultaneous equations in two unknowns using the matrix method:
     (a) $3x + 2y = 7$ (b) $x + 2y = -4$ (c) $3E + 4I = -8$
           $4x - y = 2$ $2x - 3y = 13$ $-2E - 3I = 7$
     (d)   4 lathes of type A and 10 lathes of type B turn out 450 components each week. 2 lathes of type A and 8 of type B turn out 270 components each week. Find the number of components turned out by each type of lathe.
2.   Solve the simultaneous equations in Exercise 1.2 using the matrix method.

3. Solve using the matrix method:
   (a) $4x - 3y + 2z = -7$ (b) $x + 2y + 3z = -1$
   $$2x + 2y - 3z = 27 \qquad 4x - 3y + 2z = 2$$
   $$2x - 5y - 4z = 11 \qquad 3x - 8y - 5z = 11$$
   (c) $2x + y = 4$
   $$3x + 2y = 3$$
   $$5x + 6y + 3z = 10$$

MISCELLANEOUS EXERCISE 16

1. Given the matrix

   $$M = \frac{1}{13}\begin{pmatrix} 5 & 12 \\ 12 & -5 \end{pmatrix}$$

   evaluate M^2 and the determinant of M.

2. The matrix

   $$M = \begin{pmatrix} p & q \\ r & s \end{pmatrix}$$

   is such that
   (i) all its elements are real and non-zero,
   (ii) det $M = 1$,
   (iii) $MM^T = I$, where M^T is the transpose of M and I is the unit matrix.
   Show that

   $$s(p^2 + q^2) = p$$

3. The matrices **A**, **B** and **C** are given by

   $$A = \begin{pmatrix} -2 & 3 \\ 1 & 0 \end{pmatrix}, \quad B = \begin{pmatrix} 5 & 1 \\ -1 & 2 \end{pmatrix}, \quad C = \begin{pmatrix} -5 & 9 \\ 6 & -6 \end{pmatrix}$$

   Find matrices **P**, **Q** and **R** such that
   (a) $P = 2A + B^2$,
   (b) $AQ + BQ = I$,
   (c) $RA = C$.

4. Calculate the determinant of the matrix

   $$M = \begin{pmatrix} k & 2k - 3 \\ 2 & k - 1 \end{pmatrix}$$

   Find the values of k for which M is singular.

5. (a) Given that

$$A = \begin{pmatrix} 1 & 2 \\ 3 & 4 \end{pmatrix}, \qquad B = \begin{pmatrix} x & y \\ 6 & 13 \end{pmatrix}$$

and that $AB = BA$, calculate the value of x and of y.

   (b) Given that

$$C = \begin{pmatrix} 2 & 1 \\ 3 & 0 \end{pmatrix} \quad \text{and} \quad D = \begin{pmatrix} 1 & -2 \\ 0 & 4 \end{pmatrix}$$

verify that $C^{-1}D^{-1} = (DC)^{-1}$.

6. Find the product XY where

$$X = \begin{pmatrix} a & 0 & b \\ 0 & 1 & 0 \\ b & 0 & a \end{pmatrix} \qquad Y = \begin{pmatrix} c & 0 & d \\ 0 & 1 & 0 \\ d & 0 & c \end{pmatrix}$$

Show that X is non-singular if and only if $a \neq \pm b$, and find X^{-1} when X is non-singular.

7. Express in matrix form the simultaneous equations

$$\begin{aligned} 2x - y + z &= 3 \\ x \quad + z &= 1 \\ 3x - y + 4z &= 0 \end{aligned}$$

Hence solve the equations.

8. (a) Find the inverse of the matrix

$$A = \begin{pmatrix} 2 & -1 & 1 \\ 2 & -2 & 3 \\ 1 & 1 & -2 \end{pmatrix}$$

   (b) Solve the simultaneous equations

$$\begin{aligned} 2x - y + z &= 1 \\ 2x - 2y + 3z &= -1 \\ x + y - 2z &= 3 \end{aligned}$$

   (c) Calculate the matrix

$$B = A^{-1} \begin{pmatrix} 0 & 0 & 1 \\ 0 & 1 & 0 \\ 1 & 0 & 0 \end{pmatrix} A.$$

9.   Solve the simultaneous equations

$$x + y - z = 1$$
$$3x + 4y - 2z = 3$$
$$-x + y + 4z = 2$$

Chapter 17

Vectors

17.1   Scalars and vectors

A **scalar** is a quantity which has magnitude only, that is, it has size only. Examples are temperature, energy and mass.

A **vector** is a quantity that has magnitude, but in addition it has a direction in space. Examples are force, velocity and acceleration. A vector may be represented by a line whose direction is drawn parallel to the actual vector, which has a length proportional to the size of the vector. A vector is written in heavy type such as **v** or as $\overrightarrow{PQ}$, where PQ is the line representing the vector. The magnitude of a vector **v** is written as $|\mathbf{v}|$ or as v.

17.2   Displacement or positional vectors in space

In Fig. (17.1) the line OP is a displacement vector representing a position of the point P from a fixed point O. Such a vector is anchored to a particular point in space and is called a tied vector.

P

O

Fig. 17.1

17.3   Free vectors

Vectors such as velocity, force are not tied vectors. Such a free vector can be represented by any one of a number of parallel lines, as shown in Fig. 17.2.

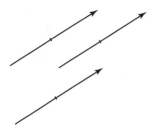

Fig. 17.2

17.4   Equal vectors

Two vectors **u** and **v** are equal if they have the same magnitude and the same direction, as shown in Fig. 17.3. As free vectors they do not have to coincide. Thus two cars with velocities 60 m/s in the same direction have equal velocity vectors.

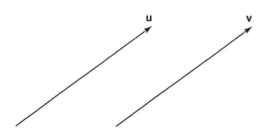

Fig. 17.3

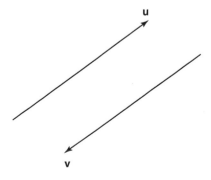

Fig. 17.4

17.5   Equal and opposite vectors

The two vectors **u** and **v** in Fig. 17.4 have equal magnitudes but act in opposite directions. They are equal and opposite, that is

$$\mathbf{v} = -\mathbf{u}$$

17.6   Addition of two vectors

In Fig. 17.5(a) two vectors **u** and **v** are shown acting at a point L. To find the single vector **w** which can replace the two vectors **u** and **v**, a vector equal to **v** is drawn from the end of the vector **u** as shown in Fig. 17.5(b). A line LN is drawn to complete the triangle, joining the free ends of the two vectors. This line LN represents the vector **w**, where

$$\mathbf{w} = \mathbf{u} + \mathbf{v}$$

w is the single vector that replaces **u** and **v**. It is called the **resultant** vector.

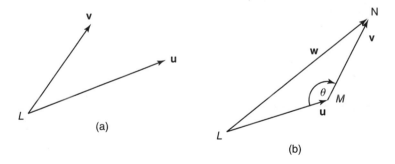

(a)

(b)

Fig. 17.5

Such a process is called the addition of two vectors.

The magnitude of **w** (written as $|\mathbf{w}|$), and its direction is obtained from the triangle LMN, either by drawing the triangle to scale or by calculation using the sine and cosine rules.

The magnitude of **w** in Fig. 17.5(a) is given by

$$|\mathbf{w}|^2 = |\mathbf{u}|^2 + |\mathbf{v}|^2 - 2|\mathbf{u}|\,|\mathbf{v}|\cos\theta$$

The direction of **w** may be obtained using the sine rule.

The calculation of the magnitude of the resultant of two vectors is shown in Example 7.1.

Example 17.1 A body has a velocity **u** of magnitude 10 m/s on a bearing 070°, and a wind velocity **v** of magnitude 20 m/s acts on it on a bearing 030°. Find the magnitude and direction of the resultant velocity.

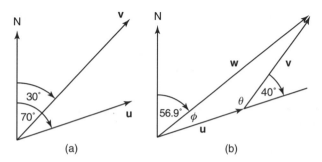

Fig. 17.6

The directions of the two velocities are shown in Fig. 17.6(a) The vector **v** is drawn end-on to vector **u** as shown in Fig. 17.6(b). The resultant velocity **w** is found from the triangle. The angle between the two vectors is 40° so that the angle θ inside the triangle is 140°. Using the cosine rule

$$|\mathbf{w}|^2 = |\mathbf{u}|^2 + |\mathbf{v}|^2 - 2|\mathbf{u}||\mathbf{v}|\cos\theta$$
$$= 10^2 + 20^2 - 2 \times 10 \times 20 \cos 140°$$
$$= 806.4$$
$$|\mathbf{w}| = 28.4\,\text{m/s}$$

The angle ϕ is obtained using the sine rule:

$$\frac{|\mathbf{w}|}{\sin 140°} = \frac{|\mathbf{v}|}{\sin\phi}$$
$$\sin\phi = \frac{10\sin 140°}{28.4} = 0.2263$$
$$\phi = 13.1°$$

The resultant vector has a magnitude of 28.4 m/s on a bearing of $(70 - 13.1)°$ to **u**, that is 056.9°.

(a) Alternating current and voltage

In Chapter 3 it was shown that an alternating current or voltage could be represented by a rotating vector. The length of the rotating vector is proportional to the amplitude of the current or voltage, and its direction determined by its phase angle. The combination of two alternating voltages, having the same frequency, can be represented by the vector addition of the two individual vectors, as shown in Example 17.2.

Example 17.2   Two alternating voltages, having the same frequency, are connected in series across a load, one with an amplitude of 200 V at a phase angle of 20°, the other with an amplitude of 150 V at a phase angle of 45°. Determine the amplitude of the combined voltage and its phase angle.

The two alternating voltages are represented as rotating vectors in Fig. 17.7.

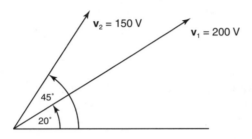

Fig. 17.7(a)

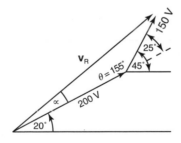

Fig. 17.7(b)

The resultant voltage $\mathbf{v_R}$ is given by

$$
\begin{aligned}
|\mathbf{v_R}|^2 &= |\mathbf{v_1}|^2 + |\mathbf{v_2}|^2 - 2|\mathbf{v_1}||\mathbf{v_2}|\cos\theta \\
&= 200^2 + 150^2 - 2 \times 200 \times 150\cos 155° \\
&= 40\,000 + 22\,500 - 60\,000(-0.9063) \\
&= 116\,878 \\
\mathbf{v_R} &= 340 \; V
\end{aligned}
$$

To find α the sine rule is used

$$\frac{|\mathbf{v_R}|}{\sin 155°} = \frac{|\mathbf{v_1}|}{\sin \alpha}$$

$$\sin \alpha = \frac{150 \sin 155°}{340}$$

$$= 0.1853$$

$$\alpha = 10.7°$$

Therefore the phase angle $\phi = \phi + 20 = 30.7°$

17.7   Addition of more than two vectors

In Fig. 17.8 the resultant of four vectors **s**, **t**, **u**, **v** is obtained, such as the resultant of four forces acting on a particle **P**.

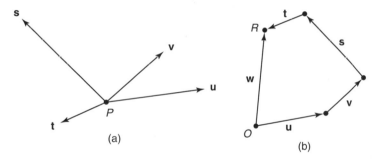

Fig. 17.8

The resultant is found in the same way as for the addition in Section 17.6. In Fig. 17.8(b) from the point O the vector **u** is drawn to scale in the correct direction. **v**, **s**, **t** are drawn in the same way in the correct order, until the point R is reached. The line OR is the resultant of the four forces. It is the single force that could replace **u**, **v**, **s**, **t** to produce the same effect on P. Fig. 17.8 is called a **polygon of vectors**.

17.8   Subtraction of vectors

Consider the two vectors **u**, **v** shown in Fig. 17.9(a). We now wish to find **u** − **v**. To do so we perform the addition **u** + (−**v**) using the method of Section 17.6.

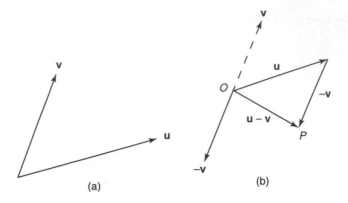

(a)   (b)

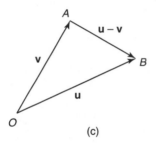

(c)

Fig. 17.9

The vector $(-\mathbf{v})$ is shown in Fig. 17.9(b). To add $\mathbf{u}$ and $-\mathbf{v}$ draw a vector parallel to $-\mathbf{v}$ at the end of vector $\mathbf{u}$.

$$\mathbf{u} + (-\mathbf{v}) = \overrightarrow{OP}$$

But $\overrightarrow{OP}$ is equal to the vector $\overrightarrow{AB}$ drawn across the ends of the two vectors, as shown in Fig. 17.9(c).

Therefore subtraction of two vectors is carried out by drawing a vector across their ends in a direction away from the subtracted vector, that is, from $\mathbf{v}$ to $\mathbf{u}$.

Example 17.3 In Fig. 17.10(a) two vectors $\mathbf{u}$ and $\mathbf{v}$ are shown acting at a point O. Show diagrammatically the vector representing (a) $\mathbf{u} + \mathbf{v}$, (b) $\mathbf{u} - \mathbf{v}$, (iii) $\mathbf{v} - \mathbf{u}$.

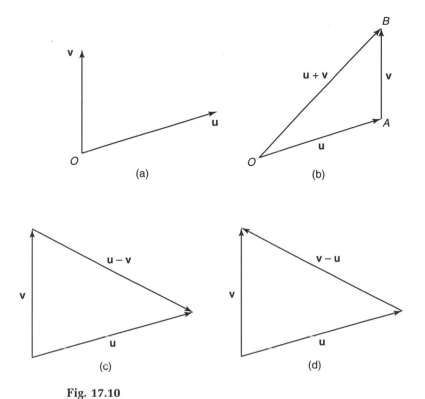

Fig. 17.10

(a)   In Fig. 17.10(b) the addition of the two vectors $\mathbf{u} + \mathbf{v}$ is shown. Vector $\mathbf{v}$ is drawn from A, the end of $\mathbf{u}$, in such a way that it follows on from $\mathbf{u}$. The resultant is $\overrightarrow{OB}$.

(b)   In Fig. 17.10(c) $\mathbf{u} - \mathbf{v}$ is the vector joining the two ends of $\mathbf{u}$ and $\mathbf{v}$, in a direction from $\mathbf{v}$ to $\mathbf{u}$.

(c)   In Fig. 17.10(d) $\mathbf{v} - \mathbf{u}$ is drawn in a similar way to (ii) above, but now in a direction from $\mathbf{u}$ to $\mathbf{v}$

17.9   Multiplication of a vector by a scalar quantity λ

A vector $\mathbf{u}$ multiplied by a scalar λ is equal to $\lambda\mathbf{u}$. The magnitude of the vector is increased by a factor λ but the direction does not change. For example, when $\lambda = 3$ in Fig. 17.11 $\overrightarrow{AC} = 3\overrightarrow{AB}$.

A vector $\overrightarrow{AB}$ will be parallel to another vector $\overrightarrow{CD}$ if one is a multiple of the other, such as, for example, $2\mathbf{u}$ and $5\mathbf{u}$. This is important when dealing with the sum or difference of two or more vectors. As shown in Fig. 17.12,

$$\mathbf{s} = 2\mathbf{u} + 3\mathbf{v}$$

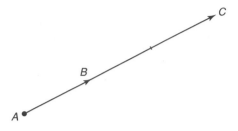

Fig. 17.11

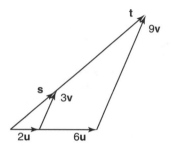

Fig. 17.12

Similarly

$$t = 6u + 9v$$
$$= 3(2u + 3v) = 3s$$

which is a vector parallel to s and three times its magnitude as discussed above. It is used in the following example.

Example 17.4   Three points A, B, C have displacement vectors with reference to the origin O of $u + v$, $2u - v$, $-u + kv$, where k is a constant. Find the value of k to ensure that ABC is a straight line.

In Fig. 17.13, if ABC is a straight line AB is parallel to BC so that

$$\overrightarrow{BC} = \lambda \overrightarrow{AB}$$

But $\overrightarrow{AB} = \overrightarrow{OB} - \overrightarrow{OA} = (2u - v) - (u + v) = u - 2v$

and $\overrightarrow{BC} = \overrightarrow{OC} - \overrightarrow{OB} = (-u + kv) - (2u - v) = -3u + (k + 1)v$

$-3u + (k + 1)v = \lambda(u - 2v) = \lambda u - \lambda 2v$

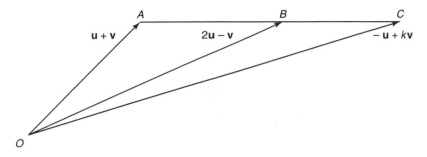

Fig. 17.13

Comparing coefficients $\mathbf{u}$: $\lambda = -3$

$\mathbf{v}$: $k + 1 = -2\lambda$

$k + 1 = 6$

$k = 5.$

EXERCISE 17.1

1. Two vectors $\mathbf{u}$ and $\mathbf{v}$ of magnitude 14 units and 26 units respectively have an angle of 35° between them. Find the magnitude and direction of the resultant.
2. Two vectors $\mathbf{u}$ and $\mathbf{v}$ have magnitudes 34 and 46 units respectively, with an angle of 42° between them. Find the magnitude and direction of
    (a) $\mathbf{u} + \mathbf{v}$,
    (b) $\mathbf{v} - \mathbf{u}$
3. Two forces $\mathbf{F}_1$ and $\mathbf{F}_2$, of magnitude 12 N and 25 N, act in a vertical plane, $\mathbf{F}_1$ making an angle of 30° with the horizontal and $\mathbf{F}_2$ making an angle of 55° with the horizontal.
    (a) Calculate the magnitude and direction of the resultant.
    (b) Calculate the magnitude and direction of $\mathbf{F}_1 - \mathbf{F}_2$.

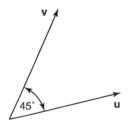

Fig. 17.14

4. In Fig. 17.14 two vectors **u** and **v** of magnitudes 16 and 19 respectively act at a point O in a plane surface. The angle between them is 45°. Calculate the magnitude and direction of (a) **u** + **v** and (b) **v** − **u**.

5. L, M, N are three points whose displacement vectors from a point O are **u** + **v**, 3**u** + 2**v**, 5**u** + n**v**. Given that LMN is a straight line find the value of n.

6. P, Q and R have positional vectors **u** − 2**v**, −3**u** + m**v**, −**u** − 3**v**. Given that PQR is a straight line find the value of m.

7. Given **s** = λ**t**, find n if **s** = **u** − 3**v**, **t** = 4**u** − n**v**.

8. An alternating current I_1 has an amplitude of 40 A and phase angle 25° and I_2 has an amplitude 50 A and a phase angle 55° both flowing through the same load. Both currents have the same frequency. Calculate the resultant amplitude and its phase angle.

17.10 Unit vector

A unit vector is a vector whose magnitude is 1. In the Cartesian axes
    the unit vector along the x-direction is called **i**
    the unit vector along the y-direction is called **j**
    the unit vector along the z-direction is called **k**
These are shown in Fig. 17.15.

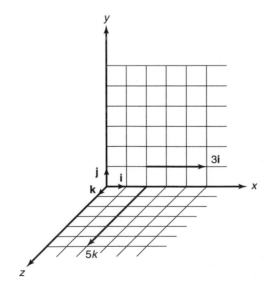

Fig. 17.15

A vector of magnitude 3 along the x-direction will be 3**i** and likewise a vector of magnitude 5 in the z-direction will be 5**k**. In general
  a vector along the x-direction of magnitude x will be x**i**
  a vector along the y-direction of magnitude y will be y**j**
  a vector along the z-direction of magnitude z will be z**k**

17.11   Resolution of a vector into its components in two dimensions

In Section 17.5 it was shown that two vectors can be combined into a single resultant vector. Conversely, a single vector **w** may be expressed in terms of components, such as **u** and **v** as shown in Fig. 17.16(a).

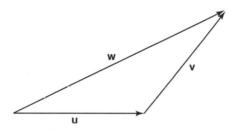

Fig. 17.16(a)

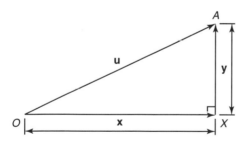

Fig. 17.16(b)

In practice it will be always more convenient to resolve a vector into two perpendicular components along the x, y directions in the Cartesian system as shown in Fig. 17.16(b). The vector $\overrightarrow{OA} = \mathbf{u}$ is resolved into two components $\overrightarrow{OX}$ and $\overrightarrow{XA}$

$$\overrightarrow{OA} = \mathbf{u} = \overrightarrow{OX} + \overrightarrow{XA}$$

Writing $\overrightarrow{OX} = x\mathbf{i}$ and $\overrightarrow{XA} = y\mathbf{j}$

$\mathbf{u} = x\mathbf{i} + y\mathbf{j}$

Using Pythagoras' theorem we can obtain the magnitude of $\mathbf{u}$ in terms of x and y.

$$|\mathbf{u}|^2 = x^2 + y^2$$

$$|\mathbf{u}| = \sqrt{x^2 + y^2}$$

The unit vector along the direction of $\mathbf{u}$ will be

$$\frac{\mathbf{u}}{|\mathbf{u}|} = \frac{\mathbf{u}}{\sqrt{x^2 + y^2}}$$

Example 17.5   Find the magnitude and direction with reference to the x-axis of

$\mathbf{u} = 2\mathbf{i} + 5\mathbf{j}.$

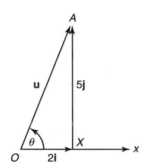

Fig. 17.17

Referring to Fig. 17.17,

$$|\mathbf{u}| = \sqrt{2^2 + 5^2}$$

$$= \sqrt{29} \qquad = \quad 5.39$$

In triangle OAX $\qquad AX = y = 5$

$$OA = |\mathbf{u}| \qquad = \quad 5.39$$

$$\tan \theta = \frac{5}{2}$$

$$\theta = 6.82°$$

Example 17.6   A force **F** of magnitude 40 N acts along a direction of 30° to the horizontal. Calculate its vertical and horizontal components and hence express **F** in terms of the unit vectors **i** and **j**.

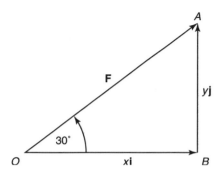

Fig. 17.18

The force **F** is shown in Fig. 17.18. Let its horizontal component be $x\mathbf{i}$ and the vertical component be $y\mathbf{j}$. Therefore

$$\mathbf{F} = x\mathbf{i} + y\mathbf{j}$$

The triangle OAB is right-angled, so that the magnitude of

$$|\overrightarrow{OB}| = x = |\mathbf{F}|\cos 30° = 40 \times 0.866 = 34.6\,\text{N}$$

$$|\overrightarrow{AB}| = y = |\mathbf{F}|\sin 30° = 40 \times 0.5 = 20\,\text{N}$$

Therefore

$$\mathbf{F} = 34.6\mathbf{i} + 20\mathbf{j}$$

17.12 Addition and subtraction of two vectors in Cartesian form

Consider two vectors **u** and **v** in Fig. 17.19. The resultant $\mathbf{w} = \mathbf{u} + \mathbf{v}$. Writing these in terms of Cartesian components:

$$\mathbf{u} = x_1\mathbf{i} + y_1\mathbf{j}$$

$$\mathbf{v} = x_2\cdot\mathbf{i} + y_2\mathbf{j}$$

$$\mathbf{w} = \mathbf{u} + \mathbf{v} = (x_1 + x_2)\mathbf{i} + (y_1 + y_2)\mathbf{j}$$

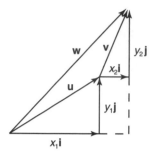

Fig. 17.19

One can see therefore that expressing vectors in Cartesian forms makes the addition of vectors an easy matter. In a similar manner

$$\mathbf{u} - \mathbf{v} = (x_1 - x_2)\mathbf{i} + (y_1 - y_2)\mathbf{j}$$

Example 17.7   Find the resultant of

$$\mathbf{u} = 3\mathbf{i} + 4\mathbf{j}$$
$$\mathbf{v} = 4\mathbf{i} - 6\mathbf{j}$$

Using the above result

$$\mathbf{w} = \mathbf{u} + \mathbf{v} = (3 + 4)\mathbf{i} + (4 - 6)\mathbf{j}$$
$$= 7\mathbf{i} - 2\mathbf{j}$$

Example 17.8   *ABC* is a triangle with *A, B, C* having position vectors **u, v, w**, respectively which are, in terms of **i** and **j**,

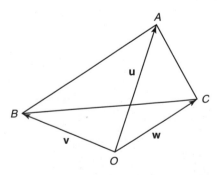

Fig. 17.20

$\mathbf{u} = 2\mathbf{i} + 4\mathbf{j}$

$\mathbf{v} = -3\mathbf{i} + \mathbf{j}$

$\mathbf{w} = 3\mathbf{i} + 2\mathbf{j}$

Express the vectors $\overrightarrow{AB}$, $\overrightarrow{BC}$, $\overrightarrow{CA}$ in terms of $\mathbf{i}$ and $\mathbf{j}$.

In Fig. 17.20

$\overrightarrow{AB} = \mathbf{v} - \mathbf{u} = (-3\mathbf{i} + \mathbf{j}) - (2\mathbf{i} + 4\mathbf{j}) = -5\mathbf{i} - 3\mathbf{j}$

$\overrightarrow{BC} = \mathbf{w} - \mathbf{v} = (3\mathbf{i} + 2\mathbf{j}) - (-3\mathbf{i} + \mathbf{j}) = 6\mathbf{i} + \mathbf{j}$

$\overrightarrow{CA} = \mathbf{u} - \mathbf{w} = (2\mathbf{i} + 4\mathbf{j}) - (3\mathbf{i} + 2\mathbf{j}) = -\mathbf{i} + 2\mathbf{j}$

Example 17.9 $OABC$ is a square with O the origin of the Cartesian axes. The position vector of A is $4\mathbf{i} + 3\mathbf{j}$, and the position vector of C is $3\mathbf{i} - 4\mathbf{j}$. Calculate the position vector of B.

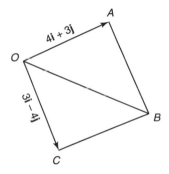

Fig. 17.21

Because OABC is a square $\overrightarrow{CB} = \overrightarrow{OA}$ (parallel and the same magnitude)

$$\overrightarrow{OB} = \overrightarrow{OC} + \overrightarrow{CB}$$

$$= 3\mathbf{i} - 4\mathbf{j} + 4\mathbf{i} + 3\mathbf{j}$$

that is

$$\overrightarrow{OB} = 7\mathbf{i} - \mathbf{j}$$

17.13   Vectors in three dimensions

In Section 17.10 we considered vectors in two dimensions only. Now we extend this treatment to three dimensions. In Fig. 17.22(a) $\mathbf{i}$, $\mathbf{j}$, $\mathbf{k}$, are three unit vectors along the axes OX, OY, OZ.

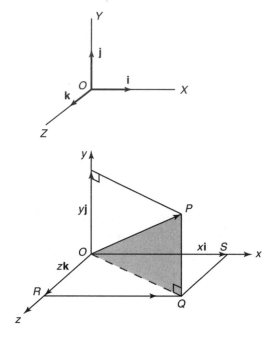

Fig. 17.22

In Fig. 17.22(b) a vector $\overrightarrow{OP}$ is resolved into three components. First of all a perpendicular PQ is dropped on the x–z plane, so that

$$\overrightarrow{OP} = \overrightarrow{OQ} + \overrightarrow{QP}$$

Now OQ is resolved along the x- and z-axes:

$$\overrightarrow{OQ} = \overrightarrow{OR} + \overrightarrow{RQ}$$

Therefore

$$\overrightarrow{OP} = \overrightarrow{OR} + \overrightarrow{RQ} + \overrightarrow{QP}$$

But

$$\overrightarrow{OR} = z\mathbf{k}$$
$$\overrightarrow{RQ} = \overrightarrow{OS} = x\mathbf{i}$$
$$\overrightarrow{QP} = y\mathbf{j}$$

Therefore

$$\overrightarrow{OP} = x\mathbf{i} + y\mathbf{j} + z\mathbf{k}$$

The magnitude of $\overrightarrow{OP}$ is given by

$$|\overrightarrow{OP}|^2 = |\overrightarrow{OQ}|^2 + |\overrightarrow{QP}|^2$$
$$= |\overrightarrow{OR}|^2 + |\overrightarrow{OS}|^2 + |\overrightarrow{QP}|^2$$
$$= x^2 + y^2 + z^2$$
$$|\overrightarrow{OP}| = \sqrt{x^2 + y^2 + z^2}$$

EXERCISE 17.2

1. Find the magnitude and direction of

    $$\mathbf{u} = 6\mathbf{i} + 7\mathbf{j}, \qquad \mathbf{v} = 8\mathbf{i} - 2\mathbf{j}$$

    Find the magnitude and direction of $\mathbf{u} + \mathbf{v}$ and $\mathbf{v} - \mathbf{u}$.

2. $\mathbf{u} = 2\mathbf{i} + \mathbf{j} + 2\mathbf{k}$, $\mathbf{v} = -\mathbf{i} + 3\mathbf{j} + 4\mathbf{k}$, $\mathbf{w} = 5\mathbf{i} - 2\mathbf{k}$, find
    (a) $\mathbf{u} + \mathbf{v} + \mathbf{w}$
    (b) $\mathbf{u} + \mathbf{v} - 2\mathbf{w}$

3. Given that $\mathbf{u} = 3\mathbf{i} + 4\mathbf{j}$ and $\mathbf{v} = -6\mathbf{i} + 2\mathbf{j}$, find m and n such that

    $$m\mathbf{u} + n\mathbf{v} = 12\mathbf{i} + 11\mathbf{j}.$$

4. Show that the points A, B, C with position vectors

    $$\mathbf{u} = \mathbf{i} + 2\mathbf{j}$$
    $$\mathbf{v} = 3\mathbf{i} + 5\mathbf{j}$$
    $$\mathbf{w} = 5\mathbf{i} + 8\mathbf{j}$$

    lie on a straight line.

5. In Question 2 find the magnitude of $\mathbf{u}$, $\mathbf{v}$, and $\mathbf{w}$ and hence determine the unit vector corresponding to each of them.

6. Find the resultant of the three vectors

    $$\mathbf{u} = 4\mathbf{i} + 3\mathbf{j} + 2\mathbf{k}$$
    $$\mathbf{v} = 3\mathbf{i} - 5\mathbf{j} - 3\mathbf{k}$$
    $$\mathbf{w} = 5\mathbf{i} + 2\mathbf{j} - 4\mathbf{k}$$

    Determine also the vector $\mathbf{u} + 2\mathbf{v} - 3\mathbf{w}$ in terms of $\mathbf{i}$, $\mathbf{i}$, $\mathbf{k}$.

7. Three points A, B, C have positional vectors $\mathbf{i} + 3\mathbf{j}$, $4\mathbf{i} + 2\mathbf{j}$, $8\mathbf{i} + 9\mathbf{j}$. Given that

    $$\overrightarrow{OC} = m\overrightarrow{OA} + n\overrightarrow{OB}$$

    find the values of m and n.

8. $OABC$ is a square with O as the origin of the Cartesian axes. The positional vector of C is $12\mathbf{i} - 5\mathbf{j}$, and the positional vector of A is $5\mathbf{i} + 12\mathbf{j}$. Find the positional vector of the point B.

9. A force $\mathbf{F} = 120\,\mathrm{N}$ makes an angle of $35°$ with the horizontal. Determine its vertical and horizontal components.

10. The velocity of an aeroplane is given by $180\mathbf{i} + 200\mathbf{j}$. Given that it flies in a cross-wind given by $40\mathbf{i} + 60\mathbf{j}$, find the resultant velocity and its direction.

17.14 The scalar product of two vectors (dot product)

Consider two vectors $\mathbf{u}$, $\mathbf{v}$ with an angle θ between them as shown in Fig. 17.23. The scalar product is written as $(\mathbf{u} \cdot \mathbf{v})$ and is defined as

$$(\mathbf{u} \cdot \mathbf{v}) = |\mathbf{u}|\,|\mathbf{v}|\cos\theta \qquad (17.1)$$

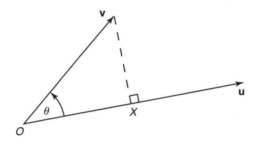

Fig. 17.23

Similarly

$$(\mathbf{v} \cdot \mathbf{u}) = |\mathbf{v}|\,|\mathbf{u}|\cos\theta$$

Both results are scalar quantities, which are independent of the order of multiplication, so that

$$(\mathbf{u} \cdot \mathbf{v}) = (\mathbf{v} \cdot \mathbf{u})$$

We see from Fig. 17.23 that $|\mathbf{v}|\cos\theta$ is the projection of the vector $\mathbf{v}$ on $\mathbf{u}$ is of length OX. Therefore the scalar product of two vectors is the magnitude of one vector multiplied by the projection of the other vector on it. It is a scalar quantity. Of particular importance are two cases:

(i) When the two vectors are perpendicular, $\theta = 90°$, so that $\cos\theta = 0$. This means that the scalar product

$$(\mathbf{u} \cdot \mathbf{v}) = 0 \qquad (17.2)$$

(ii) When the two vectors are parallel, $\theta = 0$ so that $\cos\theta = 1$. This means that that the scalar product

$$(\mathbf{u} \cdot \mathbf{v}) = |\mathbf{u}|\,|\mathbf{v}| \qquad (17.3)$$

We now apply the definition of a scalar product to the unit vectors **i**, **j**, **k**. These vectors are all perpendicular to one another. Therefore

$$(\mathbf{i}\cdot\mathbf{j}) = (\mathbf{j}\cdot\mathbf{k}) = (\mathbf{k}\cdot\mathbf{i}) = 0 \qquad (17.4)$$

Also $(\mathbf{i}\cdot\mathbf{i}) = (\mathbf{j}\cdot\mathbf{j}) = (\mathbf{k}\cdot\mathbf{k}) = 1 \qquad (17.5)$

because they are unit vectors

We shall now consider the scalar product of two vectors **u**, **v** expressed in Cartesian form.

Let $\mathbf{u} = x_1\mathbf{i} + y_1\mathbf{j} + z_1\mathbf{k}$

and $\mathbf{v} = x_2\mathbf{i} + y_2\mathbf{j} + z_2\mathbf{k}$

Therefore $(\mathbf{u}\cdot\mathbf{v}) = (x_1\mathbf{i} + y_1\mathbf{j} + z_1\mathbf{k})\cdot(x_2\mathbf{i} + y_2\mathbf{j} + z_2\mathbf{k})$

On multiplying out completely, and using the results of equations (17.4) and (17.5), this reduces to

$$(\mathbf{u}\cdot\mathbf{v}) \quad = \quad x_1x_2 + y_1y_2 + z_1z_2 \qquad (17.6)$$

because the multiplication produces such terms as $x_1y_2(\mathbf{i}\cdot\mathbf{j})$ which is 0 and $x_1x_2(\mathbf{i}\cdot\mathbf{i})$ which is $x_1\cdot x_2$.

17.15   The scalar product in engineering

(a) The work done in moving against a force

In Fig. 17.24 a force **F** acts at an angle θ to the line AB. A particle P moves a distance BA along the distance vector **s**. To find the work done against the force **F** must be resolved along AB, that is $|\mathbf{F}|\cos\theta$

work done is $|\mathbf{F}|\cos\theta \times |\mathbf{s}| = |\mathbf{F}||\mathbf{s}|\cos\theta$

Therefore work done is the scalar product $(\mathbf{F}\cdot\mathbf{s})$

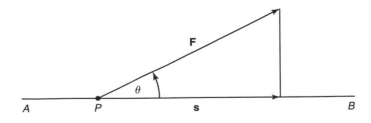

Fig. 17.24

(b) Power in an AC circuit

In an electrical circuit the phase angle between the current and voltage is $\phi°$. If the r.m.s. values are **I** and **V** the power in the circuit is $|\mathbf{I}|\,|\mathbf{V}|\cos\phi$. Therefore the power is the scalar product $(\mathbf{I}\cdot\mathbf{V})$.

Example 17.10   A vector **u** of magnitude 30 makes an angle of 25° with a vector **v** of magnitude 18. What is the scalar product of the two vectors?

The scalar product is
$(\mathbf{u}\cdot\mathbf{v}) = |\mathbf{u}|\,|\mathbf{v}|\cos 25 = 30 \times 18 \times 0.906 = 490$

Example 17.11
(a)   A force **F** of magnitude 98 N acts at 41° to a path AB. Given that a body moves along the direction BA for a distance of 11 m, calculate the work done by the body.
(b)   A force **F** is given by $(2\mathbf{i} + 3\mathbf{j})$ N. A body moves in a direction BA given by $(5\mathbf{i} - 6\mathbf{j})$ m. Find the work done by the body in moving against the force.

(a)   The work done is the scalar product
$(\mathbf{F}\cdot\mathbf{s}) = |\mathbf{F}|\,|\mathbf{s}|\cos 41 = 98 \times 11 \times 0.755 = 810\,\text{J}$
(b)   The work done against the force is the scalar product $(\mathbf{F}\cdot\overrightarrow{BA})$ as shown in Fig. 17.25(a)

$$\begin{aligned}
(\mathbf{F}\cdot\mathbf{BA}) &= (2\mathbf{i} + 3\mathbf{j})\cdot(5\mathbf{i} - 6\mathbf{j}) \\
&= 2 \times 5(\mathbf{i}\cdot\mathbf{i}) - 2 \times 6(\mathbf{i}\cdot\mathbf{j}) + 3 \times 5(\mathbf{j}\cdot\mathbf{i}) - 3 \times 6(\mathbf{j}\cdot\mathbf{j}) \\
&= 10 - 0 + 0 - 18 \\
&= -8\,\text{J}
\end{aligned}$$

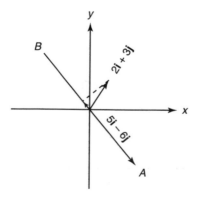

Fig. 17.25(a)

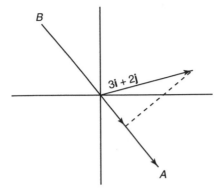

Fig. 17.25(b)

As shown in Fig. 17.25(a) the body moves against the direction of the resolved force. The scalar product has a negative sign showing that the body uses up energy in overcoming the force.

If the force is $3\mathbf{i} + 2\mathbf{j}$ the work done

$$(\mathbf{F} \cdot \mathbf{BA}) = (3\mathbf{i} + 2\mathbf{j}) \cdot (5\mathbf{i} - 6\mathbf{j}) = 15 - 12 = 3\,\text{J}$$

In this case the body moves along the resolved force and does work on the body, which gains 3 J. This is shown in Fig. 17.25(b).

Example 17.12   The vector $\mathbf{u} = 5\mathbf{i} + 6\mathbf{j}$ is perpendicular to the vector $\mathbf{v} = k\mathbf{i} + 10\mathbf{j}$. Find the value of k.

The scalar product $(\mathbf{u} \cdot \mathbf{v}) = 0$ since the two vectors are perpendicular. Therefore

$$(5\mathbf{i} + 6\mathbf{j}) \cdot (k\mathbf{i} + 10\mathbf{j}) = 5k + 60 = 0$$

$$k = -12$$

Example 17.13   Find the angle between the two vectors

$$\mathbf{u} = 2\mathbf{i} + 3\mathbf{j} - 4\mathbf{k}$$

$$\mathbf{v} = 5\mathbf{i} - 2\mathbf{j} - 3\mathbf{k}$$

Using the definition of a scalar product we obtain

$$(\mathbf{u} \cdot \mathbf{v}) = (2\mathbf{i} + 3\mathbf{j} - 4\mathbf{k}) \cdot (5\mathbf{i} - 2\mathbf{j} - 3\mathbf{k})$$

$$= 2 \times 5 + 3 \times (-2) + (-4)(-3) = 16$$

But

$$(\mathbf{u} \cdot \mathbf{v}) = |\mathbf{u}||\mathbf{v}| \cos \theta$$

The magnitude of **u** is

$$|\mathbf{u}| = \sqrt{2^2 + 3^2 + (-4)^2} = \sqrt{29}$$

The magnitude of **v**

$$|\mathbf{v}| = \sqrt{5^2 + (-2)^2 + (-3)^2} = \sqrt{38}$$

Therefore

$$\sqrt{29}\,\sqrt{38} \cos \theta = 16$$
$$\cos \theta = 0.4820$$
$$\theta = 61.2°$$

Example 17.14 $ABCD$ is a parallelogram in which $\overrightarrow{AB} = \mathbf{p}$ and $\overrightarrow{AD} = \mathbf{q}$. Express $(\overrightarrow{BD} \cdot \overrightarrow{AC})$ in terms of **p** and **q**. Hence show that if $ABCD$ is a rhombus its diagonals are at right angles.

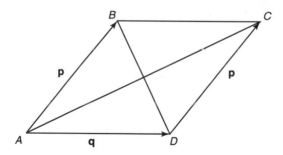

Fig. 17.26

In Fig. 17.26 we can see that $\overrightarrow{BD}$ is the difference between two vectors, and $\overrightarrow{AC}$ the resultant of two vectors, that is

$$\overrightarrow{BD} = (\mathbf{q} - \mathbf{p})$$
$$\overrightarrow{AC} = (\mathbf{p} + \mathbf{q})$$

Therefore the scalar product

$$(\overrightarrow{BD} \cdot \overrightarrow{AC}) = (\mathbf{q} - \mathbf{p}) \cdot (\mathbf{p} + \mathbf{q})$$
$$= (\mathbf{q} \cdot \mathbf{p}) + (\mathbf{q} \cdot \mathbf{q}) - (\mathbf{p} \cdot \mathbf{p}) - (\mathbf{p} \cdot \mathbf{q})$$
$$= |\mathbf{q}|^2 - |\mathbf{p}|^2$$

since $(\mathbf{q} \cdot \mathbf{p}) = (\mathbf{p} \cdot \mathbf{q})$ from equation (17.3)

If $ABCD$ is a rhombus then the sides are of equal magnitude, so that $|\mathbf{p}|^2 = |\mathbf{q}|^2$ Therefore

$$(\overrightarrow{BD} \cdot \overrightarrow{AC}) = 0$$

which by definition means that the two vectors are at right angles.

17.6 The vector product of two vectors (cross product)

Fig. 17.27 shows two vectors $\mathbf{u}$ and $\mathbf{v}$ at an angle θ to one another. The vector product of the two vectors, written as $(\mathbf{u} \times \mathbf{v})$ is defined as

$$(\mathbf{u} \times \mathbf{v}) = \mathbf{n}|\mathbf{u}||\mathbf{v}|\sin\theta \qquad (17.7)$$

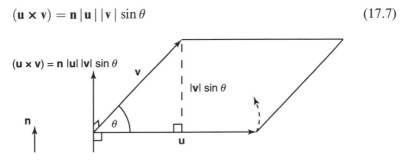

Fig. 17.27

Now $|\mathbf{u}||\mathbf{v}|\sin\theta$ is a scalar quantity equal in magnitude to the area of the parallelogram whose sides are the vectors $\mathbf{u}$ and $\mathbf{v}$, as shown in Fig. 17.27. $\mathbf{n}$ is a unit vector perpendicular to the parallelogram so formed, in a direction given by the corkscrew rule. If a corkscrew is made to rotate in the same way as $\mathbf{u}$ rotates towards $\mathbf{v}$, the direction of $\mathbf{n}$ is the direction that the tip of the

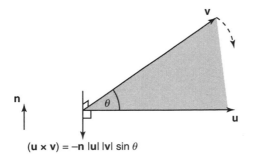

Fig. 17.28

corkscrew moves. Therefore the vector product of two vectors is a vector, whose magnitude is the area of the parallelogram formed by the two vectors as sides, in a direction given by the corkscrew rule.

The vector product $(\mathbf{v} \times \mathbf{u})$ is a vector of the same magnitude, but now acts in the opposite direction, as shown in Fig. 17.28. Therefore

$$(\mathbf{v} \times \mathbf{u}) = -(\mathbf{u} \times \mathbf{v}) \tag{17.8}$$

Example 17.15   Two vectors $\mathbf{u}$, $\mathbf{v}$, with $|\mathbf{u}| = 25$, $|\mathbf{v}| = 20$ are inclined at 22° to one another. Calculate the vector product $(\mathbf{u} \times \mathbf{v})$ and show the direction. On the same diagram show the direction of the vector product $(\mathbf{v} \times \mathbf{u})$.

The vector product $(\mathbf{u} \times \mathbf{v}) = \mathbf{n}\, 25 \times 20 \sin 22° = \mathbf{n}\, 190$ in the direction shown in Fig. 17.29.

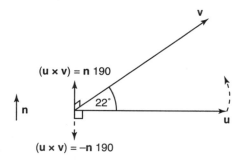

Fig. 17.29

The vector product $(\mathbf{v} \times \mathbf{u})$ has the same magnitude in the opposite direction.

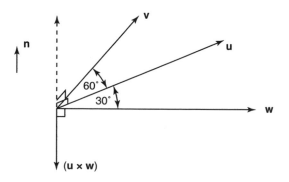

Fig. 17.30

Example 17.16   In Fig. 17.30 **u**, **v** and **w** are three vectors in the same plane with magnitude 10, 15, 20 units respectively. The angle between **u** and **v** is 60° and the angle between **u** and **w** is 30°. Find

$(\mathbf{u} \times \mathbf{v}) + (\mathbf{u} \times \mathbf{w})$

Let the unit vector **n** act upwards as shown.

In Fig. 17.30 **u** rotating towards **v** produces a vector product $(\mathbf{u} \times \mathbf{v}) = \mathbf{n}\, uv \,\sin 60°$ acting upwards. Again **u** rotating towards **w** produces a vector product $(\mathbf{u} \times \mathbf{w}) = \mathbf{n}\, uw \, \sin 30°$ acting downwards.

$$(\mathbf{u} \times \mathbf{v}) + (\mathbf{u} \times \mathbf{w}) = \mathbf{n}\, uv \sin 60° - \mathbf{n}\, uw \sin 30°$$

$$= \mathbf{n}(10 \times 15 \times \sin 60° - 15 \times 20 \times \sin 30°)$$

$$= \mathbf{n}(150 \times 0.866 - 300 \times 0.5)$$

$$= -20.1\mathbf{n} \qquad \text{which acts downwards}$$

Example 17.17   Simplify $(\mathbf{u} - \mathbf{v}) \times (\mathbf{u} + \mathbf{v})$ and show that if **u** is perpendicular to **v** then

$$|(\mathbf{u} - \mathbf{v}) \times (\mathbf{u} + \mathbf{v})| = 2\,|\mathbf{u}\mathbf{v}|$$

$$(\mathbf{u} - \mathbf{v}) \times (\mathbf{u} + \mathbf{v}) = \mathbf{u} \times (\mathbf{u} + \mathbf{v}) - \mathbf{v} \times (\mathbf{u} + \mathbf{v})$$

$$= (\mathbf{u} \times \mathbf{u}) + (\mathbf{u} \times \mathbf{v}) - (\mathbf{v} \times \mathbf{u}) - (\mathbf{v} \times \mathbf{v})$$

$$= \mathbf{n}\, u^2 \sin(0) + (\mathbf{u} \times \mathbf{v}) - (\mathbf{v} \times \mathbf{u}) - \mathbf{n}\, v^2 \sin(0)$$

$$= (\mathbf{u} \times \mathbf{v}) - (\mathbf{v} \times \mathbf{u}) \quad \text{because} \quad \sin 0 = 0$$

But

$$(\mathbf{v} \times \mathbf{u}) = -(\mathbf{u} \times \mathbf{v})$$

Therefore $(\mathbf{u} - \mathbf{v}) \times (\mathbf{u} + \mathbf{v}) = 2(\mathbf{u} \times \mathbf{v})$

If **u** and **v** are perpendicular

$(\mathbf{u} \times \mathbf{v}) = \mathbf{n}\, uv \sin(90°)$

$= \mathbf{n}\, uv$

Therefore the magnitude

$|(\mathbf{u} - \mathbf{v}) \times (\mathbf{u} + \mathbf{v})| = 2uv$

17.17 Vector products in engineering

(a) The force on a conductor carrying a current in a magnetic field

In Fig. 17.31 **B** is the vector representing the intensity of the magnetic flux, and **I** is the current in amperes flowing along a conductor at an angle θ to **B**. The force/metre on the conductor is given by the vector product $(\mathbf{I} \times \mathbf{B})$,

$$\mathbf{F} = (\mathbf{I} \times \mathbf{B}) = \mathbf{n}\,|\mathbf{I}|\,|\mathbf{B}|\,\sin\theta$$

in a direction determined by the corkscrew rule, as shown in the diagram.

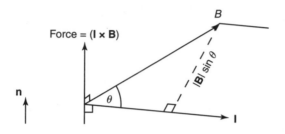

Fig. 17.31

(b) The moment of a force about a point

In Fig. 17.32 a force **F** acts at a point A, distance vector **a** from a point O, at an angle θ to OA. The moment of the force **F** about the point O is given by the vector product $(\mathbf{F} \times \mathbf{a})$, that is,

$$\text{Moment about } O = (\mathbf{F}) = \mathbf{n}\,|\mathbf{F} \times \mathbf{a}|\,|\mathbf{a}|\,\sin\theta$$

in the direction determined by the corkscrew rule.

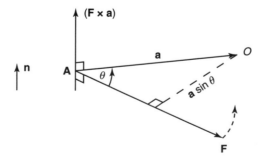

Fig. 17.32

17.18   The vector product of unit vectors i, j, k

In Fig. 17.33 **i**, **j**, **k** are unit vectors acting along the x-, y-, z-axes respectively. Using the definition of vector product we see immediately that

$$\left.\begin{array}{l} (\mathbf{i} \times \mathbf{i}) = 0 \\ (\mathbf{j} \times \mathbf{j}) = 0 \\ (\mathbf{k} \times \mathbf{k}) = 0 \end{array}\right\} \tag{17.9}$$

because the angle between each pair is $0°$ and $\sin 0° = 0$
Again using the corkscrew rule, and $\cos 90° = 1$

$$\left.\begin{array}{llll} (\mathbf{i} \times \mathbf{j}) = \mathbf{k} & & (\mathbf{j} \times \mathbf{i}) = -k \\ (\mathbf{j} \times \mathbf{k}) = \mathbf{i} & \text{and} & (\mathbf{k} \times \mathbf{j}) = -i \\ (\mathbf{k} \times \mathbf{i}) = \mathbf{j} & & (\mathbf{i} \times \mathbf{k}) = -j \end{array}\right\} \tag{17.10}$$

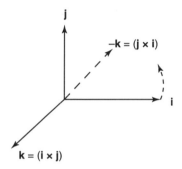

Fig. 17.33

Example 17.18   Find the vector product of $\mathbf{u} = 3\mathbf{i} + 4\mathbf{j}$ and $\mathbf{v} = 2\mathbf{i} - 5\mathbf{j}$ and determine its direction.

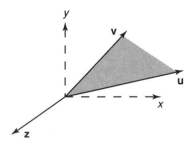

Fig. 17.34

In Fig. 17.34 **u** and **v** are the two vectors in the x–y plane.

$$(\mathbf{u} \times \mathbf{v}) = (3\mathbf{i} + 4\mathbf{j}) \times (2\mathbf{i} - 5\mathbf{j})$$
$$= -15(\mathbf{i} \times \mathbf{j}) + 8(\mathbf{j} \times \mathbf{i})$$
$$= -15\mathbf{k} - 8\mathbf{k} \qquad \text{using equations (17.10)}$$
$$= -23\mathbf{k}$$

which act in the $-z$ direction.

Example 17.19   The position vectors of two points A and B with respect to a point O are $3\mathbf{i} + 4\mathbf{j}$ and $5\mathbf{i} + 12\mathbf{j}$. Find the unit vectors along $\overrightarrow{OA}$ and $\overrightarrow{OB}$.

A force $\mathbf{F}_1$ of magnitude 75 N acting along OA, and a force $\mathbf{F}_2$ of magnitude 130 N acts along OB. Express $\mathbf{F}_1$ and $\mathbf{F}_2$ in terms of $\mathbf{i}$ and $\mathbf{j}$ and find the resultant force in vector form.

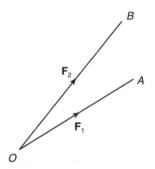

Fig. 17.35

$$|\overrightarrow{OA}|^2 = 3^2 + 4^2 = 25$$
$$|\overrightarrow{OA}| = 5$$

Therefore the unit vector along OA is $\frac{3}{5}\mathbf{i} + \frac{4}{5}\mathbf{j}$

$$|\overrightarrow{OB}|^2 = 5^2 + 12^2 = 169$$
$$|\overrightarrow{OB}| = 13$$

Therefore the unit vector along OB is $\frac{5}{13}\mathbf{i} + \frac{12}{13}\mathbf{j}$
The force $\mathbf{F}_1$ acts along OA so that it can be expressed as 75 × unit vector along OA.

$$F_1 = 75 \left(\tfrac{3}{5}\mathbf{i} + \tfrac{4}{5}\mathbf{j}\right)$$

Hence

$$F_1 = 45i + 60j$$

Similarly

$$F_2 = 130 \times \text{unit vector along OB}$$
$$= 130 \left(\tfrac{5}{13}i + \tfrac{12}{13}j\right)$$
$$= 50i + 120j$$

The resultant of F_1 and F_2 is $(45 + 50)i + (60 + 130)j = 95i + 190j$

EXERCISE 17.3

1. Find the magnitude and direction of $u = 6i + 7j$ and $v = 8i - 2j$. Hence find the magnitude and direction of (a) $u + v$, (b) $(u \times v)$

2. Given

    $$u = 3i + 2j - 5k$$
    $$v = 2i - 3j + 2k$$

    find (a) $(u \cdot v)$, (b) $(u \times v)$

3. Find the angle between $2i - 3j - 4k$ and $-3i + 4j \; 5k$.

4. Simplify $(u - v) \times (u + v) + (-v + 2v) \times (2u - v)$

    and show that if the angle between the two vectors u and v is $90°$ the above expression reduces to the vector area of the rectangle whose sides are u and v.

5. Simplify

    $$(-u + v) \cdot (u - v) - (-u + 2v) \cdot (2u - v)$$

    and show that if the angle between the two vectors u and v is $90°$ the above expression reduces to the magnitude of the diagonal of the rectangle whose sides are u and v.

6. Show that the following expression reduces to 0 and explain this on a vector diagram

    $$(-u + v) \times (u - v)$$

7. Find the unit vector (a) along, (b) perpendicular and coplanar, to each of the following vectors:

    (i) $3i + 4j$ (ii) $5i + 12j$ (iii) $4i + 7j$

MISCELLANEOUS EXERCISE 17

1. In the regular hexagon $ABCDEF$, $\overrightarrow{AB} = a$ and $\overrightarrow{BC} = b$. Express, in terms of a and b, the vectors

    (a) $\overrightarrow{AC}$ (b) $\overrightarrow{AD}$ (c) $\overrightarrow{AE}$ (d) $\overrightarrow{AF}$.

2.   Given that

$$\mathbf{a} = \mathbf{x} + \mathbf{y} \qquad \text{and} \qquad \mathbf{b} = \mathbf{x} - \mathbf{y}$$

where $\mathbf{x} \neq 0$ and $\mathbf{y} \neq 0$, show that

$\mathbf{a} \cdot \mathbf{b} = 0$ if and only if $|\mathbf{x}| = |\mathbf{y}|$.

3.   Find a unit vector which is in the opposite direction to the sum of the vectors $(3\mathbf{i} + 2\mathbf{j} + \mathbf{k})$ and $(-5\mathbf{i} - 3\mathbf{j} + 6\mathbf{k})$.
     Prove that this unit vector is perpendicular to the vector $(9\mathbf{i} - 4\mathbf{j} + 2\mathbf{k})$.

4.   Show that $\mathbf{a} \cdot \mathbf{a} = |\mathbf{a}|^2$.
     $ABCD$ is a parallelogram whose diagonals intersect at E. Given that $\overrightarrow{AE} = \mathbf{a}$ and $\overrightarrow{BE} = \mathbf{b}$, express $\overrightarrow{AB} \cdot \overrightarrow{AD}$ in terms of $|\mathbf{a}|$ and $|\mathbf{b}|$.
     Hence show that if $ABCD$ is a rectangle, its diagonals are equal in length.

5.   Referred to an origin O, the position vectors of points A and B are given respectively by

$$\overrightarrow{OA} = 3\mathbf{i} + \mathbf{j} + 3\mathbf{k}$$
$$\overrightarrow{OB} = 5\mathbf{i} - 4\mathbf{j} + 3\mathbf{k}$$

Show that the cosine of angle AOB is equal to $4/\sqrt{(38)}$.

6.   The position vectors $\mathbf{a}$, $\mathbf{b}$, $\mathbf{c}$, of the points A, B, C, respectively, are given by

$$\mathbf{a} = \qquad -\mathbf{j} + \mathbf{k}$$
$$\mathbf{b} = 2\mathbf{i} \quad +\mathbf{j} \quad +4\mathbf{k}$$
$$\mathbf{c} = -2\mathbf{i} +5\mathbf{j} -4\mathbf{k}$$

The point X on BC is such that $BX = \frac{1}{3}XC$ and the point Y, on AX produced, is such that $AX = \frac{1}{4}XY$.
     Find the position vectors of X and Y.

7.   Three points have coordinates $A(1, 3)$, $B(4, 2)$ and $C(8, 9)$. O is the origin.
     (a)   Given that $\overrightarrow{OC} = m\overrightarrow{OA} + n\overrightarrow{OB}$ where m and n are scalar constants, find the value of m and of n.
     (b)   Evaluate the scalar product $\overrightarrow{AC} \cdot \overrightarrow{BC}$ and *hence* calculate angle $A\hat{C}B$.

8.   The magnitudes of the vectors $\mathbf{a}$ and $\mathbf{b}$ are 8 and 3, respectively, and the angle between the vectors is $60°$. Sketch a diagram showing these vectors and the vector $\mathbf{a} - \mathbf{b}$. Calculate
     (a)   the magnitude of $\mathbf{a} - \mathbf{b}$,
     (b)   the resolved part of $\mathbf{a}$ in the direction of $\mathbf{b}$.

9. (a) The position vectors of the points A, B and C are $\mathbf{a}$, $\mathbf{b}$ and $\mathbf{c}$ respectively, referred to the point O as origin. Given that $3\mathbf{a} + \mathbf{b} = 4\mathbf{c}$, prove that the points A, B and C are collinear and find the ratio $AB : AC$.

   (b) Three forces $7\mathbf{i} + 5\mathbf{j}$, $2\mathbf{i} + 3\mathbf{j}$ and $\lambda\mathbf{i}$ act at the origin O, where $\mathbf{i}$ and $\mathbf{j}$ are unit vectors parallel to the x-axis and the y-axis respectively. The unit of force is the newton. Given that the magnitude of the resultant of the three forces is 17 N, calculate the two possible values of λ.

10. Referred to O as origin, the position of the points A and B are $(5\mathbf{i} + 12\mathbf{j})$ and $(16\mathbf{i} + 12\mathbf{j})$ respectively.

   (a) Find, in terms of $\mathbf{i}$ and $\mathbf{j}$, the unit vector along $\overrightarrow{OA}$ and $\overrightarrow{OB}$.

   (b) Force $\mathbf{F}_1$ of magnitude 91 units acts along OA and force $\mathbf{F}_2$ of magnitude 80 units acts along OB. Express $\mathbf{F}_1$ and $\mathbf{F}_2$ in terms of $\mathbf{i}$ and $\mathbf{j}$ and hence calculate the magnitude of their resultant.

Appendix 1

Area of a sector of a circle

In Fig. A.1 the shaded area is a sector of a circle of radius r, in which the angle subtended at the centre is θ rad. Because of the symmetry of a circle

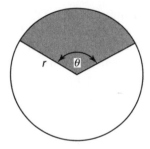

Fig. A.1

the area of any sector is proportional to the angle subtended at the centre. In this respect a circle can be regarded as a sector, in which the angle at the centre is $360°$ or 2π rad. Hence

$$\frac{\text{area of sector}}{\text{area of circle}} = \frac{\theta}{2\pi}$$

so that $\quad$ area of sector $= \dfrac{\theta}{2\pi} \times$ area of circle

$$\frac{\theta}{2\pi} \times \pi r^2$$

Area of sector $= \frac{1}{2}r^2\theta$

Appendix 2

The sum of the series
$$1^2 + 2^2 + 3^2 + \cdots + (N-1)^2$$

The binomial $(n+1)^3$ expands as follows:

$$(n+1)^3 = n^3 + 3n^2 + 3n + 1$$

so that

$$(n+1)^3 - n^3 = 3n^2 + 3n + 1 \tag{i}$$

Identity (i) is now examined for various integer values of n.

When $n = 1$: $2^3 - 1^3$ $=$ $3(1)^2$ $+$ $3(1)$ $+$ 1
$\quad\quad\;\; n = 2$: $3^3 - 2^3$ $=$ $3(2)^2$ $+$ $3(2)$ $+$ 1
$\quad\quad\;\; n = 3$: $4^3 - 3^3$ $=$ $3(3)^2$ $+$ $3(3)$ $+$ 1

$$\vdots$$

$n = (N-1)$: $N^3 - (N-1)^3 = 3(N-1)^2 + 3(N-1) +\quad 1$

Adding: $N^3 - 1^3 \qquad = 3\{1^2 + 2^2 \ldots (N-1)^2$

$$+ 3\{1 + 2 + 3 \ldots (N-1)\} + (N-1)$$

Note that on the left-hand side the numbers cancel out as shown. The last column on the right-hand side contains $(N-1)$ units.

The series $1 + 2 + 3 + 4 + \cdots + (N-1)$ is a simple arithmetic series, whose sum is $\frac{1}{2}(N-1)N$. Therefore, transposing the above result,

$$3\{1^2 + 2^2 + 3^2 \ldots (N-1)^2\} = N^3 - 1 - \tfrac{3}{2}N(N-1) - (N-1)$$

$$= \tfrac{1}{2}(2N^3 - 2 - 3N^2 + 3N - 2N + 2)$$

$$= 2N(N-1)(2N-1)$$

that is

$$1^2 + 2^2 + 3^2 + \cdots + (N-1)^2 = \tfrac{1}{6}(N(N-1)(2N-1))$$

Answers to exercises

Chapter 1

Exercise 1.1

1. $x = 5$, $x = \frac{1}{2}$,   2. $x = -\frac{7}{2}$, $x = \frac{1}{2}$ 3. $x = 0.76$, $x = -0.19$
4. $x = 1.24$, $x = -0.64$ 5. $x = \pm 3$, $x = \pm 2$
6. $x = \pm \frac{1}{3}$, $x = \pm \frac{3}{2}$ 7. $y = \frac{1}{3}$, $y = 2$
8. $y = 0.53$, $y = -2.53$ 9. $x = 1$, $x = 2$
10. $x = -1$, $x = 1$ 11. $x = 2$, $x = -2$ 12. $x = 0$, $x = 2$
13. $x = \frac{1}{3}$, -3, $\frac{3}{2}$, $-\frac{2}{3}$ 14. $x = -1$, 2, -2.17, 0.92
15. $x = -1$, 2

Exercise 1.2

1. $x = 1$, $y = 2$, $z = 3$ 2. $x = 1$, $y = -1$, $z = -2$
3. $x = -1$, $y = 1$, $z = 2$ 4. $i_1 = 2$, $i_2 = -1$, $i_3 = 3$
5. $u = -2$, $v = -3$, $w = 1$ 6. $a = -6$, $b = 3$, $c = 4$
7. $i_1 = 3$, $i_2 = 4$, $i_3 = 2$ 8. $a = 2$, $b = -1$, $c = 1$
9. $F_1 = 7$, $F_2 = -3$, $F_3 = 3$

Exercise 1.3

1. $(\frac{2}{9}, -\frac{4}{3})$, $(2, 4)$ 2. $(-1, -1)$, $(\frac{3}{13}, \frac{19}{13})$
3. $(4, -2)$, $(-2, 4)$ 4. $(-2, 1)$
5. $(\frac{1}{2} -2)$, $(2, 4)$ 6. $(6, -3)$, $(-6, 3)$
7. $(-5.35, -3.14)$, $(4.1, 0.64)$ 8. $(4, 1)$, $(1, 4)$

Exercise 1.4

1. $\frac{9}{4}$ 2. $\frac{2}{5}$ 3. $\frac{4}{5}$ 4. $27x^4$ 5. 4 6. 16 7. $25y^2$
8. 9 9. 7 10. 1 11. 1 12. $t^{\frac{4}{3}}$

Exercise 1.5

1. $3 = \log_3 27$ 2. $2 = \log_9 81$ 3. $2 = \log_{10} 100$
4. $\frac{1}{2} = \log_4 2$ 5. $t = \log_n N$ 6. $32 = 2^5$ 7. $6 = 6^1$
8. $343 = 7^3$ 9. $3 = 27^{\frac{1}{3}}$ 10. $M = x^m$

Exercise 1.6

1. (a) 1.871, (b) 2.183, (c) 0.4898, (d) 6.166, (e) –5.164,
   (f) –0.1892, (g) –3.430
2. (a) 1644, (b) 5.039, (c) 2.041, (d) 0.1172, (e) 0.2557,
   (f) 0.000 9961

Exercise 1.7

1. 39.22 2. ± 4.26 3. ± 3.008 4. 2.935 5. 4 6. $+5, -5$
7. 2, 1 8. 1110 9. 2.204 10. 14.04

Exercise 1.8

1. $x = 3$ 2. 0.541, -5.541 3. 5, -1 4. 2 log 3
5. log 5 6. $\frac{3}{2}$

Exercise 1.9

1. 2.377 2. 1.913 3. -1.530 4. -0.208 5. 1.905
6. 3.169×10^6 7. -2.423 8. $-4.512, 6.130$

Exercise 1.10

1. $\theta = \dfrac{1}{ur} \ln \dfrac{T_1}{T_2}$ 2. $b = ae^{(L/2k - 1)}$

3. $t = \dfrac{1}{R} \ln \dfrac{N}{N_0}$, (a) $\dfrac{1}{R} \ln 2$, (b) $\dfrac{1}{R} \ln 10$, (c) $\dfrac{1}{R} \ln 100$

4. $E = W - kT \ln \dfrac{j}{kT^2}$ 5. $r = e^k (A - V)2q$

6. $b = ae^{2\pi\kappa\theta/Q}$ 7. $V_2 = V_1 e^{W/RT}$

8. $t = \dfrac{1}{\lambda} \ln \dfrac{N_0}{N_t}$, 0.74 m

9. $T = \dfrac{b}{\lambda \ln\left(\dfrac{a}{E_\lambda \lambda^5} + 1\right)}$

10. $t = T \ln\left(\dfrac{8}{8-x}\right)$

Exercise 1.11

1. $\dfrac{7}{3(x-4)} + \dfrac{2}{3(x+2)}$ **2.** $\dfrac{1}{3(x+1)} - \dfrac{2}{3(2x+5)}$

3. $\dfrac{6}{5(x-3)} + \dfrac{2}{5(3x+1)}$ **4.** $-\dfrac{1}{6(x+1)} + \dfrac{17}{6(5x-1)}$

5. $\dfrac{1}{(x+1)} - \dfrac{4}{(x+1)^2}$ **6.** $\dfrac{1}{(2x-1)} + \dfrac{1}{2(2x-1)^2}$

7. $\dfrac{1}{2(x-1)} + \dfrac{-x+3}{2(x^2+2x-1)}$ **8.** $\dfrac{-11}{17(x+2)} + \dfrac{33x-20}{17(3x^2-2x+1)}$

9. $1 + \dfrac{2}{(x-2)} - \dfrac{2}{(x+1)}$ **10.** $4x + 1 + \dfrac{53}{5(x+2)} + \dfrac{82}{5(x-3)}$

Miscellaneous Exercise 1

1. (a) $(1, 1)$, $\left(4, \dfrac{4}{7}\right)$ (b) $(0, 3)$ **2.** $x = 3$, $x = 2$

3. (a) $a = 2$, $b = 3$, $c = -1$ (b)$\frac{2}{3}$, 3

4. (a) $t = -\dfrac{l}{k} \ln \dfrac{T-A}{B-A}$ (b) 12, 16, 20

5. (a) 0.865 (b) 1, 1.86 (c) 0.136

6. $x = 2.50$, $y = 0.80$ **7.** (a) 0.8342 (b) 13

8. (a) $\dfrac{1}{108} a^{12} b^{10}$ (b) $\dfrac{1}{6} \dfrac{b^3 c^{\frac{1}{12}}}{a^2}$ **9.** $-\dfrac{3}{2(x-1)} + \dfrac{2}{(x-2)} - \dfrac{1}{2(x-3)}$

10. (a) 0.5922 (b) 2.270

Chapter 2

Exercise 2.1

1. (a) 54 (b) 4.360 (c) 148.6 (d) 0.038 (e) 632000

2. 292 000, 0.313, 0.00916, 0.000501

3. 0.1390 N   **4.** 1910000 mm^3, 74500 mm^2, 18700 mm^3, 485 mm^2

5. (a) 43 (b) 0.93%   **6.** 0.283, 0.280, 0.00146, 0.52%

7. (a) 0.4540, 0.4226, 3.6% (b) −0.3256, −0.2924, 5.4% (c) 1.2913, 1.2656, 1.01%

8. 0.048, 3.0%   **9.** 1074-1173   **10.** 31 550 mm^2, 30 850 mm^2, 1.1%

Exercise 2.2

1. 1°, 1.1% **2.** 6.0%, 154 cm^2 **3.** (a) 0.28% (b) 0.14% (c) 0.56%
4. 0.41%, 0.187 to 0.189
5. (a) 0.1, 0.1, 2.27 (b) 1.08%, 1.75% **6.** ±1.68%
7. ±0.4% **8.** +5% or −3% **9.** −7.3 Hz
10. −5% or +3.8% **11.** (a) 0.00004, 0.024% (b) 0.0515, 0.0655%

Exercise 2.3

1. (a) 1.98 (b) 4.44 (e) 10 700 (d) 0.158
2. (a) 0.237 5 (b) 0.336 0 (c) 493 (d) 0.154 (e) 4.10 × 10^{-7} (f) 14.65
    (g) 0.1932 (h) 0.768 (j) 0.031 49 (k) 0.201 9
3. (a) $c = 247.6$ (b) $p = 53.92$ (c) $v = 2.10$
4. $i = 0.89$ **5.** $t = 0.622$
6. 450 N **7.** −0.27 A **8.** 0.0046 n
9. 13 × 10^6 **10.** 0.0029 J **11.** 66 × 10^6

Chapter 3

Exercise 3.1

1. 1.3 m, 77° **2.** 44°, 4.3 m
3. 114 V, 35° **4.** 30.62 mm, 9.518°

Exercise 3.2

1. 40°, 140° **2.** 198°, 342° **3.** 73°, 287°
4. 123°, 237°, **5.** 12°, 192° **6.** 115°, 295°

Exercise 3.3

1. 2.4 m, 86° **2.** 1788 m, 1654 m **3.** 1.4 V, 134°
4. 230, 17° **5.** 1.5 m, 1.2 m

Exercise 3.4

1. 59.0 cm^2 **2.** 22° **3.** 4.1 m^2 **4.** 34 cm^2

Exercise 3.5

1. (a) 30°, 270°, 900°, 420° (b) $\dfrac{\pi}{4}, \dfrac{\pi}{10}, 2\pi, \dfrac{5\pi}{6}, \dfrac{3\pi}{4}$

2. (a) 35.5°, 78.5°, 224.5° (b) 0.279 3, 0.880 5, 5.396 8

5. 51.5°, 88.5°, 141.5°, 178.5°
6. (a) 3, π rad, 0 (b) 60°, 27° (c) 10, $\frac{1}{3}$ s, −0.4 rad (d) 250, 0.025 s, 0

(e) 16, 1 rad, $\dfrac{\pi}{3}$ rad

9. 2.4 m, 0.334 s, 1.88   **10.**   0.02 s, 126 sin $(100\pi + 18.5)$
11. 47.1 sin$(\theta - 24.2°)$ **12.**   9.27 sin$(\theta - 27.2°)$

Exercise 3.6

1. (a) −1.730, −0.709, 1.225 6 (b) −2.028, 0.567 0, −1.149 6

(c) 1.031 3, −3.971, −4.096 (d) −1, ∞, ∞ (e) 2, $\dfrac{1}{\sqrt{3}}$, $\dfrac{2}{\sqrt{3}}$

3. $\cot^2 A$

4. $\sin\theta = \dfrac{a}{\sqrt{(a^2 + b^2)}}$, $\sec^2\theta = \dfrac{a^2 + b^2}{a^2}$

Exercise 3.7

3. $2 - \sqrt{3}$

Exercise 3.8

1. (a) 2 sin $\frac{5}{2}\theta$ cos $\frac{3}{2}\theta$ (b) 2 cos 7θ sin θ (c) 2 cos 11θ cos θ
(d) 2 sin $\frac{3}{2}$ sin $\frac{1}{2}\theta$ (e) 2 sin $\frac{7}{24}\pi$ cos $\frac{1}{24}\pi$ (f) -2 sin $50°$ sin $20°$
2. (a) sin 8θ + sin 4θ (b) cos 8θ + cos 2θ (c) sin 5θ − sin 3θ

(d) cos 2θ − cos 12θ (e) $\frac{1}{2}\left(\sin\dfrac{\pi}{2} + \sin\dfrac{\pi}{6}\right)$

(f) $\frac{1}{2}$ (sin $220°$ − sin $80°$) (g) cos $2A$ + cos $2B$ (h) $\frac{1}{2}$ (cos $\frac{3}{4}\pi$ − cos $\frac{5}{4}\pi$)
3. (a) cot θ (b) 2 sin$(A + B)$ (c) $-\frac{1}{2}$ sec $\frac{1}{2}\theta$

Exercise 3.9

1. (a) 61.1°, 118.9° (b) 54.3°, 234.3° (c) 77.3°, 282.7°
(d) 193.1°, 346.9°(e) 49.6°, 130.4° (f) 109.9°, 289.9°
(g) 58.6°, 301.5° (h) 117.3°, 152.7°, 297.3 , 332.7
(j) 64.1° (k) 39.9°, 80.1°, 159.9°, 200.1°, 279.9°, 320.1°
(l) 0°, 120°, 360° (m) 101.7°, 178.8°, 281.7°, 358.8°
(n) 149.4°, 328.4° (o) 41.1°, 97.5°, 161.1°, 217.5°. 281.1°, 337.5°
2. (a) 30°, 60° (b) 40°, 80°,160° (c) 30°, 150°, 210°, 330°(d) 40°, 160°
3. (a) 30°, 150°, 270° (b) 36.9°, 216.9°
(c) 60°, 300°, 126.9°, 233.1° (d) 48.2°, 138.6 , 221.4°, 311.8°
(e) 0°, 36°,°180°, 300°, 360° (f) 0°, 120°, 240°
(g) 0°, 36°, 72°, 108°, 144° (h) 0°, 90°, 180°

(i) $36°$, $60°$, $108°$, $180°$ (j) $0°$, $22.5°$, $67.5°$, $90°$, $112.5°$, $157.5°$, 180
(k) $10.5°$, $144.2°$ (l) $32.6°$, $82.7°$ (m) $106.7°$, $170.9°$
(n) $136.2°$, $358.6°$

Exercise 3.10

1. (a) $0.378\ 4$ (b) $1.175\ 9$ (c) $1.070\ 6$ (d) $0.696\ 5$ (e) $1.856\ 9$
    (f) $\dfrac{\pi}{2}$ (g) $\dfrac{\pi}{6}$ (h) $\dfrac{\pi}{4}$ (i) 0
3. (a) $\dfrac{\pi}{6}$ (b) $0.965\ 5$

Miscellaneous Exercise 3

1. (b) $69.6°$, $169.5°$ **2.** 0.159, 8.9, $146.6°$
3. (a) 6.40 at $\theta = 141.3°$, -6.40 at $\theta = 321.3°$
    (b) $79.3°$, $203.4°$
4. 918 m
5. (a) $-14.9°$, (b) $32.2°$, $237.8°$
6. (b) $161.6°$, $341.6°$ **7.** 0.46 rad
8. $79.22°$, $92.12°$, $139.6°$, $49.07°$, 14.2 m
9. (a) $77.3°$
10. (a) $111.7°$ (b) $\dfrac{\pi}{8}, \dfrac{\pi}{3}, \dfrac{3\pi}{8}, \dfrac{2\pi}{3}, \dfrac{5\pi}{8}, \dfrac{7\pi}{8}$ **11.**   (b) $49.8°$
12. (a) $2\left(\sin 2t + \dfrac{\pi}{4} \right)$
    (b) (i) $t - \tfrac{3}{8}\pi$ (ii) 2 when $t = \dfrac{\pi}{8}$, -2 when $t = \dfrac{5\pi}{8}$
13. (a) $35.7°$, $144.3°$, $270°$ (b) $23.1°$, $263.1°$

Chapter 4

Exercise 4.1

1. -8, 14, -10, 0, 24 **2.** 42, $18x^2h + 18xh^2 + 6h^3$, $6h^2$
3. 81, 4^6

Exercise 4.2

1. 4 **2.** $0, -1$ **3.** $1, -1$

Exercise 4.3

1. $6x^5$ **2.** $12x^3$ **3.** $-2x^{-2}$ **4.** $\frac{21}{4}x^{-\frac{1}{4}}$ **5.** $-4x^{-3}$
6. $-\frac{3}{2}x^{-\frac{3}{2}}$

Exercise 4.4

1. $42x^6$ **2.** $-21x^{-4}$ **3.** $-4x^{-2}$ **4.** $\frac{3}{2}x^{-\frac{1}{2}}$ **5.** $\frac{49}{2}x^{\frac{5}{2}}$
6. $-\frac{15}{2}x^{-\frac{5}{2}}$ **7.** $x^{-\frac{3}{2}}$ **8.** $-6x^{-\frac{5}{3}}$ **9.** $-x^{\frac{1}{2}}$
10. $28, 756$ **11.** (a) $(0, 0)$ (b) $(\pm 3, \pm 108)$

Exercise 4.5

1. $8x + 8x^3$ **2.** $12x^3 + 4x^{-3} + \frac{3}{2}x^{-\frac{1}{2}}$ **3.** $15x^2 + \frac{7}{2}x^{-\frac{1}{2}} - 3x^{-\frac{5}{2}}$
4. $-21x^{-8} + 12x^{-5}$ **5.** $-14t^{-3} - 16t^3 + 9t^{-4}$ **6.** $\frac{7}{2}\theta^{-\frac{1}{2}} - 100\theta^{\frac{3}{2}}$
7. $8x + 4$ **8.** $-2x^{-2} - 10x^{-3} + 9x^{-4}$ **9.** $3, 2, 1$
10. (a) $(1, 1)$ (b) $(2, 4)$

Exercise 4.6

1. $36, 42$ **4.** $3, 1$

Exercise 4.7

1. $8(2x + 3)^3$ **2.** $84x(7x^2 - 3)^5$ **3.** $-\frac{9}{2}x^{-\frac{1}{2}}(-3x^{\frac{1}{2}} + 4)^2$

4. $2(4t - 3)^{-\frac{1}{2}}$ **5.** $9\left(t + \dfrac{1}{t^3}\right)\left(6t^2 - \dfrac{6}{t^2}\right)^{-\frac{1}{4}}$

6. $2\left(\theta - \dfrac{1}{\theta}\right)\left(\theta + \dfrac{1}{\theta^2}\right)$ **7.** $-\frac{2}{3}x(x^2 - 3)^{-\frac{4}{3}}$

8. $-\frac{1}{2}(2t - 3)(t^2 - 3t + 4)^{-\frac{3}{2}}$

Exercise 4.8

1. $6x(2x - 1)$ **2.** $2x^2(x - 3)^{-\frac{1}{2}}(7x - 18)$ **3.** $14x(x^2 - 1)^3(5x^2 - 1)$
4. $\frac{1}{2}x^{-\frac{1}{2}}(7x^2 - 4)(63x^2 - 4)$ **5.** $\frac{1}{2}x^{-\frac{1}{2}}(1 + x^2)^{-\frac{1}{2}}(1 + x^2)$
6. $\frac{1}{2}(x + 1)^{-\frac{1}{2}}(5x^2 + 4x - 4)$

Exercise 4.9

1. $\dfrac{2x(x + 1)}{(2x + 1)^2}$ **2.** $\dfrac{2x^2(2x^2 + 3)}{(2x^2 + 1)^2}$ **3.** $\dfrac{4x}{(x^2 + 1)^2}$

4. $\dfrac{4}{(x^2 + 1)^{\frac{3}{2}}}$ **5.** $\dfrac{x(x^2 + 7)}{(x^2 + 4)^{\frac{3}{2}}}$ **6.** $\dfrac{x^2 + 2x - 4}{(x + 1)^2}$ **7.** $\dfrac{-x + 2}{2x^{\frac{1}{2}}(x + 2)^2}$

Exercise 4.10

1. $7\cos(7x-4)$ **2.** $-8x\sin(4x^2-3)$
3. $6x\tan(3x^2-1)\sec(3x^2-1)$ **4.** $3(x-1)\sec^2(\frac{3}{2}x^2-3x+1)$
5. $-3\cot 3x\operatorname{cosec} 3x$ **6.** $-5\operatorname{cosec}^2(5x-3)$
7. $-6\cos^5 x\sin x$ **8.** $-\frac{2}{3}\cot^{-\frac{1}{3}}x\operatorname{cosec}^2 x$ **9.** $2\sec^2 x\tan x$
10. $-\frac{5}{2}\operatorname{cosec}^{\frac{1}{2}}x\cot x$ **11.** $4\tan^3 x\sec^2 x$ **12.** $\frac{5}{8}\sin^{-\frac{1}{6}}x\cos x$
13. $24x\cos(3x^2)\sin^3(3x^2)$ **14.** $-18\cos^{\frac{1}{2}}(3x-2)\sin(3x-2)$
15. $24x\tan^{-\frac{1}{2}}(4x^2-1)\sec^2(4x^2-1)$

Exercise 4.11

1. $\dfrac{12x^2}{(2y+3y^2)}$ **2.** $\dfrac{6x}{12y^3}$ **3.** $\dfrac{3x^2}{\cos y}$ **4.** $\dfrac{3x}{y\cos y^2}$ **5.** $\dfrac{\frac{3}{2}x^{-\frac{1}{2}}+4}{\frac{3}{2}y^{-\frac{1}{2}}+4}$

6. $\dfrac{2x\sec^2 x}{1-\sec^2 y}$ **7.** $\dfrac{y\cos x-\sin y}{x\cos y-\sin x}$ **8.** $-\dfrac{y}{x}$

Exercise 4.12

1. $5e^{5x}$ **2.** $14xe^{x^2}$ **3.** $e^{\frac{1}{2}x}-x$ **4.** $24xe^{(4x^2-6)}+3x^2$
5. $12t^2e^{4t^3}$ **6.** $-7\sin xe^{\cos x}+12\cos xe^{4\sin x}$

7. $4\sec^2 xe^{\tan x}-6xe^{-3x^2}$ **8.** $\dfrac{8x}{e^y}$ **9.** $\dfrac{7}{6e^{3y}x^{\frac{1}{2}}}$ **10.** $-\dfrac{3x^2}{(e^y-2y)}$

Exercise 4.13

1. $\dfrac{1}{x}$ **2.** $\dfrac{16x}{(8x^2-1)}$ **3.** $\dfrac{8x}{(4x^2-7)}$ **4.** $\dfrac{6t-7}{(3t^2-7t+1)}$ **5.** $\dfrac{1}{2s}$

6. $\dfrac{1}{x}-\dfrac{1}{(x-2)}$ **7.** $\dfrac{6x}{3x^2-2}-\dfrac{6x}{3x^2-4}$ **8.** $-\dfrac{2}{x}$ **9.** $\dfrac{\cos x}{\sin x}$

10. $\dfrac{\sec^2 x}{\tan x}$ **11.** $\dfrac{0.4343}{x}$ **12.** $\dfrac{4x(y^2+2)}{y}$ **13.** $\dfrac{0.4343}{x}$

14. $\dfrac{2x(y-2)}{(1+4y-2y^2)}$

Exercise 4.14

1. $-\dfrac{(2x^3+3x^2+1)}{(x^3-1)^2}$ **2.** $e^{3x}(3x^2+2x+3)$ **3.** $\frac{1}{2}x^{-\frac{1}{2}}(5x^2-1)$

4. $2x\ln x+\dfrac{1+x^2}{x}$ **5.** $-\dfrac{1}{2(1-x)}$ **6.** $e^{2x}(2\cos x-\sin x)$

7. $e^{2x}(3x^2+2x^3)$ **8.** $\dfrac{3}{x}-\dfrac{\sin x}{\cos x}$ **9.** $e^{\frac{2}{3}x}(\frac{2}{3}\tan 4x+4\sec^2 4x)$

10. $\dfrac{5 \cos 10t \cos 5t + 10 \sin 10t \sin 5t}{\cos^2 10t}$ **11.** $27x^2 e^{3x}$

12. $\sin 2\theta + 2\theta \cos 2\theta$ **13.** $\dfrac{6 \ln 5y \cos 5y - \dfrac{1}{y} \sin 5y}{(\ln 5y)^2}$

14. $\ln x + 1$ **15.** $\dfrac{e^{2x}}{\ln 10}\left(2 \ln 2x + \dfrac{1}{x}\right)$

Exercise 4.15

5. 3

Miscellaneous Exercise 4

1. (a) $-\dfrac{3}{(4\pi)^3 x^4}$ (b) $-0.16x^{-\frac{1}{2}} + 1.2x^{-\frac{1}{2}}$ (c) $\frac{2}{9}x^{-3} + 18x$

2. (a) $-\frac{1}{2}(1-x)^{-\frac{1}{2}}$ (b) $-\frac{3}{2}x^{-4}$ (c) $\dfrac{4}{(1-3x)^2}$ **3.** $11\frac{1}{8}$, $14\frac{3}{16}$

4. (a) $6x + 4 + \dfrac{6}{x^3}$ (b) 138 (c) 5

5. (a) $6x + \dfrac{1}{x^2}$ (b) $3(t^2 - 2t)^{\frac{1}{2}}(t-1)$ (c) $1 - \dfrac{1}{p^2} - \dfrac{2}{p^3}$

6. (a) $\dfrac{x^2(x^2+3)}{(x^2+1)^2}$ (b) $\frac{1}{4}$, $\dfrac{\pi}{2}$ (c) $7 - 9x^2$, $-18x$, $4\frac{3}{4}$

7. (a) (i) $1 - \dfrac{4}{x^2}$ (ii) $-3x^{-4}$ (c) $6.72x^{0.6} - 11.61x^{3.3}$

(b) $3 - 4x^{-3}$, 3

8. (a) (i) $12x^2$ (ii) $\dfrac{x-4}{y}$ (iii) $\frac{3}{2}y^{\frac{1}{3}}$ (b) $\dfrac{14}{x^3} - \dfrac{12}{x^4}$

9. (a) (i) $\dfrac{2x5}{x^2 - 5x + 3}$ (ii) $\dfrac{15 - 3x^2}{(x^2+5)^2}$ (iii) $-12 \cos^3 3x \sin 3x$

(b) 1

10. (a) (i) $-(1-2x)^{-\frac{1}{2}}$ (ii) $4x \cos \frac{1}{2}x - x^2 \sin \frac{1}{2}x$ (b) 2.598

11. (a) (i) $-x^{-\frac{3}{2}} + 3x^{-4} + \frac{3}{2}x^{\frac{1}{2}}$ (ii) $\dfrac{3(\cos 3x - \sin 3x)}{e^{3x}}$

(b) $\dfrac{2x}{(x^2+1)}$, $\dfrac{2 - 2x^2}{(x^2+1)^2}$, $\dfrac{4}{(x^2+1)^2}$

12. (a) (i) $\dfrac{-4 \sin 2x - 2 \sin x \sin 2x - \cos 2x \cos x}{(2 + \sin x)^2}$

(ii) $\frac{3}{2}x^2(2 - x^3)^{-\frac{3}{2}}$ (iii) $3 \cot 3x$

13. (a) (i) $\dfrac{-x}{(1-x^2)^{\frac{1}{2}}}$ (ii) $-e^{-x}(\cos x + \sin x)$ (iii) $-\dfrac{1}{x}$ (c) $-\dfrac{2}{3}$

14. (a) $x^2\left[3\cos^{-1}x - \dfrac{x}{\sqrt{1-x^2}}\right]$, (b) $4x^4\left[5\sin^{-1}2x + \dfrac{2x}{\sqrt{1-4x^2}}\right]$

(c) $e^x\left[\tan^{-1}z + \dfrac{1}{1+x^2}\right]$, (d) $\dfrac{x^2(3\ln x - 1)}{(\ln x)^2}$

Chapter 5

Exercise 5.1

1. $(4 - 128t)$, -128, -252 **2.**   (a) 104 m/s (b) $+5$ s
3. 28×10^{-6} **4.**   (a) 16 m/s (b) 3 s or $\frac{4}{3}s$ (c) 3 s
5. $\frac{1}{2}e^{=t/10}(t - 10)$, $\frac{1}{20}e^{-t/10}(20 - t)$, $t = 10$ **6.**   (a) 19.2 (b) 14.4

Exercise 5.2

1. Max $\frac{19}{27}$ at $x = -\frac{1}{3}$, min -12 at $x = 2$
2. Max $2\frac{14}{27}$ at $x = -\frac{1}{3}$, min -6 at $x = 3$
3. Max at $(-\frac{1}{2}, -4)$, min at $(\frac{1}{2}, 4)$
4. Max at $(1, 4)$, min at $(3, 0)$ **5.** $x = 2, -3$ **6.**   min

Exercise 5.3

1. $125\,\text{mm}^3$ **2.** $\frac{32}{3}\,\text{m}^3$ **3.**   (a) $\frac{2}{3}\,\text{m}$ (b) $8\,\text{m}^2$ **5.** $\frac{4}{3}, \dfrac{2\sqrt{2}}{3}$
6. $16\pi\,\text{mm}^3$ **7.**   £8.02

Exercise 5.4

1. $0.21\,m^3$ **2.** $2\,880\pi\,\text{mm}^3/\text{s}$ **3.** $\dfrac{300}{1\,000 + 200t}$, -0.031
4. (a) $\dfrac{1}{12}\pi x^3$ (b) $\dfrac{80}{\pi}$ mm/s   **5.**   100 mm/s, $10^5\,\pi\,\text{mm}^2/\text{s}$
6. 16 W/s   **7.** $-0.1\,\mu\text{F}/\text{s}$

Exercise 5.5

1. $6xy + 5y^2$, $6y$, $10x$ **2.** $\dfrac{1}{y}\cos\dfrac{x}{y}$, $-\dfrac{x}{y^2}\cos\dfrac{x}{y}$
3. ye^{xy}, xe^{xy}, $xye^{xy} + e^{xy}$ **5.** $6x\{\cos(2x + 3y) - x\sin(2x + 3y)\}$
7. -1% **8.** -1%

Miscellaneous Exercise 5

1. $\dfrac{15}{\pi}$ **2.** (a) Min at $x = 4$ max at $x = -2$ (b) $\dfrac{8\pi}{27}\,\mathrm{m}^3$

3. (a) Max at $(-3, 102)$, min at $(2, -23)$ (b) $\dfrac{3}{8\pi}\,\mathrm{mm/s}$

4. 122 mm   **5.** (a) $0.000.6\ \mathrm{m}^3/\mathrm{s}$ (b) 0.25 mm   **6.**   8 km
7. Max at $(-7, 398)$, min at $(+3, -102)$ **8.**   6 m
9. $t = 1$ sec, 2 m/s   **10.**   (a) 2 sec, 1 sec (b) 6, -6 (c) $+1.5$
11.   Min at $x = +\sqrt{\tfrac{2}{3}}$, max at $x = -\sqrt{\tfrac{2}{3}}$

12. $\dfrac{1}{y}\cos\left(\dfrac{x}{y}\right),\ -\dfrac{x}{y^2}\cos\left(\dfrac{x}{y}\right),\ -\dfrac{1}{y^2}\sin\left(\dfrac{x}{y}\right),$

$-\dfrac{x^2}{y^4}\sin\left(\dfrac{x}{y}\right) + \dfrac{2x}{y^3}\cos\left(\dfrac{x}{y}\right),\ \dfrac{x}{y^3}\sin\left(\dfrac{x}{y}\right) - \dfrac{1}{y^2}\cos\left(\dfrac{x}{y}\right)$

13. $-\dfrac{1}{\sqrt{2}}$ **15.**   0.2%

Chapter 6

Exercise 6.1

1. $13\frac{1}{2}$ **2.** 12   **3.** -27 **4.** 0   **5.** -614.4 **6.** $\frac{1}{2}$
7. 16   **8.** $-58\frac{1}{3}$ **9.** $\frac{1}{3}b^{30} - \frac{1}{3}a^{30}$ **10.** $30a^4$

Exercise 6.2

1. $\dfrac{x^4}{4} + C$ **2.** $\dfrac{x^{11}}{11} + C$ **3.** $\dfrac{7x^2}{2} + C$ **4.** $-\frac{3}{7}x^7 + C$
5. $-\frac{1}{18}x^6 + C$ **6.** $\frac{1}{20}x^5 + C$ **7.** $\dfrac{x^6}{36} + C$ **8.** $\dfrac{x^4}{40} + C$
9. $-1\frac{3}{7}$

Exercise 6.3

1. $-\dfrac{x^{-5}}{5} + C$ **2.** $\dfrac{765}{2\,048}$ **3.** $\frac{4}{7}t^{\frac{7}{4}} + C$ **4.** $\dfrac{x^{\frac{3}{4}}}{3} + C$

5. $18x^{\frac{1}{3}} + C$ **6.** $t^{\frac{2}{3}} + C$ **7.** $\frac{7}{3}$ **8.** $\frac{14}{3}$ **9.** $\dfrac{x^4}{4} - 2x^3 + 6x^2 - 8x + C$

10. $-49\frac{1}{6}$ **11.** $\dfrac{x^3}{3} - 9x + C$ **12.** $\dfrac{x^5}{5} - \dfrac{2x^3}{3} + x + C$

13. $\frac{2}{3}x^{\frac{3}{2}} + C$ **14.** $\frac{2}{3}x^{\frac{3}{2}} + 2x^{\frac{1}{2}} + C$ **15.** $-x^{-1} - 2x^{-2} + C$
16. $-2x^{-1} + \frac{3}{2}x^{-2} - 2x^{-3} + C$ **17.** $\frac{1}{6}x^{-2} - \frac{1}{12}x^{-4} + C$
18. $\frac{2}{7}x^{\frac{7}{2}} - \frac{6}{5}x^{\frac{5}{2}} - \frac{4}{3}x^{\frac{3}{2}} + C$ **19.** $\frac{4}{3}x^{\frac{3}{4}} - \frac{4}{5}x^{\frac{5}{4}} + C$ **20.** $\frac{3}{2}x^{\frac{2}{3}} + \frac{3}{4}x + C$

Exercise 6.4

1. $-20\frac{5}{6}$ 2. $156\frac{1}{4}$ 3. 21 4. $10\frac{2}{3}$ 5. $21\frac{1}{3}$ 6. $\frac{1}{6}$
7. -18 8. $\frac{1}{6}$ 9. $4\frac{1}{6}$

Exercise 6.5

1. $\dfrac{256\pi}{3}$ 2. $\dfrac{92}{15}\pi$ 3. 30π 4. $563.7,\ 8\pi$

Exercise 6.6

1. $\dfrac{(2x-5)^4}{4}+C$ 2. $\frac{3}{16}(2x^2-3)^{\frac{4}{3}}$ 3. $\frac{1}{4}(2x^2-2x+1)^4+C$

4. $\frac{2}{27}(3x^3-2)^{\frac{3}{2}}$ 5. $-\frac{1}{8}(6t^2-4t+2)^{-2}+C$
6. $\frac{1}{56}(4x^2-1)^7+C$ 7. $\frac{2}{3}(x^2+2x+2)^{\frac{3}{4}}+C$ 8. $\frac{1}{9}$
9. $\frac{3}{4}(2x-4)^{\frac{2}{3}}+C$ 10. $2(\frac{1}{3}x^3+1)^{\frac{1}{3}}+C$

Exercise 6.7

1. $\ln(x^2-3)+C$ 2. $\frac{1}{2}\ln(3x^2+4x+7)+C$ 3. $\frac{1}{2}\ln\frac{3}{2}$
4. $\ln(x^2+2x+6)+C$ 5. $\frac{2}{3}\ln(x^{\frac{3}{2}}+4)+C$ 6. $\frac{7}{8}\ln 33$
7. $-2\ln(\frac{1}{s}+2)+C$ 8. $\frac{1}{4}\ln(y^4-1)+C$

Exercise 6.8

1. $-\frac{1}{4}\cos 4x+C$ 2. $\frac{1}{3}$ 3. $\frac{2}{7}\sin(\frac{7}{2}x-3)+C$
4. $\frac{1}{2}\ln[\operatorname{cosec}(2x+4)-\cot(2x+4)]+C$ 5. 0.4230
6. $\frac{4}{3}\ln\cos(-\frac{3}{4}t+\frac{1}{2})+C$ 7. -0.0312
8. $-2\ln[\operatorname{cosec}(-\frac{1}{2}x)-\cot(-\frac{1}{2}x)]+C$

Exercise 6.9

1. $-\frac{1}{18}\sin 9x+\frac{1}{2}\sin x+C$ 2. $\frac{1}{16}\sin 8x+\frac{1}{4}\sin 2x+C$
3. $-\frac{1}{20}\cos 10x-\frac{1}{8}\cos 4x+C$ 4. $-0.103\,6$ 5. 1
6. $+\frac{1}{4}\sin 2x-\frac{1}{20}\sin 10x+C$ 7. 0
8. $-\frac{1}{4}\cos 2x-\frac{1}{16}\cos 8x+C$

Exercise 6.10

1. $-\frac{1}{3}\sin^{-3}x+C$ 2. $-\frac{4}{7}\cos^{\frac{7}{4}}x+C$ 3. $\frac{3}{2}\tan^{\frac{2}{3}}x+C$
4. $-\frac{1}{6}\cot^6 x+C$ 5. $\frac{1}{16}\sin^8 2x+C$ 6. $-\frac{6}{5}\tan^{\frac{5}{4}}(-\frac{2}{3}x)$
7. $\frac{4}{15}\cot^{\frac{3}{4}}5x+C$ 8. $-\frac{1}{5}\cos^{\frac{5}{2}}(2x-1)$

Exercise 6.11

1. $\frac{1}{3}e^{3x} + C$ 2. $-\frac{1}{2}e^{-2x-1} + C$ 3. $-2e^{-\frac{1}{2}x}$ 4. $-\frac{1}{5}e^{-5\theta + \frac{1}{3}}$
5. $0.632\,1$ 6. 0.282 7. $0.129\,7$ 8. $\frac{1}{3}(e^{3x} - e^{-3x}) + C$
9. $-\frac{4}{3}e^{-\frac{3}{4}v} - \frac{1}{2}e^{2v} + C$

Exercise 6.12

1. $6\ln(x-3) - 2\ln(x-1) + C$ 2. $\frac{5}{8}\ln(x-4) + \frac{11}{8}\ln(x+4)$

3. $\frac{4}{3}\ln\left(\dfrac{x-4}{x-1}\right) + C$ 4. $0.174\,3$ 5. $0.182\,3$

6. $\frac{1}{2}\ln\left(\dfrac{x-5}{x+1}\right)$ 7. $\frac{1}{2}\ln(x+1) - 3\ln(x+2) + \frac{7}{2}\ln(x+3) + C$

8. $\frac{1}{4}\ln(2u-1) - \dfrac{1}{4(2u-1)} + C$ 9. $\frac{1}{3}\ln(3t-4) - \dfrac{1}{3(3t-4)} + C$

10. 1.763

11. $\frac{1}{6}\ln(x-1) + \frac{1}{2}\ln(x+1) - \frac{2}{3}\ln(x+2) + C$

Exercise 6.13

1. $e^x(x-1) + C$ 2. $\frac{1}{25}x^5(5\ln x - 1) + C$
3. $-\frac{1}{4}x^2\cos 4x + \frac{1}{8}x\sin 4x + \frac{1}{32}\cos 4x + C$ 4. $0.594\,1$

5. $-\dfrac{1}{x}(\ln x + 1)$ 6. $x\tan x - \frac{1}{2}x^2 + \ln(\cos x) + C$

7. $-\frac{4}{25}(e^{3\pi/2} + 1)$ 8. $\dfrac{e^{5x}}{29}(5\sin 2x - 2\cos 2x) + C$

9. 11.18 10. $\frac{1}{32}e^{4x}(8x^2 - 52x + 85)$

Miscellaneous Exercise 6

1. $x - \frac{4}{3}x^{\frac{3}{2}} + \frac{1}{2}x^2 + C$ 2. $\frac{16}{45}$ 3. $1\frac{3}{4}$ 4. $\frac{1}{3}x^3 + 2x - \dfrac{1}{x} + C$

5. $24\frac{2}{3}$ 6. $\frac{2}{7}x^{\frac{7}{2}} - \frac{5}{6}x^{\frac{6}{5}} + C$ 7. $31\frac{13}{21}$

8. (a) (i) $\dfrac{r^7}{56}(7r - 8)$ (ii) $\dfrac{y^2}{4}(y^2 - 2)$ (b) $-1\frac{1}{3}$

9. (a) $-\dfrac{1}{6x} + C$ (b) $\frac{3}{2}x^2 + 2x^{\frac{3}{2}} + C$ (c) $\dfrac{x^3}{3} + ax^2 + a^2x + C$ (d) $108\frac{6}{7}$

   (e) $\frac{8}{3}p^2q$

10. $11\frac{1}{3}$ 11. (a) 1 (b) $\frac{9}{10}\pi$ 12. $6\frac{1}{3}$, $\frac{18}{3}\pi$ 13. 5.27
14. $\pi h(\frac{1}{3}a^2h^2 + aRh + R^2)$, $\frac{26}{3}\pi$

15. (a) $1 - \dfrac{\pi}{4}$ (b) $\frac{3}{4}$ (c) 1.11

17. (a) $\frac{1}{2}x^2 + 2\ln x + C$ (b) $\frac{1}{4}$ (c) 0.432
18. (a) $\sqrt{2x} + \frac{2}{5}x^{\frac{5}{2}} + C$ (b) $\frac{3}{2}x\sin 2x + \frac{3}{4}\cos 2x + C$ (c) 0.3465 (d) 0.632 1
19. (a) $\frac{2}{3}x^{\frac{3}{2}} + 4x^{\frac{1}{2}} + C$ (b) $\frac{1}{3}x^3\, n_e x - \frac{1}{9}x^3 + C$ (c) $\frac{3}{128}$ (d) $\frac{2}{9}(x^3+1)^{\frac{3}{2}} + C$

20. (a) $\ln(6x^2 - x - 1) + C$ (b) $-\dfrac{\sqrt{2}}{3}$

21. (a) 6.67 (b) 0.549 (c) 3.33
22. (a) $\frac{1}{3}(e^{3x} - e^{-3x}) + C$ (b) 7.443 (c) $\frac{1}{4}$
25. (a) $\frac{1}{4}(2x - 1)e^{2x} + C$ (b) $\frac{1}{5}e^{2x}(2\cos x + \sin x) + C$

26. (a) 0.470 (b) $-\dfrac{v^{-0.2}}{0.2} + C$ (c) -0.00318 (d) $\tan x - x + C$ (e) 2.89

Exercise 7.1

1. $y = \frac{1}{5}\sin(5x - 3) + C$ **2.** $y = \frac{5}{3}e^{3x}$ **3.** $y = \frac{1}{3}x^4 + x + 3$
4. $y = \frac{-1}{4}\sin 2x + Cx + D$ **5.** $y = x\, n_e x + 0.9x + 3.1$

6. $y = e^{(\frac{3x^4}{4} - 10)}$ **7.** $\dfrac{e^{4y}}{4} = \dfrac{4e^{3x}}{3} + C$ **8.** $y = Ce^{(-\cos x)}$

9. $y^3 = \frac{3}{2}x^2 + 9x + C$ **10.** $y = C(x - 2) + \frac{5}{2}$

11. $i = \dfrac{E}{R}(1 - e^{-Rt/L})$ **12.** $n_e y = -\frac{1}{2}\sin 2x + x + 1$

13. $y = \frac{3}{2}e^{x^2} - \frac{1}{2}$

Exercise 7.2

1. $y = 1 + 2e^{-\frac{1}{2}x^2}$
2. $y = x\ln x + 10x$
3. $y(x + 2) = \frac{1}{2}x^2 + 10$
4. $y\cos^2 x - \frac{1}{2}x - \frac{1}{4}\sin 2x$
5. $y = \frac{1}{2}x(\ln x)^2 + 2x$

Exercise 7.3

1. $y = -\frac{1}{4}e^{-5x} + \frac{1}{4}e^{-x}$
2. $y = Ce^{-x}\sin(2x + \alpha)$
3. $y = 3e^{-x} + 3e^{-2x}$
4. $y = Ae^{3x} + Be^{-2x}$
5. $y = \frac{5}{2} - \frac{1}{2}e^{-2x}$
6. $y = 24e^{-2.5t}(\sin \frac{3}{2}t + 0.12)$

Miscellaneous Exercise 7

1. (a) $y = 3.5 + \ln 2$ (b) 4
2. $y = \frac{1}{160}(\frac{1}{6}x^3 - \frac{1}{2}x^2 + \frac{2}{3})$, $y = \frac{1}{240}$
3. (a) $y = \frac{1}{3}x^3 + \frac{5}{2}x^2 + \frac{2}{3}x + 3$ (b) $y = \frac{1}{2}\ln(x^2 + 1)$
4. (a) $y = \frac{16}{3}x^{\frac{3}{2}} + x + 7$ (b) $\theta = 30e^{-0.05t} + 10$, 13.9 min
5. $EIy = \frac{5}{6}x^3 - \dfrac{x^5}{900} - \dfrac{525}{4}x$
6. $y = y_c e^{kt}$, 2^{12}
7. $i = 0.02e^{-50t}$, 13.8 ms
8. $i = \dfrac{E}{R}(1 - e^{-Rt/L})$
9. $\theta - \theta_0 = Ce^{-0.2t}$, 5.3 min
10. $x = Ce^{-3t}\sin(t + \propto)$, $\dfrac{1}{2\pi}$ Hz

Chapter 8

Exercise 8.1

1. $\frac{4}{3}$ 2. 0 3. 0.159 4. 102.2 5. 0.316 6. 1.313

7. $1.6\,V_0$ 8. 5.33 9. 2.18 10. $0.432N_0$

Exercise 8.2

1. (a) 5.47 (b) 5.77 (c) 4.375 (d) 4.71 (e) 5.2 (f) 4.79

2. 150 mm from apex 3. 1 308 mm 4. 3.59 5. 1.69 6. $\dfrac{\pi}{4}$

Exercise 8.3

1. 2.31, 11.16 2. 0.402, 2.66 3. $-1, \frac{2}{5}$ 4. $1, \dfrac{\pi}{8}$

5. 4, 35.7 6. $-0.9, 0$ 7. $\frac{1}{2}, \frac{4}{5}$ 8. 0.6, 0.34

Exercise 8.4

1. 1.125×10^8 mm^3 2. 1.961×10^4 mm^2, 1.70×10^5 mm^3

3. 1.15×10^8 mm^3 4. $\dfrac{4r}{3\pi}$

5. (a) 1.62×10^4 mm^2 (b) 1.67×10^5 mm^3

6. (a) 65.78 (b) 3.59×10^4 mm^2 (c) 4.10×10^6 mm^3
7. 11.52×10^5 m^3 **8.** (a) 8.75 m (b) 7 065 m^3

Exercise 8.5

1. (a) $1\,325\frac{1}{3}$ mm^4, 308 mm^4 (b) 3 278 mm^4, 1 081 mm^4
(c) 2 166.7 mm^4, 635.4 mm^4 (d) 2 705 mm^4, 756.1 mm^4
(e) 2 425 mm^4, 725 mm^4 (f) 2 244 mm^4, 585.6
2. $\dfrac{bh^3}{12}, \dfrac{bh^3}{36}$
3. (a) $5\,973\frac{1}{3}$ mm^4, $973\frac{1}{3}$ mm^4 (b) 3.58×10^5 mm^4, 2.61×10^5 mm^4

Exercise 8.6

1. 6662, 193.6 **2.** 0.52, 18.14 **3.** 0.31, 1.6 **5.** 3.26×10^5, 2 604

Miscellaneous Exercise 8

1. (b) $3\,534 \times 10^{-6}$ m^4
2. (a) 4.54×10^8 mm^4 (b) 1.64×10^8 mm^4 (c) 26.1×10^6 mm^4
3. 228.5 mm^4 **4.** 93 mm **5.** 0.673 m **6.** 38.1
7. 1.12×10^7 mm^4, 3.36×10^7 mm^4 **8.** 5.03 kg
9. 5.88, 2 026 **10.** $\dfrac{7\pi}{15}, \dfrac{\pi}{6}$ **11.** (i) $\frac{16}{3}$ (ii) 53.6 (iii) 1.60
12. (a) $\dfrac{9\pi}{8}$ (b) $\dfrac{3}{\sqrt{2}}$ **13.** $1, \dfrac{16\pi}{15}$ **14.** $\dfrac{2a\omega}{\pi}$ **15.** 0.363

Chapter 9

Exercise 9.1

1. (a) $(26, -51°)$ (b) $(5.4, 112°)$ (c) $(6.7, 243°)$
2. (a) $(10.4, 6)$ (b) $(-12.9, -15.3)$ (c) $(-14.1, 5.13)$
3. $(40.0, 60.0°)$, $(100, 60.0°)$, 60 mm
4. A $(-26.0, 122)$, B $(108, 62.5)$, C $(92.9, -83.6)$, D $(-50.8, -114)$,
E $(-124, 13.0)$ AB= 147 mm BC= 147 mm
5. 59.8°, 62.0 mm : 34.4 mm, 264.3 mm : 95.8 mm

Exercise 9.2

1. $y = 5x - 4$ **2.** $y = -6x - 17$ **3.** $y = -2x - 1$ **4.** $y = -4x + 22$
5. $y = qx + 2q - q^2$ **6.** (a) 2, −1 (b) −1, 4 (c) 1, 1 (d) 3, 0 (e) −1, 0
(f) 0, 2 **7.** $F = 0.2W + 0.4$ **8.** $V = -3I + 17$
9. $R = 0.5\theta + 50$

Exercise 9.3

1. (a) $y = -0.25x + 2.25$ (b) $y = x - 8$ (c) $y = 0.2x + 4.6$ (d) $y = 2x + 1$
   (e) $y = x + 7$
2. (a) $(4.5, 5.5)$, -1 (b) $(0.5, -2)$, $\frac{13}{8}$ (c) $(0.5, 1.5)$, $\frac{13}{7}$ (d) $(11, 6)$, ∞
   (e) $(-4.5, -6.5)$, -0.6 (f) $(-2a, -0.5a)$, $\frac{8}{7}$
3. $(3, 0)$, $(0.5, 4)$, $(3.5, 4)$, $x = 3$, $y = -0.125x + 4.0625$, $(3, 3.6875)$
4. (a) $(-5, 1)$ (b) $y = \frac{1}{6}x + \frac{11}{6}$, $y = -0.5x - 1.5$
5. $y = -0.5x + 2.5$, $y = 6x - 8$, $\left(\frac{22}{13}, \frac{21}{13}\right)$
6. (a) $y = -1.5x + 2$ (b) $y = \frac{2}{3}x + 2$ (c) $(0, 2)$

Exercise 9.4

1. $(3, 4)$, 9 2. $(2, -6)$, 4 3. $(-3.5, 4.5)$, 5.7
4. $(-5, 0)$, 4.25 5. $(0, 4)$, 6 6. $x^2 + y^2 - 8x - 12y - 12 = 0$
7. $x^2 + y^2 + 6x + 14y + 19 = 0$ 8. $x^2 + y^2 + 8x - 10y - 59 = 0$
9. $x^2 + y^2 - 2ax - 4ay - 20a^2 = 0$

Exercise 9.5

1. (a) 5, 3 (b) 3.74, 3.16 (c) 2.65, 2 (d) $2p^2$, $3m$ (e) $\frac{1}{5}$, $\frac{1}{6}$
2. $y = -4.94x^2$ 3. $y^2 = 2000x$ 4. $4x^2 + 5y^2 = 84$ 5. $\dfrac{x^2}{16} + \dfrac{y^2}{9} = 1$

Exercise 9.6

1. (a) $y = -0.75x + 6.25$, $y = -\frac{4}{3}x$ (b) $y = -2.4x + 33.8$, $y = \frac{5}{12}x$
   (c) $y = -\frac{7}{6}x + \frac{85}{6}$, $y = \frac{6}{7}x$ (d) $y = -\frac{6}{17}x - 5$, $y = \frac{17}{6}x - 5$
   (e) $y = -0.29x - 23.8$, $y = 3.4x - 12.8$
2. (a) $y = -x + 5$, $y = x + 1$ (b) $y = -\frac{2}{15}x + \frac{47}{5}$, $y = \frac{15}{2}x - \frac{9}{2}$
   (c) $y = -0.7x + 6.4$, $y = 1.43x + 2.14$ (d) $y = 2x - 4$, $y = -0.5x + 3.5$
   (e) $y = 3x - 2.5$, $y = -\frac{1}{3}x + \frac{5}{2}$
3. $(2.571, -5.143)$ 4. $(36, -24)$

Exercise 9.7

1. -3.6, 2.1 min $y = -16.1$ at $x = -0.75$
2. (a) $6.25 \, \text{MN/m}^2$ at $x = 4.5 \, \text{cm}$ (b) $2.0 \, \text{cm}$, $7.0 \, \text{cm}$

Exercise 9.8

1. $2.15 \, \text{V/s}$

Exercise 9.10

1. $\sigma = 3.1x^2 - 7.3x$ 2. $R = \dfrac{42}{I} + 45.4$

3. $I = 0.94 L^2 + 2.12$, 10.58 kN m
4. $A = 50 r^{-1.5}$, 3.80 cm^2 5. $\theta = 500(2.0^{-t})$
6. $i = 3\mathrm{e}^{-0.14t}$

Exercise 9.11

1. 0, 0, 9   2.   3, -2 3.   (a) -3.45, 1.45 (b) -4.19, 1.19
4. -3.1, -0.6, 2.68   5.   2.61, 5.88   6. -3.9, 0.13, 3.8
7. -1.42, 0.54   8.   0
9. $(-0.3, 3.9)$, $(2.1, -3.3)$
10. $(0.67, 2.99)$, $(4.95, 0.40)$, $y = -0.60x + 3.40$

Miscellaneous Exercise 9

1. 1.39   2.   3.41 or 0.58
4. $(5, 36.9°)$, $(5, 126.9°)$, $(7.07, 261.9°)$
5. $y = -2x + 2$ 6.   (a) $7y = 4x + 5$, $(5, 36.9°)$ (b) -1, -3
7. $y = 4x + 8$ 8. -2.56, 1, 1.56
9. (a) 1, -1, -2 (b) 0, -2.41, 0.41 (c) -2.62, -0.38, 1.0
10. $m = 4.2 \times 10^{-3}$, $c = 2.7 \times 10^{-4}$ 11. $a = \frac{7}{8}$, $b = 8$, $t = \frac{20}{7}$
12. $k = 0.01$, $M = 200$ mg, 300 min   13. $a = -10$, $b = 320$
14. $n = 2.34$, $c = 178$, 0.175 mm/rev
15. $n = 0.63$, $K = 5.78$, $d = 21.1$ mm
16. $y = \frac{1}{3}x + 2$, $y = -\frac{1}{3}x + 3$, $(1.5, 2.5)$, $(1, 0)$, $(2, 0)$
18.   (a) $y = -5x + 4$, $y = \frac{1}{5}x - \frac{6}{5}$ (c) 52
19.   (a) $y = x^2 - 2x + 2.75$ (b) $y = 4x - 6.25$
20. $y = 12x$, $y = -12x$, $(3, 36)$, $(-3, 36)$
21.   5, 10, 4, 3   23.   (a) $x^2 + y^2 - 10x - 14y + 2 = 0$ (b) $(5, 7)$

Chapter 10

Exercise 10.1

1. 5, 18   2. -22, -60 3.   31   4.   840, 1080, 1320, 1560

Exercise 10.2

1.   (a) -128, -341 (b) 6561, 1093 (c) $\dfrac{1024}{19\,683}$, $\dfrac{422}{81}$ (d) x^{11}, $\dfrac{x(x^{18} - 1)}{x - 1}$

2.   2, 1536, 3069   3.   10   4.   100, 135, 181, 244, 328, 442, 596, 800

5. 11, 206   **6.**   4374, 1458, 486, 162, 54, 18
7. (a) £218.30 (b) 15 years

Exercise 10.3

1.   (a) $1\frac{1}{3}$ (b) $\frac{2}{9}$ (c) $\frac{5}{6}$ (d) $6\frac{3}{4}$ (e) $\frac{2}{3}$ **2.**   10   **3.**   £12 500
4.   300 mm   **5.**   (a) $\frac{1}{3}$ (b) $\frac{35}{99}$

Exercise 10.4

1.   0.479 4, 0.980 1
2. $1 - 2x^2 + \frac{2}{3}x^4 - \frac{4}{45}x^6 + \cdots$, $\frac{1}{2}x - \frac{1}{48}x^3 + \frac{1}{3840}x^5 - \frac{1}{645120}x^7 + \cdots$,
       $ax + \frac{1}{3}a^3x^3 + \frac{2}{15}a^5x^5 + \frac{17}{35}a^7x^7 + \cdots$
3.   (a) $3x - 10.5x^3 + 11.03x^5 - \cdots$
        (b) $x^2 + x^3 - \frac{1}{6}x^6 - \frac{1}{6}x^7 + \frac{1}{120}x^{10} + \cdots$

Exercise 10.5

1.   0.0953, −0.1054
2.   0.405 465   **3.**   1.791 8, 1.945 9
4.   (a) $-3x - \frac{9}{2}x^2 - 9x^3 - \frac{81}{4}x^4 - \cdots |x| < \frac{1}{3}$
        (b) In $3 - x - \frac{1}{2}x^2 - \frac{1}{3}x^3 - \cdots |x| < 1$
        (c) $-2x - x^2 - \frac{2}{3}x^3 - \frac{1}{2}x^4 \cdots |x| < 1$
        (d) $-4x - 4x^2 - \frac{28}{3}x^3 - 20x^4 \cdots |x| < \frac{1}{3}$
        (e) $x - \frac{1}{2}x^2 - \frac{1}{6}x^3 + \frac{3}{40}x^4 \cdots |x| < 1$
        (f) In $4 + \frac{5}{4}x - \frac{17}{32}x^2 + \frac{65}{192}x^3 \cdots |x| < 1$
        (g) $\frac{5}{2}x - \frac{25}{4}x^2 + \frac{125}{6}x^3 - \frac{625}{8}x^4 \cdots |x| < \frac{1}{5}$

Exercise 10.6

1.   (a) $1 + 7x + 21x^2 + 35x^3 + 35x^4 + 21x^5 + 7x^6 + x^7$
        (b) $1 - 4x + 6x^2 - 4x^3 + x^4$
        (c) $1 - 6x^2 + 15x^4 - 20x^6 + 15x^8 - 6x^{10} + x^{12}$
        (d) $1 - 2x + \frac{3}{2}x^2 - \frac{1}{2}x^3 + \frac{1}{16}x^4$
2.   (a) $1 - 3x + 6x^2 - 10x^3 + \cdots |x| < 1$
        (b) $1 + 10x + 60x^2 + 280x^3 + \cdots |x| < \frac{1}{2}$
        (c) $1 + 3x + 9x^2 + 27x^3 + \cdots |x| < \frac{1}{3}$
        (d) $1 + \frac{2}{5}x - \frac{3}{25}x^2 + \frac{8}{125}x^3 - \cdots |x| < 1$
        (e) $1 - 6x + 27x^2 - 108x^3 + \cdots |x| < \frac{1}{3}$
        (f) $1 + \frac{1}{3}x^2 - \frac{1}{9}x^4 + \frac{5}{81}x^6 - \cdots |x| < 1$

(g) $1 - x + \frac{3}{2}x^2 - 9\frac{5}{2}x^3 \ldots |x| < \frac{1}{2}$

(h) $x^4 + 4x^5 + 10x^6 + 20x^7 \cdots |x| < 1$

3. (a) $32 - 240x + 720x^2 - 1080x^3 + 810x^4 + 243x^5$
    (b) $81a^4 + 216a^3b + 216a^2b^2 + 96ab^3 + 16b^4$
4. (a) $2 - \frac{1}{4}x - \frac{1}{64}x^2 - \frac{1}{512}x^3 - \cdots |x| < 4$
    (b) $\frac{1}{2} - \frac{1}{16}x + \frac{3}{256}x^2 - \frac{5}{2048}x^3 + \cdots |x| < 4$
    (c) $\frac{1}{2} - \frac{5}{4}x^2 + \frac{25}{8}x^4 - \frac{125}{16}x^6 + \cdots |x| < \frac{2}{5}$
    (d) $x^{\frac{1}{3}} - \frac{1}{3}x^{\frac{7}{3}} + \frac{2}{9}x^{\frac{13}{3}} - \frac{14}{81}x^{\frac{19}{3}} + \cdots |x| > 1$
5. (a) $10\,376/a^3$ (b) $12\,870x^{16}y^{16}$ **6.** $1 + x$
7. (a) $1 + x - \frac{1}{2}x^2 + \cdots, \; |x| < \frac{1}{2}$ (b) $1 + \frac{1}{2}x + \frac{3}{8}x^2 + \cdots, \; |x| < 1$
8. (a) $1 + 4x + 12x^2 + \cdots, \; |x| < \frac{1}{3}$ (b) $1 + 10x + 55x^2 + \cdots, \; |x| < \frac{1}{3}$
    (c) $1 + x + \frac{1}{2}x^2 + \cdots, \; |x| < 1$ (d) $1 - x - 5x^2 + \cdots, \; |x| < \frac{1}{3}$

Exercise 10.7

1. 1.08, 0.968, 1.002 5, 0.997, 1.002. 1.015
2. (a) 0.995 53 (b) 0.995 2   **3.**   2%, 3%
4. (a) 5.916 08 (b) 0.1414 (c) 3.009 2   **5.**   1.5% low
6. 4.5% increase   **7.**   63 106   **8.**   0.5% low

Exercise 10.8

1. 1.094 2, 0.895 83
2. (a) $1 + 4x + \frac{9}{2}x^2 + \frac{8}{3}x^3 + \cdots$ (b) $1 - 3x + \frac{5}{2}x^2 - \frac{7}{6}x^3 + \cdots$
3. $x + x^2 + \frac{1}{3}x^3 + \cdots$
4. (a) $x + \frac{1}{6}x^3 + \cdots$ (b) $1 + \frac{1}{2}x^2 + \frac{1}{24}x^4 + \cdots$
    (c) $1 - x^2 + \frac{1}{2}x^4 - \frac{1}{6}x^6 + \cdots$ (d) $-x - \frac{5}{2}x^2 - \frac{10}{3}x^3 - \frac{13}{4}x^4 + \cdots$
    (e) $x - \frac{3}{2}x^2 + \frac{4}{3}x^3 - x^4 + \cdots$

Miscellaneous Exercise 10

1. 120, 150, 180   **2.**   £18 500   **3.**   £3 130, £6 000
4. (a) 3, 1 458, 728 (b) 0.332 8   **5.**   1.051 6
6. (a) $\dfrac{1}{9} - \dfrac{4}{27}x + \dfrac{4x^2}{27} - \dfrac{32}{243}x^3 + \cdots, \; -\frac{3}{2} < x < \frac{3}{2}$
7. $1 + x + x^2 + \frac{2}{3}x^3 + \frac{2}{3}x^4 + \cdots$
8. $256 + 512x + 448x^2 + 224x^3 + 70x^4 + 14x^5, \; 311.912$
9. (a) 2.015 94
10. (a) $x^5 + 5x^6 + 10x^7 + 10x^8 + 5x^9 + x^{10}$ (b) 6% (d) 1.02

12. 0.75% high

13. $\dfrac{c_1}{c_2} = \dfrac{\gamma}{\gamma - 1}$

Chapter 11

Exercise 11.1

1. 1.893 3 **2.** −0.567 **3.** 4.536
4. (a) 0.523 8 (b) 1.936, −2.062 (c) 0.251
5. (a) 38.24 (b) 1.526 4 (c) 0.687 33 (d) 5.393 2 (e) 6.149 57
6. (a) 0.287 (b) 0.007 194 24 (c) 468.6

Exercise 11.2

1. $x = -1.4$, $y = 2.5$, $z = 1.6$
2. $x = 0.688$, $y = 0.362$, $z = 0.204$
3. $x = 1.38$, $y = 2.45$, $z = 3.13$
4. $x = 1.360$, $y = 1.764$, $z = 1.158$
5. $x = -1.037\,0$, $y = 0.875\,8$, $z = 1.521\,8$

Exercise 11.3

1. (a) 46.3 (b) −0.625
3. (a) $2x^2 + 3x - 7$ (b) $x^3 - 2x^2 + x - 3$
     (c) $-\theta^3 + \theta$: $f(8) = 145$, $f(-4) = -103$, $f(-0.1) = -0.099$:
     $f(1.5) = 2$, $f(-0.6) = -4.54$, $f(0.48) = 0.37$

Exercise 11.4

1. (a) 0.3407 3 (b) 0.493 11 (c) 1.905 (d) 0.881 (e) 5.75
2. (a) 0.113 (b) 2.34 **3.** 15.290 096 65, 0.004% low **4.** 132 J

Exercise 11.5

1. (a) 2.5 (b) 0.368 (c) 1.17 **2.** 1.095

Miscellaneous Exercise 11

1. 0.57 **2.** 4.472 **3.** 0.682 **4.** $p = 1.38$, $q = 0.01$
5. 0.178 57 **6.** $x = -1.8$, $y = 3.2$, $z = -4.9$
7. $a = \frac{3}{4}$, $b = -\frac{7}{4}$, $c = -\frac{9}{2}$ **8.** 9.518
9. $x^2 + x + 2$, 16.96, 0.24 high **10.** 12.50, 3.12 **11.** 4.047
12. 3.15 **13.** 0.347 **14.** (a) 5.436 (b) −0.23

Chapter 12

Exercise 12.2

1. 16.1 kN   **2.** 6.5   **3.** 37.6 years   **4.** 1.282 m^2 **5.** 0.878 m^2
6. medians 1.282 m^2, 0.878 m: modes 1.28 m^2, 0.88 m

Exercise 12.3

1. $\bar{x} = 8$, $s = 4$ **2.** $\bar{x} = 5.640 \text{ m}$, $s = 0.0144 \text{ m}$
3. $\bar{x} = 78.49 \text{ mm}^2$, $s = 0.233 \text{ mm}^2$ **4.** $\bar{x} = 0.339 \text{ m}$, $s = 0.015 \text{ m}$
5. $\bar{x} = 1.718 \text{ m}$, $s = 0.046 \text{ m}$ **6.** $\bar{x} = 30.85 \text{ kN}$, $s = 0.238 \text{ kN}$
7. (a) medians: 5.64 m, 78.48 mm^2, 0.339 m, 1.720 m, 30.84 kN
   (b) semi-interquartile ranges: 0.010 m. 0.14 mm^2, 0.01 m, 0.034 m, 0.15 m
8. $\bar{x} = 58.92$, mode 59, median 59.0, $s = 1.567$
9. 78.5p, 6.4p   **10.** 62.98 kg, 62.9 kg

Exercise 12.4

1. (a) $f = 21.3i - 11.4$ (b) $i = 0.0450f + 0.693$
2. (a) $\theta_1 = 0.837\theta_2 - 10.7$ (b) $\theta_2 = 1.00\,\theta_1 + 14.4$, 26.4 °C, 7.24 °C
3. (a) 0.915 (b) 0.76   **4.** 0.83

Miscellaneous Exercise 12

1. $\bar{x} = 27.52 \text{ mm}$, $s = 0.159 \text{ mm}$ **2.** $\bar{x} = 38.6$, $s = 5.53$
3. 26 h: 18.35 h   **4.** 30.8 km/h   **6.** 69%
7. $\bar{x} = 11.52 \text{ min}$, 11.0 min, 1.97 min   **8.** $x = 35.84$, $s = 2.62$
9. $\bar{x} = 11$, $s = 5.94$ **10.** $\bar{x} = 15.2$, $s = 1.13$ **11.** 9.732, 9.787
12. $\bar{x} = 6.66 \text{ N/mm}^2$, $s = 1.51 \text{ N/mm}^2$, 94.8%
13. -0.96 **14.** $t = -0.087\theta + 32.3$, $\theta = -7.34t + 255$, 80, 0.37

Chapter 13

Exercise 13.1

1. $\frac{1}{30}$ **2.** $\frac{1}{2}$ **3.** (a) $\frac{4}{9}$ (b) $\frac{4}{9}$
4. (i) (a) $\frac{4}{25}$ (b) $\frac{9}{25}$ (c) $\frac{12}{25}$ (ii) (a) $\frac{2}{15}$ (b) $\frac{1}{3}$ (c) $\frac{8}{15}$ **5.** 6/5525
6. (a) $\frac{1}{30}$ (b) $\frac{1}{2}$ (c) $\frac{5}{6}$ **7.** (a) $\frac{1}{12}$ (b) $\frac{1}{12}$ (c) $\frac{1}{6}$
8. $\frac{1}{216}$, $\frac{15}{216}$, $\frac{72}{216}$, $\frac{125}{216}$ **9.** (a) $\frac{1}{144}$ (b) $\frac{121}{144}$.
10. (a) $\frac{8}{125}$ (b) $\frac{54}{125}$ (c) $\frac{27}{125}$

Exercise 13.2

1. (a) $\frac{1}{64}, \frac{6}{64}, \frac{15}{64}, \frac{20}{64}, \frac{15}{64}, \frac{6}{64}, \frac{1}{64}$ (b) $\frac{21}{32}$ 2.   (a) 0.014 2 (b) 0.815 3
3. 0.07 7   4.   0.078 1   5.   250, 11.2
6. (a) 0.259 (b) 0.010 2 (c) 0.077 8   7.   (a) 100 (b) 300   8.   56

Exercise 13.3

1. (a) 8 (b) 69 (c) 340   2.   703   3.   0.344 6, 250 ± 20, 250 ± 26
4. (a) 5457 (b) 96 (c) 2590 hours   5.   (a) 80.3% (b) 38.3%
6. (a) 0.022 8 (b) 0.954 4 (c) 0.977 2   7.   18.2%
8. (a) 44.0 kN, 2.0 kN (b) 54.1 kg, 1.4 kg

Exercise 13.4

1. (a) 0.606 5 (b) 0.303 3 (c) 0.090 2   2.   (a) 0.047 (b) 0.442 2
3. (a) 0.270 6 (b) 0.270 6 (c) 0.180 4 (d) 0.143 1
4. (a) 0.049 8, 0.149 4, 0.224 1 (b) 0.199 2 (c) 0.083 6
5. (a) 0.449 3 (b) 0.550 7 (c) 0.047 5
6. 18.4, 18.4, 9.2, 3.1, 0.7, 0.2
7. (a) 27.3% (b) 35.4% (c) 23.0% (d) 14.3% (e) 85.7%
8. 110, 90, 35, 10, 2, 0

Exercise 13.5

1. 120 ± 1.33 kN/m^2 2,   67   3.   0.961
4. (a) 0.242 (b) 0.013 4   5. $27.73 - 28.76$ kg
6. $60.10 - 60.50$ ohms. sample standard deviation high

Miscellaneous Exercise 13

1. (a) (i) 0.154 (ii) 0.025 6 (iii) 0.590 (b) 8.2%
2. (a) $\frac{1}{75}$ (b) $\frac{73}{75}$ (c) $\frac{1}{9}$ 3.   (b) (i) 0.599 (ii) 0.075 (iii) 0.011 6
4. (a) 5.5% (b) 1 219.6 g (c) 95.5%   5.   (c) 0.407
6. (a) (i) 0.279 (ii) 0.0573 (b) 0.062   7.   (a) 0.965
8. (a) 0.366 1, 0.369 6, 0.184 9 (b) 0.367 9, 0.367 9, 0.184 0
9. (a) $\frac{19}{500}$ (b) $\frac{24}{500}$ (c) $\frac{1}{500}$ (d) $\frac{44}{500}$ (e) $\frac{436}{500}$
10.   0.216 7, 0.204 8

Chapter 14

Exercise 14.1

1. (a) $-3 \pm j2$ (b) $0.33 \pm j1.25$ 2.   (a) $-j$ (b) $-j$ (c) -2 (d) $-j2$

3.   (a) $-3 - j2$ (b) $1 + j6$ (c) $4 + j5$ (d) $-14 + j8$ (e) $35 + j12$
    (f) $-2 - j4$ (g) $-6 + j9$ (h) $-\frac{20}{37} - j\frac{9}{37}$ (i) $-1.4 + j4.7$
    (j) $\frac{5}{169} - j\frac{12}{169}$ (k) $14 + j12$ (l) $-9 + j46$ (m) $-\frac{1}{10} + j\frac{1}{5}$
    (n) $\dfrac{-j}{3}$ (o) $-\frac{1}{9}$

Exercise 14.2

2.   (a) $12 + j9$ (b) $-2 + j7$ (c) $5 - j8$ (d) $7 - j$ (e) $-2 - j9$
3.   Anticlockwise rotations of $90°$, $180°$, $270°$ and $360°$ respectively

Exercise 14.3

1.   (a) $3.61(\cos 33.7° + j \sin 33.7°)$
    (b) $4.47(\cos 116.6° + j \sin 116.6°)$
    (c) $6.08(\cos -170.5° + j \sin -170.5)$ (d) $3(\cos 90° + j \sin 90°)$
    (e) $1.41(\cos -45° + j \sin -45)$ (f) $1.41(\cos 135° + j \sin 135°)$
2.   (a) $1.73 + j$ (b) $1.5 - j2.6$ (c) $-2.5 + j4.3$ (d) $7.07 + j7.07$
    (e) $1.69 - j1.41$
4.   (a) $5.60°$ (b) $13, -112.6°$
5.   (a) $5.65\underline{/47°}$ (b) $1.6\underline{/122.5°}$ (c) $36.5\underline{/100°}$
6.   (a) $14\underline{/101°}$ (b) $6.0\underline{/-150°}$ (c) $1.2\underline{/11°}$ (d) $4.3\underline{/-44°}$
    (e) $18\underline{/90°}$ (f) $4.5\underline{/90°}$
7.   (a) $3125(\cos 94.3° + j \sin 94.3°)$
    (b) $2.24(\cos -26.6° + j \sin -26.6°)$
    (c) $0.45(\cos 26.6° + j \sin 26.6°)$ (d) $0.53\underline{/-40.9°}$
    (e) $0.2\underline{/-1.26.9°}$
9.   (i) $2.23e^{-j0.463}$ (ii) $5e^{j0.698}$ (a) $11.2e^{j0.235}$ (b) $5e^{-j0.926}$
    (c) $0.446e^{-j1.46}$ (d) $2.63\underline{/24°}$ (e) $0.802 - j0.463$ (f) $2.41 - j1.39$
10. (a) $1.08 + j1.68$ (b) $-0.909 - j3.90$ **11.** $63°$ Anticlock rotation

Exercise 14.4

1. $33 + j7$ **2.** 24.2 kN at $-85.7°$ to positive x-axis
3. 4.12 amps lagging voltage by $59°$
4.   (a) $53.8°$ leading (b) 1.03 ohms
5.   (a) $Z = 56 + j37$ ohms (b) $I = 2.5 - j1.64$

Miscellaneous Exercise 14

1. $-5 - j10$, $11.2\underline{/-111.6°})$, $-26.6°$, $+j$
2.   (a) $-8.5, -3.5$ (b) $-2.12 + j10.6$
3. $r = 29.2, \theta = 39.7°$

4.   (a) $4 + j6$, $5 + j$ (b) $5.66\underline{/105°}$, $4\underline{/150°}$
5.   (a) (i) $\cos 10° - j \sin 10°$ (ii) $\cos 2\theta - j \sin 2\theta$ (iii) $\cos 3\theta + j \sin 3\theta$
     (b) (i) $\pm j$ (ii) $-1.5 \pm j1.32$
6.   (a) (i) $-12 - j5$ (ii)$-\frac{54}{13} - j\frac{23}{13}$ (iii) $\frac{27}{26} + j\frac{47}{26}$
     (b) $\cos \theta - j \sin \theta$ (c)(i) $4.47\underline{/-116.6°}$ (ii) $5\underline{/-126.9°}$
7.   (a) $13^6\underline{/43.8°}$
     (c) (i) $26.3\underline{/40.4°}$ (ii) $2\underline{/120°}$ (iii) $1.41\underline{/45°}$ (iv) $3\underline{/-90°}$
     (v) $0.28\underline{/33.7°}$
8.   (a) 131 volts, $6.58°$ lagging (b) 2.52 ohms (c) 260 watts

Chapter 15

Exercise 15.1

1.   (a) 6 (b) 9 (c) 27(d) 38 (e)117 (f) 127 (g) 665 (h) 2670
2.   (a) 10 110 (b) 110 101 (c) 1000 1100 (d) 110 1110 (e) 1 000 000 000
     (f) 111 0101 111 (g) 111 111 000 01

Exercise 15.2

1.   (a) 1 011 (b) 11 000 (c) 100 000 (d) 110 011 (e) 101 1110 (f) 100 111
     (g) 100 001 (h) 111 001
2.   (a) 11 (b) 11 (c) 1 001 (d) 1 (e) 111 010 (f) 1 110 (g) 100 (h) 11 011
3.   (a) 1 111 (b) 111 100 (c) 110 11100 (d) 111 011101 (e) 110 11000
     (f) 111 00111001
4.   (a) 10 (b) 101 (c) 111 (d) 1 010 (e) 1 100 (f) 11 000

Exercise 15.4

1.   {2, 4, 8}, {1}, {1, 2, 4, 6, 8, 10}, {1, 2, 4, 8} {0}
2.   (a) $(\pm 4, 0), (\pm 3, \sqrt{7}), (\pm 3, -\sqrt{7}), (\pm 2, \sqrt{12}), (\pm 2, -\sqrt{12}),$
     $(\pm 1, \sqrt{15}), (\pm 1, -\sqrt{15}), (0, \pm 4)$
     (b) $(-4, 8), (-3, 7), (-2, 6), (-1, 5), (0, 4), (1, 3), (2, 2), (3, 1), (4, 0)$
     (c) $(A \cap B) = \{(0, 4), (4, 0)\}$ i.e. coordinates of points of intersection.
3.   {1, 2, 4, 8, 16}, {1, 4, 9, 16}, {1, 4, 16}, {1, 2, 4, 8, 9, 16}

Exercise 15.5

1. 6 **2.**   (a) 14 (b) 6 **3.**   (a) 7 (b) 15 (c) 19 **4.** 9

Exercise 15.6

1. 0 **2.** $a + \bar{c}$ **3.** a **4.** ab **5.** abc **6.** $a + bcd$ **7.** a
8. $a + bc$ **9.** $a + b$ **10.** a **11.** a **12.** b **13.** $a\bar{c} + \bar{b}\bar{c}$
14. $a + b$ **15.** $y\bar{z}$

Exercise 15.7

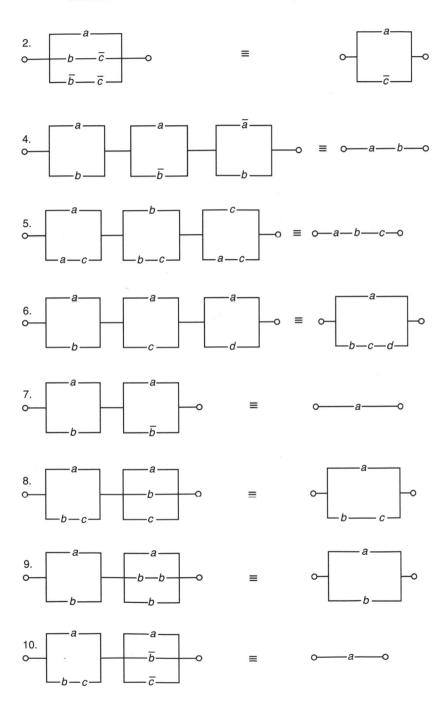

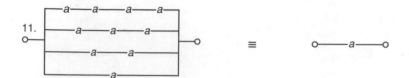

Exercise 15.8

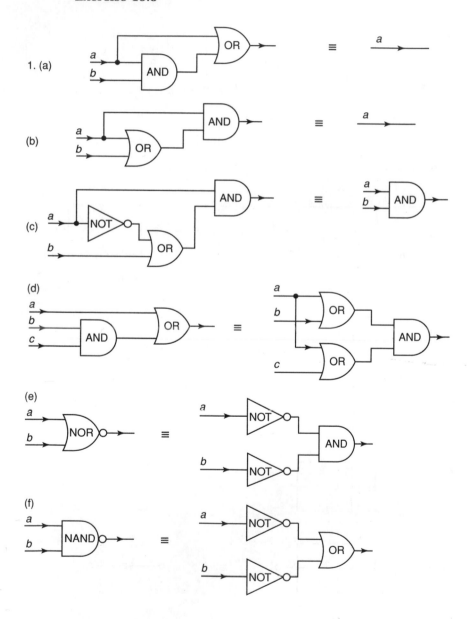

2. (a) $a + bcd$

(b) $a(b + c)$

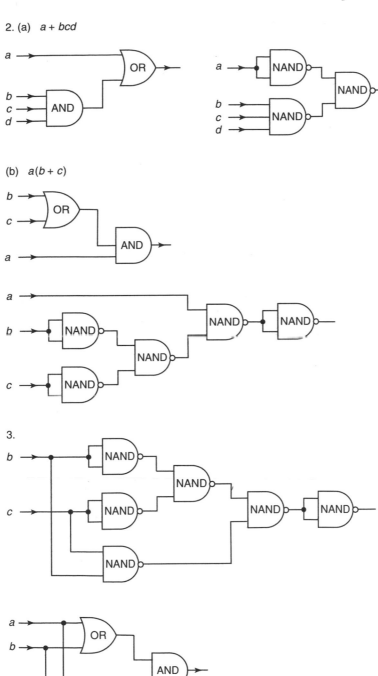

3.

4.

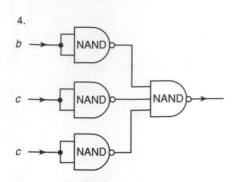

5. $\overline{a}b + a\overline{b}$

6.

a	b	s	c
0	0	0	0
0	1	1	0
1	0	1	0
1	1	0	1

7. $\overline{a}\overline{b} + a\overline{b}$

8. $\overline{a}\overline{b}c + bc$

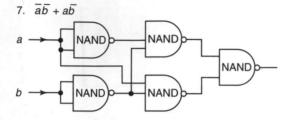

9. $\overline{a}b + a\overline{b}$

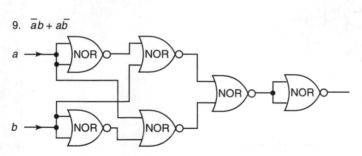

Chapter 16

Exercise 16.1

1. -1 **2.** 10 **3.** $-5xy$
4. -21 **5.** 1 **6.** -11
7. 23 **8.** 1 **9.** $18x$

Exercise 16.2

1. 5, −6, −4, −5
2. 4, 0
3. $+7xz, -10xy$

Exercise 16.3

2. (a) $abc(a-b)(b-c)(c-a)$ (b) $-(a-b)(b-c)(c-a)$
3. (a) 102 (b) –1200 (c) $(x-a)(x-b)(x+a+b)$

Exercise 16.4

1. 2, 3 **2.** 2, 0 **3.** $\frac{7}{3}, \frac{3}{2}$
4. 1, 2, 3 **5.** –3, –4, 5 **6.** 2, –2, 1

Exercise 16.5

1. $\begin{pmatrix} 7 & 7 \\ 21 & 0 \end{pmatrix}$

2. $\begin{pmatrix} 7 & 3 \\ -7 & 1 \\ 20 & 17 \end{pmatrix}$

3. $\begin{pmatrix} -9 & -28 \\ 37 & 0 \end{pmatrix}$

4. $\begin{pmatrix} 4 & -7 & -5 \\ 3 & 16 & -5 \\ 8 & 10 & -16 \end{pmatrix}$

Exercise 16.6

1. $-\dfrac{1}{13}\begin{pmatrix} -5 & -1 \\ 2 & 3 \end{pmatrix}$

2. $\dfrac{1}{4}\begin{pmatrix} -2 & -2 \\ 3 & -4 \end{pmatrix}$

3. $-\dfrac{1}{30}\begin{pmatrix} -16 & 13 & 2 \\ 4 & -7 & -8 \\ 2 & -11 & -4 \end{pmatrix}$

4. $\dfrac{1}{15}\begin{pmatrix} -4 & 3 & -2 \\ 5 & 0 & -5 \\ 12 & -9 & -9 \end{pmatrix}$

Exercise 16.7

1. (a) 1, 2 (b) 2, –3, (c) 4, –5 (d) 75, 15
2. As for Exercise 1.2
3. (a) 3, 3, –5, (b) 6, 4, –5, (c) 5, –6, 7

Miscellaneous Exercise 16

1. $\begin{pmatrix} 1 & 0 \\ 0 & 1 \end{pmatrix}, -13$

3. (a) $\begin{pmatrix} 20 & 13 \\ -5 & 3 \end{pmatrix}$ (b) $\frac{1}{6}\begin{pmatrix} 2 & -4 \\ 0 & 3 \end{pmatrix}$

   (c) $\begin{pmatrix} 3 & 1 \\ -2 & -2 \end{pmatrix}$

4. $k = 2, k = 3$
5. (a) $x = 7, y = 4$

6. $\begin{pmatrix} ac+bd & 0 & ad+bc \\ 0 & 1 & 0 \\ ad+bc & 0 & ac+bd \end{pmatrix}, \dfrac{1}{(a^2-b^2)^2}\begin{pmatrix} a & 0 & -b \\ 0 & 1 & 0 \\ -b & 0 & a \end{pmatrix}$

7. 3, 1, –2

8. (a) $-\begin{pmatrix} 1 & -1 & -1 \\ 7 & -5 & -4 \\ 4 & -3 & -2 \end{pmatrix}$ (b) 1, 0, –1 (c) $\begin{pmatrix} 3 & -4 & 6 \\ 11 & -21 & 33 \\ 6 & -12 & 19 \end{pmatrix}$

9. 7, 3, –3

Chapter 17

Exercise 17.1

1. 38, 23° to vector **u**
2. (a) 75, 24° (b) 30, 91° to **u**
3. (a) 36N, 47° (b) 15N, 20° to F_2
4. (a) 32, 24° to u (b) 13.7, 54° to **v**
5. 3
6. –4
7. 12
8. 87A, 42°

Exercise 17.2

1. 15, 19°; 9.2, –77°
2. (a) $6\mathbf{i} + 4\mathbf{j} + 4\mathbf{k}$, (b) $-9\mathbf{i} + 4\mathbf{j} + 10\mathbf{k}$
3. $m = 3, n = -\frac{1}{2}$

5. $3, \sqrt{26}, \sqrt{29}, ; \dfrac{\mathbf{u}}{3}, \dfrac{\mathbf{v}}{\sqrt{26}}, \dfrac{\mathbf{w}}{\sqrt{29}}$

6. $12\mathbf{i} - 5\mathbf{k} - 5\mathbf{i} - 13\mathbf{j} + 8\mathbf{k}$
7. $m = 2, n = \frac{3}{2}$
8. $17ic + 7j$

9.   98.3 N, 68.8 N

10. $220\mathbf{i} + 260\mathbf{j}$, 49.8

Exercise 17.3

1.   9.2, 49° ; 8.2, −14°
       (a) 14.8, 19° (b) −68**k**
2.   −10 (b) −11**i** − 16**j** − 3**k**
3.   86°
4.   −(**u** × **v**)
5. $|\mathbf{u}|^2 + (\mathbf{v}|^2 - 3(\mathbf{u} \cdot \mathbf{v})$
7.   (i)   (a) $(\frac{3}{5}\mathbf{i} + \frac{4}{5}\mathbf{j})$ (b) $(-\frac{4}{5}\mathbf{i} + \frac{3}{5}\mathbf{j})$
       (ii)   (a) $(\frac{5}{13}\mathbf{i} + \frac{12}{13}\mathbf{j})$ (b) $(-\frac{12}{13}\mathbf{i} + \frac{5}{13}\mathbf{j})$
       (iii) (a) $(\frac{4}{\sqrt{65}}\mathbf{i} + \frac{7}{\sqrt{65}}\mathbf{j})$ (c) $(-\frac{7}{\sqrt{65}}\mathbf{i} + \frac{4}{\sqrt{65}}\mathbf{j})$

Miscellaneous Exercise 17

1.   (a) $\mathbf{a} + \mathbf{b}$ (b) $2\mathbf{b}$ (c) $2\mathbf{b} - \mathbf{a}$ (d) $\mathbf{b} - \mathbf{a}$
2. $|\mathbf{x}|^2 - |\mathbf{y}|^2$
3. $2\mathbf{i} + \mathbf{j} - 7\mathbf{k}$
4. $a^2 - b^2$
6. $(\mathbf{i} + 2\mathbf{j} + 2\mathbf{k}), (5\mathbf{i} + 14\mathbf{j} + 6\mathbf{k})$
7.   (a) $m = 2, n = \frac{3}{2}$; (b) 70, 20°
8.   (a) 7 (b) 4
9.   (a) 4:1, (b) 6, −24
10.   (a) $(\frac{5}{13}\mathbf{i} + \frac{12}{13}\mathbf{j}), (\frac{4}{5}\mathbf{i} + \frac{3}{5}\mathbf{j})$
        (b) $\mathbf{F}_1 = 7(5\mathbf{i} + 12\mathbf{j}); \mathbf{F}_2 = 16(4\mathbf{i} + 3\mathbf{j}); 165$

Study guide

GNVQ Engineering:

Further, Additional and Extended Mathematics

GNVQ Engineering students who wish to proceed to higher education are recommended to take the following BTEC units:

- Optional unit 17: Further Mathematics
- Additional unit 23: Additional Mathematics in Engineering
- Additional unit 40: Extended Mathematics for Engineering

This study guide shows how these BTEC units are covered by this book.

Unit 17 Further mathematics

Index